教育部—邦飞产学合作协同育人项目
大数据与人工智能技术丛书

Hadoop+Spark 大数据技术 微课版

◎ 刘彬斌 主编 李柏章 周磊 李永富 编著

清华大学出版社
北京

内 容 简 介

本书从初学者角度出发，通过丰富的实例，详细介绍了大数据开发环境和基本知识点的应用。全书内容包括：大数据系统基础篇、Hadoop 技术篇、Spark 技术篇和项目实战篇。大数据系统基础篇讲解 Linux 的安装、Linux 的使用和在 Linux 系统上安装并使用 MySQL；Hadoop 技术篇讲解 Hadoop 集群的搭建、Hadoop 两大核心的原理与使用、Hadoop 生态圈的工具原理与使用（Hive、HBase、Sqoop、Flume 等）；Spark 技术篇讲解 Spark 集群的搭建、Scala 语言、RDD、Spark SQL、Spark streaming 和机器学习；项目实战篇将真实的电力能源大数据分析项目作为实战解读，帮助初学者快速入门。

本书所有知识点都结合具体实例和程序讲解，便于读者理解和掌握。本书适合作为高等院校计算机应用、大数据技术及相关专业的教材；也适合作为大数据开发入门者的自学用书，可快速提高开发技能。

本书封面贴有清华大学出版社防伪标签，无标签者不得销售。
版权所有，侵权必究。举报：010-62782989，beiqinquan@tup.tsinghua.edu.cn。

图书在版编目（CIP）数据

Hadoop+Spark 大数据技术：微课版/刘彬斌主编. —北京：清华大学出版社，2018（2023.1 重印）
（大数据与人工智能技术丛书）
ISBN 978-7-302-51427-5

Ⅰ. ①H… Ⅱ. ①刘… Ⅲ. ①数据处理软件–高等学校–教材 Ⅳ. ①TP274

中国版本图书馆 CIP 数据核字（2018）第 242160 号

责任编辑：付弘宇　薛　阳
封面设计：刘　键
责任校对：焦丽丽
责任印制：沈　露

出版发行：清华大学出版社
网　　址：http://www.tup.com.cn，http://www.wqbook.com
地　　址：北京清华大学学研大厦 A 座　　邮　编：100084
社 总 机：010-83470000　　邮　购：010-62786544
投稿与读者服务：010-62776969，c-service@tup.tsinghua.edu.cn
质 量 反 馈：010-62772015，zhiliang@tup.tsinghua.edu.cn
课 件 下 载：http://www.tup.com.cn，010-62795954

印 装 者：三河市龙大印装有限公司
经　　销：全国新华书店
开　　本：185mm×260mm　　印　张：22.5　　字　数：537 千字
版　　次：2018 年 11 月第 1 版　　印　次：2023 年 1 月第 9 次印刷
印　　数：9901～10900
定　　价：69.00 元

产品编号：079663-01

前 言

随着信息技术的不断发展,以及物联网、社交网络、移动终端等新兴技术与服务的不断涌现和广泛应用,数据种类日益增多,数据的规模急剧增大,大数据时代已悄然来临。由于大数据对政府决策、商业规划和危险预防等方面所起的重大作用,大数据逐渐成为一种重要的国家战略性资源,受到政府、能源及信息领域的普遍关注。大数据的多样性(Variety)、规模性(Volume)和高速性(Velocity)等特点,使得传统的数据存储、管理、分析技术已经无法满足大数据的处理要求。

时至今日,无论你是来自互联网、通信行业,还是来自金融业、服务业或零售业,相信你都不会对大数据感到陌生。调查显示,32.5%的公司正在搭建大数据平台,29.5%的公司已经在生产环境实践大数据技术,并有成功的用例/产品;24.5%的公司已经做了足够的了解,开发准备就绪;基本不了解的只占调查对象的 13.5%。根据某知名数据公司的调查数据,目前国内市场的 IT 人才缺口已经高达几十万,到 2025 年,这一数字还会增加至 200 万,"尤其是大数据技术方面的人才"。在智联、58 同城等大型招聘网站最新发布的招聘职位中,大数据相关岗位占比已经超过 50%,薪酬比软件工程师高 10%以上。由此可见,大数据人才的培养是一份重大的责任和使命。

1. 高校大数据人才培养的背景

(1)高校教育中,大数据人才培养存在起步晚、规模化不足的问题,而且高校学生从大学入学到研究生毕业需要相当长的一段时间。本书从实用的角度出发,为高校快速培养大数据人才提供可行性。

(2)如前文所述,大数据人才紧缺的现象在全球越来越突出。在此背景下,本书旨在弥补高校大数据教材的不足,以模拟真实生产环境为教学目标,为企业培养"到岗就能用"的大数据实用型人才。

(3)经济社会的高速发展,对 IT 产业(尤其是软件产业)提出了更高的要求,对大数据开发人才从数量和质量方面提出了更高的要求。

(4)教育技术的进步和移动互联网时代的到来,打破了高校进行知识传播的技术壁垒。大量的资本和风险投资涌进 IT 培训产业。达内、传智播客等实体 IT 培训机构,开课吧、慕课网、极客网等在线 IT 培养机构纷纷引入先进的教学理念、强大的技术支持,再加上商业化运作,对高校 IT 人才培养带来巨大的挑战和竞争压力。

(5)教学环境的变化。教室、实验室硬件配置齐全,实现了高速稳定的互联网接入,笔记本电脑和手机等互联网接入设备日渐普及,这些都为先进教学理念和教学模式(如微课)的实施提供了硬件和软件上的准备。

(6)教育参与者。教师应该树立"教育就是服务"的教育观念,贯彻工程教育的教育理念,从注重"教师教什么"转移到"学生学到了什么"。学生作为"数字原住民",对新鲜事物、新技术、新教学方式(人性化学习、泛在学习等)有着天然的渴望,教师应尽量多利用新的教学手段,提升课程的吸引力。

综上所述，IT 产业、软件技术以及软件人才培养中的教学理念、教学模式、教学环境、教学对象等因素的发展变化倒逼着高校进行教学改革，教师必须围绕以上因素进行教学创新，传统教材形式的革新也势在必行。

2．本书内容

全书内容分为大数据系统基础、Hadoop 技术、Spark 技术和项目实战 4 部分。其中，Linux 是学习大数据技术的基础，先从 Linux 入手，打下坚实的基础，之后才能更好地学习 Hadoop 和 Spark。4 部分内容分别介绍如下。

大数据系统基础篇通过大数据概述、Linux 系统安装、Linux 系统基础命令、Shell 编程和 MySQL 数据操作，为以后编程奠定坚实的基础。

Hadoop 技术篇以 Hadoop 生态圈为中心，详细介绍 Hadoop 高可用集群搭建、HDFS 技术、MapReduce 技术、Hive 技术，为读者学习大数据开发技术提供便利，并以实用的方式简单介绍 HBase、Sqoop、Flume 工具的使用，使读者在精通一门技术的前提下，能扩展了解相关知识，真正成为一专多能的专业型人才。

Spark 技术篇从 Spark 概述、Scala 语言、环境搭建、RDD 核心技术、Spark SQL 和机器学习等多方面讲解 Spark 大数据的开发，从基础的 Scala 语言开始学习，并以 Hadoop 环境为基础搭建 Spark 大数据集群，从最基础、最常用、最容易理解的思路出发，帮助读者逐步掌握 Spark 大数据技术。

项目实战篇从真实项目"电力能源大数据分析"中抽取一部分业务作为实战解读，通过简洁的流程讲解，使读者了解大数据项目开发的整个过程。

3．本书特色

本书不是对相关原理进行纯理论的阐述，而是提供了丰富的上机实践操作和范例程序，极大地降低了读者学习大数据技术的门槛。对于需要直接上机实践的读者而言，本书更像是一本大数据学习的实践上机手册。书中首先展示了如何在单台 Windows 系统上通过 VirtualBox 虚拟机安装多台 Linux 虚拟机，而后建立 Hadoop 集群，再建立 Spark 开发环境。搭建这个上机实践的平台并不限制于单台实体计算机，主要是考虑个人读者上机实践的实际条件和环境。对于有条件的公司和学校，参照这个搭建过程，同样可以将实践平台搭建在多台实体计算机上。

搭建好大数据上机实践的软硬件环境之后，就可以在各个章节的学习中结合本书提供的范例程序逐一设置、修改、调试和运行，从中体会大数据实践应用的真谛——对大数据进行高效的"加工"，萃取大数据中蕴含的"智能和知识"，实现数据的"增值"，并最终将其应用于实际工作或者商业项目中。

4．本书的使用

第 1 篇讲解 Linux 系统和 Linux 系统上的软件应用。本篇是学习大数据技术的第一步，就如同你要学习 Java 开发，必须先学会操作 Windows 系统一样。

第 2 篇讲解 Hadoop 大数据技术。Hadoop 大数据集群要求在 CentOS 6.9 版本的系统上搭建，JDK 版本为 JDK 1.8，Hadoop 版本为 Hadoop 2.6.5，Zookeeper 版本为 Zookeeper 3.4.10。

第 3 篇讲解在 Hadoop 大数据技术的基础上搭建 Spark 环境，所以读者在学习本篇内容之前，需要熟悉第 2 篇中的 Hadoop 大数据集群搭建的内容。

第 4 篇讲解电力大数据项目，是基础 HDFS 的离线分析项目，读者需要掌握 Java 知识、

Hadoop 技术和 Web 前端知识。

5．作者与致谢

本书由刘彬斌主编。参与本书的编写、资料整理、书稿校对、课件制作等工作的还有李永富、李柏章、周磊、汪磊等。另外，感谢清华大学出版社相关编辑专业和严谨的工作，为本书的顺利出版提供了宝贵的意见，并付出了辛勤的劳动。

编 者

2018 年 3 月

目 录

第 1 篇　大数据系统基础

第 1 章　大数据概述 ... 3
1.1　数据的产生与发展 ... 3
1.2　大数据的基础知识 ... 4
1.3　大数据架构 ... 5

第 2 章　系统的安装与使用 ... 7
2.1　系统安装 ... 7
2.1.1　安装 CentOS 6.x ... 7
2.1.2　安装步骤 ... 7
2.2　基本命令 ... 18
2.2.1　cd 命令 ... 18
2.2.2　打包和解压指令 ... 19
2.2.3　其他常用命令 ... 21
2.3　权限与目录 ... 26
2.3.1　权限 ... 26
2.3.2　目录 ... 27
2.4　文件操作 ... 28
2.4.1　文件与目录管理 ... 28
2.4.2　用户和用户组管理 ... 39
2.5　习题与思考 ... 46

第 3 章　任务命令 ... 47
3.1　脚本配置 ... 47
3.1.1　Shell 脚本 ... 47
3.1.2　Shell 变量 ... 47
3.1.3　Shell 传递参数 ... 48
3.1.4　Shell 数组 ... 50
3.1.5　Shell 运算符 ... 51
3.1.6　Shell echo 命令 ... 55

3.1.7　Shell printf 命令 ·· 57
　　3.1.8　Shell test 命令 ·· 58
　　3.1.9　Shell 流程控制 ·· 60
3.2　网络配置 ·· 67
3.3　习题与思考 ·· 70

第 4 章　数据库操作 ·· 71

4.1　数据库简介 ·· 71
　　4.1.1　MySQL 数据库简介 ·· 71
　　4.1.2　安装 MySQL ·· 72
4.2　数据库基本操作 ·· 72
　　4.2.1　MySQL 的 DDL 操作 ·· 72
　　4.2.2　MySQL 的 DML 操作 ··· 80
4.3　数据库用户操作 ·· 83
　　4.3.1　创建用户 ·· 83
　　4.3.2　给用户授权 ·· 83
　　4.3.3　撤销授权 ·· 84
　　4.3.4　查看用户权限 ·· 85
　　4.3.5　删除用户 ·· 85
　　4.3.6　修改用户密码 ·· 86
4.4　数据库查询操作 ·· 86
4.5　习题与思考 ·· 90

第 2 篇　Hadoop 技术

第 5 章　Hadoop 开发环境 ··· 95

5.1　Hadoop 生态圈工具 ·· 95
5.2　环境搭建 ·· 97
　　5.2.1　步骤 1——虚拟机安装 ·· 97
　　5.2.2　步骤 2——安装 JDK 和 Hadoop ··· 97
　　5.2.3　步骤 3——复制虚拟机 ··· 113
　　5.2.4　步骤 4——设置免密 ··· 117
　　5.2.5　步骤 5——安装 Zookeeper ·· 119
　　5.2.6　步骤 6——启动 Hadoop 集群 ·· 122
　　5.2.7　正常启动顺序 ·· 125
5.3　常见问题汇总 ·· 127
5.4　习题与思考 ·· 128

第 6 章　HDFS 技术 ·· 129

6.1　HDFS 架构 ··· 129

6.2	HDFS 命令	130
	6.2.1 version 命令	131
	6.2.2 dfsadmin 命令	131
	6.2.3 jar 命令	132
	6.2.4 fs 命令	132
6.3	API 的使用	140
6.4	习题与思考	142

第 7 章 MapReduce 技术 … 143

7.1	MapReduce 工作原理	143
	7.1.1 MapReduce 作业运行流程	143
	7.1.2 早期 MapReduce 架构存在的问题	144
7.2	YARN 运行概述	144
	7.2.1 YARN 模块介绍	144
	7.2.2 YARN 工作流程	145
7.3	MapReduce 编程模型	146
7.4	MapReduce 数据流	148
	7.4.1 输入文件	150
	7.4.2 输入格式	150
	7.4.3 数据片段	151
	7.4.4 记录读取器	151
	7.4.5 Mapper	151
	7.4.6 Shuffle	152
	7.4.7 排序	153
	7.4.8 归约	153
	7.4.9 输出格式	153
7.5	MapReduce API 编程	154
	7.5.1 词频统计	154
	7.5.2 指定字段	156
	7.5.3 求平均数	158
	7.5.4 关联	160
7.6	习题与思考	163

第 8 章 Hive 数据仓库 … 165

8.1	Hive 模型	165
	8.1.1 Hive 架构与基本组成	165
	8.1.2 Hive 的数据模型	166
8.2	Hive 的安装	167
	8.2.1 Hive 的基本安装	167

8.2.2 MySQL 的安装 ········168
8.2.3 Hive 配置 ········169
8.3 HQL 详解 ········170
8.3.1 Hive 数据管理方式 ········170
8.3.2 HQL 操作 ········174
8.4 习题与思考 ········182

第 9 章 HBase 分布式数据库 ········183

9.1 HBase 工作原理 ········183
9.1.1 HBase 表结构 ········183
9.1.2 体系结构 ········184
9.1.3 物理模型 ········186
9.1.4 HBase 读写流程 ········187
9.2 HBase 完全分布式 ········189
9.2.1 安装前的准备 ········189
9.2.2 配置文件 ········189
9.2.3 集群启动 ········191
9.3 HBase Shell ········192
9.3.1 DDL 操作 ········192
9.3.2 DML 操作 ········194
9.4 习题与思考 ········197

第 10 章 Sqoop 工具 ········198

10.1 Sqoop 安装 ········199
10.2 Sqoop 的使用 ········200
10.2.1 MySQL 的导入导出 ········200
10.2.2 Oracle 的导入导出 ········201
10.3 习题与思考 ········202

第 11 章 Flume 日志收集 ········203

11.1 体系架构 ········204
11.1.1 Flume 内部结构 ········204
11.1.2 Flume 事件 ········204
11.2 Flume 的特点 ········205
11.3 Flume 集群搭建 ········206
11.4 Flume 实例 ········207
11.4.1 实例 1：实时测试客户端传输的数据 ········207
11.4.2 实例 2：监控本地文件夹并写入到 HDFS 中 ········208
11.5 习题与思考 ········210

第 3 篇　Spark 技术

第 12 章　Spark 概述 ... 213

12.1　Spark 框架原理 ... 213
12.2　Spark 大数据处理 ... 214
12.3　RDD 数据集 ... 215
12.4　Spark 子系统 ... 215

第 13 章　Scala 语言 ... 216

13.1　Scala 语法基础 ... 216
13.1.1　变量、常量与赋值 ... 216
13.1.2　运算符与表达式 ... 217
13.1.3　条件分支控制 ... 217
13.1.4　循环流程控制 ... 218
13.1.5　Scala 数据类型 ... 218
13.2　Scala 运算与函数 ... 219
13.3　Scala 闭包 ... 220
13.4　Scala 数组与字符串 ... 220
13.4.1　Scala 数组 ... 220
13.4.2　Scala 字符串 ... 221
13.5　Scala 迭代器 ... 221
13.6　Scala 类和对象 ... 222
13.7　习题与思考 ... 223

第 14 章　Spark 高可用环境 ... 224

14.1　环境搭建 ... 224
14.1.1　准备工作 ... 224
14.1.2　下载并安装 Spark ... 224
14.2　常见问题汇总 ... 226

第 15 章　RDD 技术 ... 228

15.1　RDD 的实现 ... 228
15.1.1　数据源 ... 228
15.1.2　调度器 ... 228
15.2　RDD 编程接口 ... 229
15.3　RDD 操作 ... 229
15.3.1　Spark 基于命令行的操作 ... 229
15.3.2　Spark 基于应用作业的操作 ... 231

15.3.3 Spark 操作的基础命令与开发工具介绍 ················231
15.3.4 Spark 基于 YARN 的调度模式 ······················231
15.3.5 Spark 基于 Scala 语言的本地应用开发 ··············234
15.3.6 Spark 基于 Scala 语言的集群应用开发 ··············235
15.3.7 Spark 基于 Java 语言的应用开发 ··················236
15.3.8 Spark 基于 Java 语言的本地应用开发 ···············237
15.3.9 Spark 基于 Java 语言的集群应用开发 ···············238
15.4 习题与思考 ··241

第 16 章 Spark SQL ···242

16.1 Spark SQL 架构原理 ··242
   16.1.1 Hive 的两种功能 ···242
   16.1.2 Spark SQL 的重要功能 ··································242
   16.1.3 Spark SQL 的 DataFrame 特征 ························243
16.2 Spark SQL 操作 Hive ···243
   16.2.1 添加配置文件，便于 Spark SQL 访问 Hive 仓库 ······243
   16.2.2 安装 JDBC 驱动 ···243
   16.2.3 启动 MySQL 服务及其 Hive 的元数据服务 ············243
   16.2.4 启动 HDFS 集群和 Spark 集群 ·························244
   16.2.5 启动 spark-shell 并测试 ··································244
16.3 Spark SQL 操作 HDFS ···244
   16.3.1 操作代码 ··244
   16.3.2 工程文件 ··246
   16.3.3 创建测试数据 ···246
   16.3.4 运行 Job 并提交到集群 ··································247
   16.3.5 查看运行结果 ···247
16.4 Spark SQL 操作关系数据库 ··248
   16.4.1 添加访问 MySQL 的驱动包 ·····························248
   16.4.2 添加必要的开发环境 ····································248
   16.4.3 使用 Spark SQL 操作关系数据库 ·······················248
   16.4.4 初始化 MySQL 数据库服务 ·····························250
   16.4.5 准备 Spark SQL 源数据 ·································251
   16.4.6 运行 Spark 代码 ···252
   16.4.7 创建 dist 文件夹 ···252
   16.4.8 安装数据库驱动 ···252
   16.4.9 基于集群操作 ···253
   16.4.10 打包工程代码到 dist 目录下 ····························256
   16.4.11 启动集群并提交 Job 应用 ·······························256
   16.4.12 检查关系数据库中是否已有数据 ·······················258

16.5 习题与思考 258

第17章 Spark Streaming 260

17.1 架构与原理 260
    17.1.1 Spark Streaming 中的离散流特征 260
    17.1.2 Spark Streaming 的应用场景 260

17.2 KafKa 中间件 261
    17.2.1 KafKa 的特点 261
    17.2.2 ZeroCopy 技术 261
    17.2.3 KafKa 的通信原理 261
    17.2.4 KafKa 的内部存储结构 262
    17.2.5 KafKa 的下载 262
    17.2.6 KafKa 集群搭建 262
    17.2.7 启动并使用 KafKa 集群 263
    17.2.8 停止 KafKa 集群 264
    17.2.9 KafKa 集成 Flume 264

17.3 Socket 事件流操作 265
    17.3.1 netcat 网络 Socket 控制台工具 265
    17.3.2 基于本地的 Spark Streaming 流式数据分析示例 266
    17.3.3 基于集群的 Spark Streaming 流式数据分析示例 269
    17.3.4 基于集群模式下的集群文件 I/O 流分析示例 272

17.4 KafKa 事件流操作 275
    17.4.1 基于 Receiver 模式的 KafKa 集成 275
    17.4.2 基于 Direct 模式的 KafKa 集成 278

17.5 I/O 文件事件流操作 280
    17.5.1 基于路径扫描的 Spark Streaming 281
    17.5.2 打包至工程的 dist 目录 284
    17.5.3 启动集群 284

第18章 Spark 机器学习 289

18.1 机器学习原理 289
    18.1.1 机器学习的概念 289
    18.1.2 机器学习的分类 289
    18.1.3 Spark 机器学习的版本演变 290
    18.1.4 DataFrame 数据结构 290
    18.1.5 DataSet 数据结构 290
    18.1.6 执行引擎的性能与效率 290
    18.1.7 Spark 2.x 的新特性 290

18.2 线性回归 291

18.2.1　线性回归分析过程 ················· 291
　　18.2.2　矩阵分析过程 ····················· 291
　　18.2.3　基于本地模式的线性回归分析 ········ 291
　　18.2.4　基于集群模式的线性回归分析 ········ 294
18.3　聚类分析 ································ 300
　　18.3.1　K-Means 聚类算法原理 ·············· 300
　　18.3.2　聚类分析过程 ····················· 300
　　18.3.3　基于本地模式的聚类算法分析 ········ 301
　　18.3.4　基于集群模式的聚类算法分析 ········ 305
18.4　协同过滤 ································ 312
　　18.4.1　个性化推荐算法 ··················· 312
　　18.4.2　相关性推荐算法 ··················· 312
　　18.4.3　基于本地的协同过滤算法分析 ········ 312
　　18.4.4　基于集群的协同过滤算法分析 ········ 317

第 4 篇　项目实战

第 19 章　基于电力能源的大数据实战 ············ 325

19.1　需求分析 ································ 325
19.2　项目设计 ································ 325
　　19.2.1　数据采集 ························· 325
　　19.2.2　数据处理 ························· 326
　　19.2.3　数据呈现 ························· 326
19.3　数据收集与处理 ·························· 329
　　19.3.1　数据收集 ························· 329
　　19.3.2　数据处理 ························· 329
19.4　大数据呈现 ······························ 341
　　19.4.1　数据传输 ························· 341
　　19.4.2　数据呈现 ························· 342
19.5　项目总结 ································ 343

第1篇 大数据系统基础

第 1 章　大数据概述

大数据是目前最热门的技术之一。顾名思义，大数据的"大"是指相对于传统意义上的数据量来说，它的数据体量很大，这些数据信息是海量信息，且在动态变化和快速增长。多样性：大数据是异构的，且是多样性的；价值密度大，大量的数据包含了大量的数据信息，对这些看似不相关的数据信息进行对未来趋势和模式的可预测分析具有巨"大"的价值。速度：对于这些海量数据，需要快速获取信息，这就需要奇"快"的速度。大数据的这些特征能带来预测未来、决胜千里的能力，给生活特别是企业带来巨大的价值。

要想实现大数据的这些价值，就需要大数据技术的支撑，包括数据的存储、数据的管理、数据的计算和建模、数据分析以及数据的可视化。

为了解决这些问题，Hadoop 和 Spark 应运而生，作为大数据处理系统，它们利用分布式文件存储系统、高可靠性以及高可用的分布式计算框架可以构建庞大的高可靠、高效率的集群，为快速地从大体量、庞杂的数据中获取有价值的数据提供了强有力的技术支撑。本书立足于生产实践的实际应用，从最基础的 Linux 平台的学习入手，再到 Hadoop 体系和 Spark 体系。本书包含了想要走向或者正在走向大数据之路的读者所要掌握的绝大部分基础知识。

1.1　数据的产生与发展

随着计算机和信息技术的迅猛发展，人们从工业时代迈向了互联网时代。随着人们获取信息的方式和方法的改变，各行业应用系统的规模迅速扩大，所产生的数据呈井喷式增长。很多应用产生的数据量每天达到数十 TB、数百 TB 甚至数 PB 的规模，各行业所应用的大数据已远远超出了传统计算机和信息技术的处理能力。因此，需要迫切寻求有效的大数据处理技术、方法和手段。

2003 年，Google 发布了有关 Google File System 的论文，详细阐述了一个可扩展的分布式文件系统，用于大型的、分布式对大量数据进行访问的应用。在此基础上，Google 又陆续发表了有关 MapReduce 和 Bigtable 的论文。

2005 年，基于 Google 发表的前两篇论文的理论，Hadoop 应运而生。Hadoop 最初只是雅虎公司用来解决网页搜索问题的一个项目，后因其技术的高效性，被 Apache Software Foundation 公司引入并成为开源应用，为其发展和广泛使用打下了坚实基础。

2008 年末，"大数据"得到部分美国知名计算机科学研究人员的认可，业界组织计算社区联盟（Computing Community Consortium）发表了一份有影响力的白皮书——《大数据计算：在商务、科学和社会领域创建革命性突破》。它使人们的思维不再局限于数据处理的机器，并提出"大数据真正重要的是新用途和新见解，而非数据本身"，使人们对大数据有了新的认知。

2009 年，Spark 在伯克利大学 AMPLab 产生。它最初属于伯克利大学的研究性项目，2010 年正式开源，2013 年成为 Apache 基金项目，并于 2014 年成为 Apache 基金的顶级项目，整个过程用时不到 5 年。

2011 年 5 月，全球知名咨询公司麦肯锡（McKinsey&Company）在麦肯锡全球研究院（MGI）发布了一份报告——《大数据：创新、竞争和生产力的下一个新领域》，标志着大数据开始备受关注。这也是专业机构第一次全方面介绍和展望大数据，为大数据的发展和普及带来良好机遇。2011 年 12 月，我国工信部发布物联网"十二五"规划，提出把信息处理技术作为 4 项关键技术的创新工程之一，其中包括了海量数据存储、数据挖掘、图像视频智能分析，这都是大数据的重要组成部分，为我国的大数据发展提供了新的发展机遇和强有力的政府支持。

2012 年 1 月，在瑞士达沃斯召开的世界经济论坛上，大数据作为主题之一。会上发布的报告《大数据，大影响》（*Big Data, Big Impact*）提出大数据的价值和黄金一样，使人们认识到数据的重要性。

2015 年，国务院正式印发《促进大数据发展行动纲要》，明确了大数据的发展前景和应用，标志着大数据正式上升为国家战略。

2016 年，《大数据"十三五"规划》（以下简称《规划》）出台。《规划》提出实施国家大数据战略，促进大数据产业健康发展，深化大数据在各行业的创新应用，探索与传统产业协同发展的新业态、新模式，加快完善大数据产业链，加快海量数据采集、存储、清洗、分析发掘、可视化、安全与隐私保护等领域的关键技术攻关，促进大数据软硬件产品发展，完善大数据产业公共服务支撑体系和生态体系，加强标准体系和质量技术基础建设。

从大数据的发展历程和政府对大数据的支持程度来看，大数据已成为促进时代发展的重要因素。它将带动千亿甚至更多的产业价值，依托互联网的发展和普及，获得更多的数据。如何从中提取出有价值的信息，成为急需解决的问题。现如今 Hadoop 和 Spark 技术提供了很好的技术平台和解决方案，但国内大数据发展现状并不乐观。虽然得到政府的大力支持，但由于大数据从业者的匮乏，企业难以找到合适的从业人员，阻碍了企业的发展，这就要求培养更多的大数据从业者来支撑大数据产业的健康快速发展。

目前大数据所涉及的范围非常广泛，小到日常生活的购物、医疗，大到企业未来决策和长期规划。不论是消费行业、金融行业、医疗行业，还是区块链、机器学习、人工智能等，都离不开大数据的支持。大数据已融入生活的方方面面，未来的时代是大数据的时代。大数据产业已经上升到国家战略的高度，未来将会有更广阔的市场前景和应用价值。

1.2 大数据的基础知识

初学者想要系统地学习大数据知识，必须要有一定的大数据基础知识储备，只有打下牢固的基础，才能收到事半功倍的效果。

首先，目前主流的大数据平台 Hadoop 和 Spark 都是运行在 JVM 上的。特别是 Hadoop

本身就是使用 Java 开发出来的系统，所以在学习 Hadoop 之前必须掌握一定的 Java 知识，这对后面 Hadoop 应用的开发和使用是必不可少的。另外，目前这些大数据系统基本都运行在 Linux 系统中，因此学会 Linux 的基本操作也是必需的。本书考虑到广大学习者的需求，从最基础的知识入手，在第 2 章对常用 Linux 命令进行了详细讲解，消除读者由于 Linux 系统不熟悉带来的困扰。

在大数据的操作过程中，会涉及对现在使用非常广泛的关系型数据库的使用。因此，在前面的基础章节也对 MySQL 数据库的操作进行了基本的讲解。

Spark 由 Scala 语言开发而来，虽然它可以和 Java 语言进行很好的兼容，但是它提供了很多新的特性，对 Scala 语言的学习有助于深入地了解和学习 Spark。

本书充分考虑初学者的学习需求，立足实战，从基本的运行平台开始，再到关键知识点的讲解，事无巨细，以解决大家在迈向大数据殿堂的路上所遇到的问题。

1.3　大数据架构

大数据架构体系现在主要使用的是 Hadoop 和 Spark。下面对 Hadoop 和 Spark 的体系架构进行简要介绍，具体的内容将在每一个章节进行详解。

就目前的大数据体系来说，要构建一个大数据平台，主要依靠 Linux 系统，它是工业界运行大数据平台的不二之选，所以目前很多大数据软件只支持 Linux。本书立足实际需要，以 Linux 为构建平台，它是搭建大数据平台的基础，具体常用操作需要在正式讲解开始之前对其做详细讲解。

Hadoop 的核心技术包含分布式文件处理系统 HDFS 以及分布式数据处理模型和执行环境 MapReduce。本书主要讲解的 Hadoop 项目如下。

- HDFS：分布式文件处理系统，用于对大型文件的处理和拆分，为构建大规模集群和高可用的文件处理打下基础。
- MapReduce：分布式数据处理和执行环境，用于对大规模数据集进行运算。
- Hive：基于 Hadoop 的一个数据仓库工具，可将结构化的数据文件映射为数据库表，并提供简单 SQL 查询功能，可以将 SQL 转化为 MapReduce 进行运算。
- HBase：分布式的、面向列的开源数据库，它适合于类似大数据的非结构化的数据存储的数据库。
- Sqoop：一款开源的数据传输工具，主要用于在 Hadoop 与传统的数据库间数据的传递。
- Flume：由 Cloudera 提供的一个高可用、高可靠，分布式的海量日志采集、聚合和传输的系统。

Spark 是当今最活跃且相当高效的大数据通用计算平台，是 Apache 三大顶级开源项目之一。目前，Spark 已经发展成为包含 Spark SQL、Spark Streaming、GraphX、MLlib 等众多子项目的集合。本书主要对 Spark 的一些实际应用和入门开发平台的搭建进行讲解。主要框架体系包括如下项目。

- RDD：弹性分布式数据集，是分布式内存的抽象概念，它提供了高效的数据流处理。
- Spark SQL：它是用来处理结构化数据的 Spark 组件，提供了 DataFrames 的可编程抽象模型，可视为分布式的 SQL 查询引擎。
- Spark Streaming：它是基于 Spark 核心的流式计算的扩展，具有高吞吐量和容错能力强的特点。
- MLlib：一个 Spark 的可以扩展的机器学习库，包含通用的学习算法和工具，本书主要讲线性回归算法、聚类分析和协同过滤这几种常用的算法。
- KafKa：一种高吞吐量、分布式的发布订阅消息系统，它可以处理消费者规模消息的数据。

第 2 章　系统的安装与使用

2.1　系统安装

2.1.1　安装 CentOS 6.x

本节介绍 Linux 的安装。Linux 是通过 VMware 虚拟平台进行安装的,这里以 CentOS 6.9 为例。扫描二维码可以观看 Linux 安装视频。

CentOS 最新版本的官方下载地址为:https://www.centos.org/download/。
以下针对各个版本的 ISO 镜像文件进行说明。

- CentOS-6.9-x86_64-DVD-1708.iso:标准安装版,一般推荐下载这个版本。
- CentOS-6.9-x86_64-Everything-1708.iso:对完整版安装盘的软件进行补充,集成所有软件,包含 CentOS 6.9 的一套完整软件包,可以用来安装系统或者填充本地镜像。
- CentOS-6.9-x86_64-Minimal-1708.iso:精简版,自带软件最少。

2.1.2　安装步骤

创建虚拟机安装(以 VMware 10 为例)。

(1)在 VMware 中创建一台新的虚拟机,如图 2.1 所示。

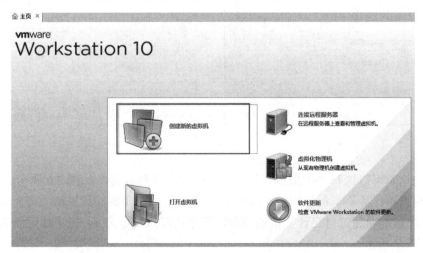

图　2.1

(2)选择"自定义(高级)",然后单击"下一步"按钮,如图 2.2 所示。
(3)单击"下一步"按钮,如图 2.3 所示。

图 2.2

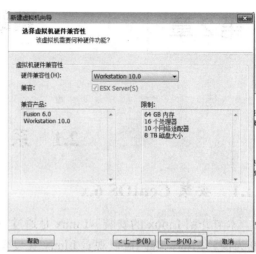

图 2.3

(4)选择"稍后安装操作系统",然后单击"下一步"按钮,如图 2.4 所示。
(5)客户机操作系统选择 Linux,版本选择"CentOS 64 位",然后单击"下一步"按钮,如图 2.5 所示。

图 2.4

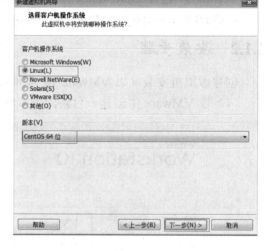

图 2.5

(6)在这里可以选择"虚拟机名称"和"位置"进行设置,如图 2.6 所示。
(7)根据自己计算机的情况给 Linux 虚拟机分配"处理器数量"和"每个处理器的核心数量"。注意不能超过自己计算机的核数,推荐处理器数量为 1,每个处理器的核心数量为 1,如图 2.7 所示。
(8)给 Linux 虚拟机分配内存。分配的内存大小不能超过自己本机的内存大小,多台运行的虚拟机的内存总和不能超过自己本机的内存大小,如图 2.8 所示。

图 2.6　　　　　　　　　　　　　　　图 2.7

（9）使用 NAT 方式为客户机操作系统提供主机 IP 地址访问主机拨号或外部以太网网络连接，如图 2.9 所示。

图 2.8　　　　　　　　　　　　　　　图 2.9

（10）选择"SCSI 控制器"类型为 LSI Logic(L)，然后单击"下一步"按钮，如图 2.10 所示。

（11）选择"虚拟磁盘类型"为 SCSI(S)，然后单击"下一步"按钮，如图 2.11 所示。

图 2.10　　　　　　　　　　　　　　　图 2.11

（12）选择"创建新虚拟磁盘"，然后单击"下一步"按钮，如图 2.12 所示。

（13）根据本机的磁盘大小给 Linux 虚拟机分配磁盘，并选择"将虚拟磁盘拆分成多个文件"，然后单击"下一步"按钮，如图 2.13 所示。

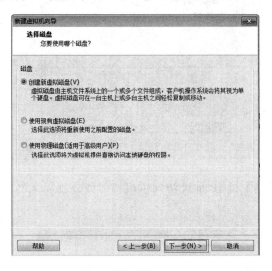

图 2.12

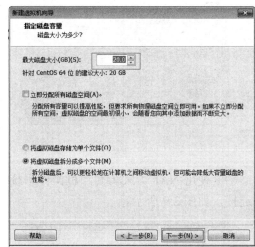

图 2.13

（14）根据需要修改存储磁盘文件的位置。如果不更改，则默认存储在 Linux 虚拟机安装文件目录，如图 2.14 所示。

（15）单击"完成"按钮，完成新建虚拟机向导，如图 2.15 所示。

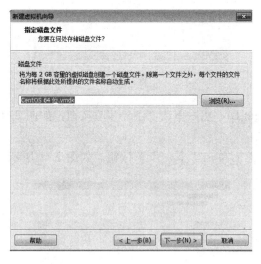

图 2.14

图 2.15

（16）在虚拟机名字上右击，选择"设置"命令来设置安装的 ISO 文件。选择 CD/DVD→"使用 ISO 映像文件"，选择自己的镜像文件，然后单击"确定"按钮，如图 2.16 所示。

（17）开启设置好的虚拟机，进行 Linux 虚拟机的安装，如图 2.17 所示。

图 2.16

（18）选择"不再显示此消息"，然后单击"取消"按钮。

（19）进入 VMware 虚拟机，选择 Install or upgrade an existing system 后按 Enter 键进行安装。或者不进行任何操作，它将倒计时 90 秒后自动安装，如图 2.18 所示。进入 VMware 后可以按 Ctrl+Alt 组合键退出 VMware。

图 2.17

图 2.18

（20）进入安装前的测试页面，通过方向键选择 Skip 按钮，跳过测试，如图 2.19 所示。

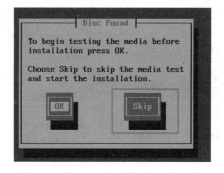

图 2.19

(21)单击 Next 按钮,进行下一步操作,如图 2.20 所示。

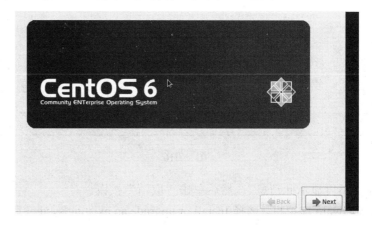

图 2.20

(22)选择"中文(简体)",然后单击 Next 按钮,如图 2.21 所示。

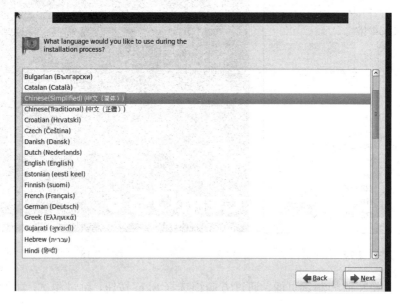

图 2.21

(23)选择"美国英语式",然后单击"下一步"按钮,如图 2.22 所示。

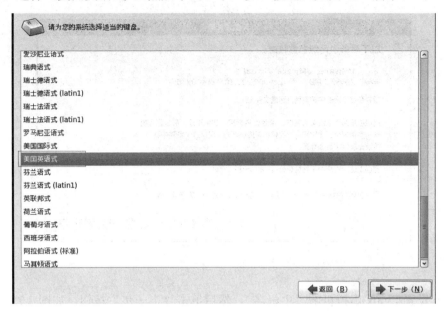

图 2.22

(24)选择"基本存储设备",然后单击"下一步"按钮,如图 2.23 所示。

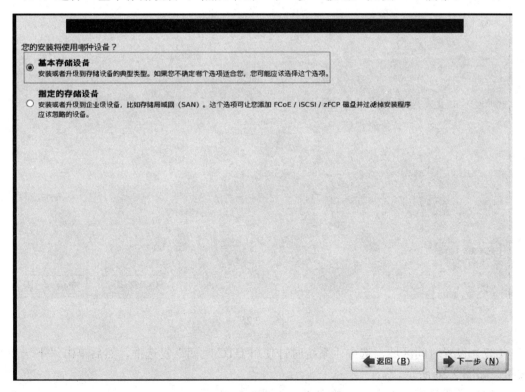

图 2.23

（25）单击"是，忽略所有数据"按钮，如图 2.24 所示。

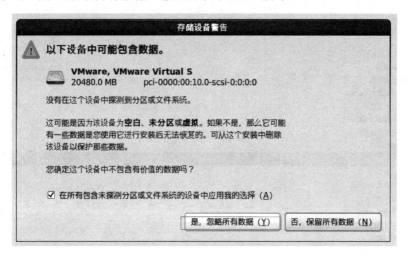

图 2.24

（26）设置 Linux 虚拟机的"主机名"，然后单击"下一步"按钮，如图 2.25 所示。

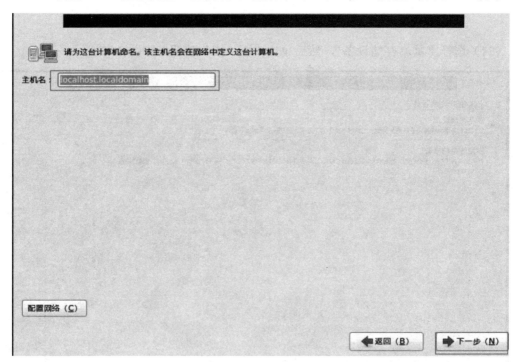

图 2.25

（27）设置时间区域，选中"系统时钟使用 UTC 时间"复选框，然后单击"下一步"按钮，如图 2.26 所示。

（28）设置管理员用户 root 的密码，如图 2.27 所示。

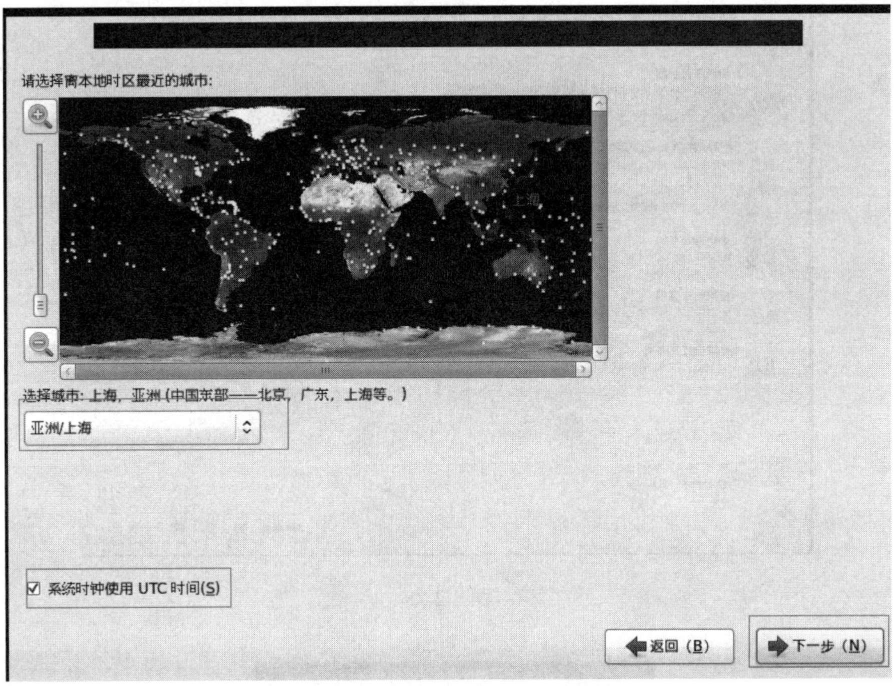

图 2.26

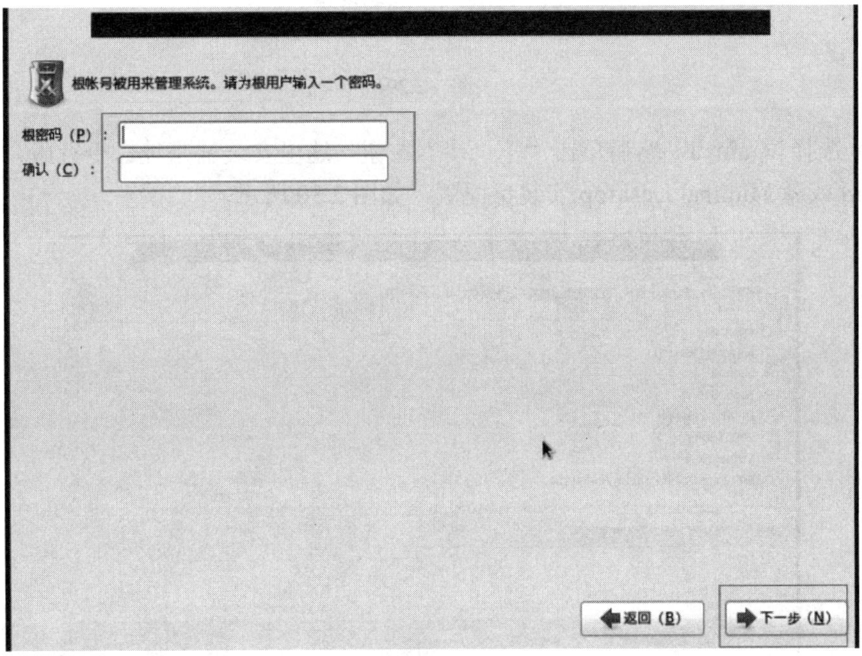

图 2.27

（29）选择"使用所有空间"，然后单击"下一步"按钮，如图 2.28 所示。
（30）单击"将修改写入磁盘"按钮，如图 2.29 所示。

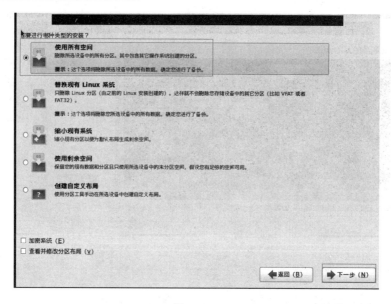

图 2.28

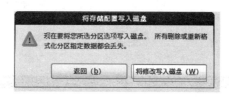

图 2.29

（31）选择 Minimal，然后单击"下一步"按钮。这里安装的是纯命令行版，也可以选择 Desktop 或者 Minimal Desktop 安装桌面版，如图 2.30 所示。

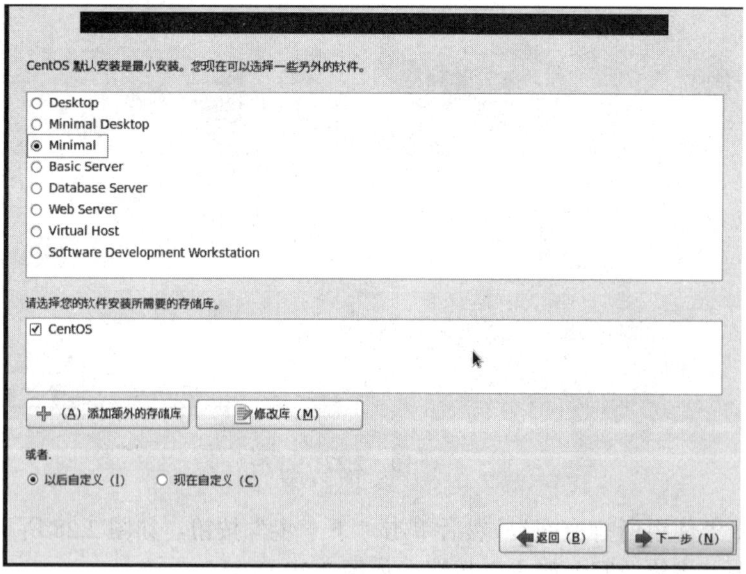

图 2.30

（32）进入安装界面，这里一共需要安装 332 个软件包，如图 2.31 所示。

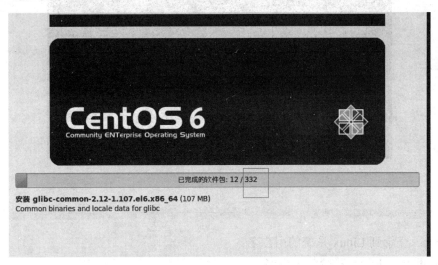

图 2.31

（33）单击"重新引导"按钮，完成安装，如图 2.32 所示。

图 2.32

（34）测试安装是否成功。输入用户 root 和用户密码，如果能进入如图 2.33 所示的界面，则说明安装成功。

图 2.33

```
[root@localhost ~]#
```

- root：登录到 Linux 系统的用户名。
- localhost：Linux 系统的主机名。
- ~：root 用户所在的位置，"~"表示 root 的家目录，root 家目录实际路径是/root。

2.2 基本命令

关于基本命令的讲解视频可扫描二维码观看。

2.2.1 cd 命令

cd 命令是在 Linux 系统中用得比较频繁的命令，因此可以使用 cd 命令在不同的目录中切换。就像是在 Windows 系统中操作前进、后退和目录树一样，在 Windows 中是通过单击的方式来实现的，在 Linux 中是通过命令输入的方式来实现的。可以通过执行 cd 命令加路径的方式，按 Enter 键实现切换目录的操作。

语法：

```
cd [相对路径/绝对路径]
```

示例如下：

（1）进入用户主目录（家目录）。

```
[root@locahost ~] # cd
[root@locahost ~] # cd ~
```

（2）返回进入此目录之前所在的目录。

```
[root@locahost ~] # cd -
```

（3）返回上级目录。

```
[root@locahost ~] # cd ..
```

（4）返回上两组目录（可以类推）。

```
[root@locahost ~] # cd ../..
```

（5）把上个命令的参数作为 cd 参数使用。

```
[root@locahost ~] # cd !$
```

（6）切换到根目录下的 tmp 目录。

```
[root@locahost ~] # cd /tmp
```

2.2.2 打包和解压指令

在 Windows 中使用最多的压缩文件就是.rar 格式，但在 Linux 下，.rar 格式并不能被识别，因为 Linux 有特有的压缩工具。有一种格式在 Windows 和 Linux 下都能使用，那就是.zip 格式的文件。本节将学习在 Linux 环境下各种文件的打包和解压的方法。

在 Linux 下最常见的压缩文件通常是以.tar 和.gz 结尾的，除此之外还有.bz2、.zip 等。在 Linux 系统中，文件可以不附带后缀名，但是压缩文件必须带后缀名，这是为了判断压缩文件是由哪种压缩工具压缩的，而后才能正确地解压这个文件。以下介绍常见的后缀名所对应的压缩工具。

- .gz：gzip 压缩工具压缩的文件。
- .bz2：bzip2 压缩工具压缩的文件。
- .tar：tar 打包程序打包的文件（tar 并没有压缩功能，只是把一个目录合并成一个文件）。
- .tar.gz：将打包过程分为两步执行，即先用 tar 打包，再用 gzip 压缩。
- .tar.bz2：过程同上，先用 tar 打包，再用 bz2 压缩。

1. gzip 命令

gzip 命令是应用最广泛的压缩命令，gzip 可以解压 zip 与 gzip 软件压缩的文件。而 gzip 创建的压缩文件后缀名为 gz（即*.gz），同时 gzip 解压或压缩都会将源文件删除。gzip 语法如下。

语法：

```
gzip [参数] [文件]
```

参数：
-d：解压。
-#：压缩等级（#取值范围为 1～9，其中 1 为压缩最差，9 为压缩最好，6 为默认等级）。
示例如下：
（1）将 bunfly 文件压缩成扩展名为.gz 的文件。将 bunfly 文件压缩后，是对原文件直接进行操作，或者可以理解为把原文件直接变成压缩文件。

```
[root@locahost tmp]# gzip bunfly        <==压缩bunfly文件
[root@locahost tmp]# ls                 <==查看解压后的当前路径下的文件/文件夹
bunfly.gz                               <==ls命令查询的结果(压缩结果)
```

（2）将 bunfly.gz 格式的压缩文件解压。解压后，同样压缩文件将直接变成解压后的文件。

```
[root@locahost tmp]# gzip -d bunfly.gz    <==解压bunfly.gz压缩包
[root@locahost tmp]# ls                   <==查看解压后的当前路径下的文件/文件夹
bunfly                                    <==ls命令查询的结果(解压结果)
```

（3）gzip 命令不可以压缩目录。先创建一个 FileDirectory 目录，并使用 gzip FileDirectory 命令对 FileDirectory 目录进行压缩，此时 gzip 命令会弹出一个提示"FileDirectory 文件是一个目录"。通过 ls 命令查看，发现刚刚的压缩操作并没有执行成功。

```
[root@locahost tmp]# mkdir -p etc/        <==创建名为etc的文件夹
[root@locahost tmp]# ls                   <==查看当前目录下的所有文件/文件夹
    etc    bunfly                         <==ls命令查询的结果
[root@locahost tmp]# gzip etc             <==压缩etc文件夹
gzip: etc/ is a directory - ignored       <==gzip命令执行结果，表示不能压缩文件夹
```

2. bzip2 命令

bzip2 是用于代替 gzip 并能提供更好的压缩功能，bzip2 的用法与 gzip 几乎相同，但压缩文件的后缀是 bz2（即*.bz2），即为不可以解压/压缩文件夹。用 bzip2 解压或压缩会删除源文件，与 gzip 一样。用法介绍如下。

语法：

```
bzip2 [参数] [文件]
```

选项或参数：
- -d：解压。
- -z：压缩。

示例如下：

（1）用 bzip2 命令打包。可以用"bzip2 –z bunfly"命令对 bunfly 文件进行压缩。-z 是压缩的参数，这个参数是可以省略的。例如"bzip2 bunfly"命令，默认情况下等于"bzip2 –z bunfly"。

```
[root@locahost tmp]# bzip2 -z bunfly      <==压缩bunfly文件，且删除bunfly文件
[root@locahost tmp]# ls                   <==查看当前目录下的所有文件/文件夹
bunfly.bz2                                <==ls命令查询的结果(压缩结果)
```

（2）用 bzip2 命令解压。bzip2 命令加上-d 参数后，压缩命令就成为解压命令。

```
[root@locahost tmp]# bzip2 -d bunfly.bz2  <==解压bunfly.bz2文件并删除
[root@locahost tmp]# ls                   <==查看当前目录下的所有文件/文件夹
bunfly                                    <==ls命令查询的结果(解压结果)
```

3. tar 命令

gzip、bzip2 不能压缩文件夹，但 tar 命令可解压或压缩 gzip、bzip2 等软件的文件，tar 可以解压或压缩文件夹（多层文件结构），压缩文件的后缀名为 gz、bz2 等。tar 解压或压缩后，不会删除源文件，用法如下。

语法：

```
tar [参数] [压缩文件] [打包文件]
```

参数：
- -z：是否同时用 gzip 压缩。
- -j：是否同时用 bzip2 压缩。
- -x：解压。
- -t：查看 tar 包里面的文件。
- -c：建立一个 tar 包。
- -v：可视化。
- -f：如果压缩时加上-f 参数，则表示压缩后的文件为 filename；如果解压时加上-f 参数，则表示解压 filename 文件；如果是多个参数组合的情况下带有-f，则需要把 f 写到最后。

示例如下。

（1）用 tar 命令压缩 bunfly 文件。注意：使用 tar 命令打包或解压，原来的文件是不会被删除的。

```
[root@locahost tmp]# tar -cvf bunfly.tar bunfly
                                          <==将bunfly文件压缩成bunfly.tar
[root@locahost tmp]# ls                   <==查看当前目录下的所有内容
bunfly  bunfly.tar                        <==查看当前目录的结果
```

（2）用 tar 命令压缩 bunfly.tar 文件。用 tar 解压后的文件名是压缩前的文件名，为了看出效果，这里执行了一步 rm 删除 bunfly 文件的步骤。

```
[root@locahost tmp]# rm -rf bunfly        <==删除当前目录中的bunfly文件
[root@locahost tmp]# ls                   <==查看当前目录下的所有内容
bunfly.tar                                <==查看当前目录的结果
[root@locahost tmp]# tar -xvf bunfly.tar  <==解压bunfly.tar压缩包
[root@locahost tmp]# ls                   <==查看当前目录下的所有内容
bunfly  bunfly.tar                        <==查看当前目录的结果
```

2.2.3 其他常用命令

1. ls 命令

ls 命令是 list 的缩写，list 字面意思是"列出"，所以 ls 命令与字面意思一样，作用是列出文件夹内所有文件和指定文件夹内的所有文件。

语法：

```
ls [参数] [目录名称]
```

参数：
- -a：列出全部的文件，包括隐藏文件（开头为 . 的文件）（常用）。
- -l：列出文件的属性、权限等（常用）。

- -n：列出 UID、GID，而非用户和用户组的名称。
- -S：以文件大小排序。
- -t：以时间排序。

示例如下：

（1）ls 未加目录名称，则表示列出当前目录下的所有文件，但隐藏文件不会列出。

```
[root@locahost ~] # ls
anaconda-ks.cfg  install.log  install.log.syslog
[root@locahost ~] #
```

（2）列出当前目录下的所有文件及其属性，包括隐藏文件。

```
[root@locahost ~] # ls -al
总用量 76
dr-xr-x---.   4 root root 4096 12月 27 00:17 .
dr-xr-xr-x.  23 root root 4096 12月 26 17:40 ..
-rw-------.   1 root root 1098 12月  6 18:09 anaconda-ks.cfg
......
-rw-r--r--.   1 root root 9458 12月  6 18:09 install.log
-rw-r--r--.   1 root root 3091 12月  6 18:07 install.log.syslog
[root@locahost ~] #
```

（3）查看指定目录/tmp 下的所有文件。

```
[root@locahost ~] # ls /tmp
bunfly  yum.log
[root@locahost ~] #
```

（4）查看指定目录/tmp 下的所有文件、属性及权限等。

```
[root@locahost ~] # ls -al /tmp
总用量 192
drwxrwxrwt.  7 root   root   4096 12月 26 22:49 .
dr-xr-xr-x. 23 root   root   4096 12月 26 17:40 ..
-rw-r--r--.  1 root   root      0 12月 25 18:06 bunfly
-rw-------.  1 root   root      0 12月  6 18:05 yum.log
```

（5）ls -l 命令可以简化成 ll，查看指定目录下所有文件的详细信息。

```
[root@locahost ~] # ll                   <==查看当前目录下文件的属性、权限等
总用量 3
-rw-------. 1 root root 1098 12月  6 18:09 anaconda-ks.cfg
-rw-r--r--. 1 root root 9458 12月  6 18:09 install.log
-rw-r--r--. 1 root root 3091 12月  6 18:07 install.log.syslog
[root@locahost ~] # ls -l                <==查看当前目录下文件的属性、权限等
总用量 3
-rw-------. 1 root root 1098 12月  6 18:09 anaconda-ks.cfg
```

```
-rw-r--r--. 1 root root 9458 12月  6 18:09 install.log
-rw-r--r--. 1 root root 3091 12月  6 18:07 install.log.syslog
```

2. pwd 命令

Linux 中 pwd 命令是 Print Working Diredctory 的缩写，其功能是打印当前的工作目录，因此在 Linux 中可以用 pwd 命令来查看当前工作目录的完整路径。在终端进行操作时，总会有一个当前工作目录，在不太确定当前位置时，可以使用 pwd 来判定当前目录在文件系统内的确切位置。简单来说，也就可理解为 pwd 命令就是显示当前所在的目录或路径。

语法：

```
pwd [参数]
```

参数：
-P：显示出实现路径，而非使用连接路径。

示例如下：

```
[root@locahost ~] # pwd           <==打印/显示当前路径
/root                              <== 显示出目录，/root表示根目录
```

3. touch 命令

touch 命令有两个功能：一是用于把已存在文件的时间标签更新为系统当前时间（默认方法），将它们的数据原封不动地保留下来；二是用来创建新的空文件。

语法：

```
touch [参数] [参考文件] [文件名]
```

参数：
- -r：把指定文件或目录的日期时间统统设置为和参考文件或目录的日期时间相同。
- -t：使用指定的日期时间，而非现在的时间。

示例如下：在 root 目录下建立一个空文件 ex2，利用 ll 命令可以发现文件 ex2 大小为 0，表示它是空文件。

```
[root@locahost ~] # touch ex2        <==创建ex2文件
[root@locahost ~] # ll               <==ll命令查看ex2文件大小
总用量 0
-rw-r--r--. 1 root root 0 12月 25 19:03 ex2
[root@locahost ~] #
```

4. ln 命令

ln 命令在 Linux 中是一个非常重要的命令，它的功能是为某个文件在另一个位置建立一个同步的链接。这个链接又分为软件链接和硬链接，通过加参数-s 区分。当加上-s 参数时创建一个软件链接，如果不加-s 参数，则创建出来的链接是硬链接。

语法：

```
ln [-s] [源文件] [目标文件]
```

选项或参数:
-s: 创建软件链接。
示例如下:

(1) 创建文件/etc/issue 的软链接 issue.soft。

```
[root@locahost tmp] # ln -s /etc/issue /issue.soft
```

(2) 创建文件/etc/issue 的硬链接 issue.soft。

```
[root@locahost tmp] # ln /etc/issue /issue.soft
```

5. cp 命令

cp 命令是 copy 的简称,copy 字面意思是"复制",所以 cp 命令的作用是复制文件命令,不仅可以复制文件,还可以创建连接文件(类似快捷方式)。cp 命令的参数选项可以不设置,源文件也可以有多个,并以空格隔开。

语法:

```
cp [参数] [源文件] [目标文件]
```

参数:
- -l: 进行硬链接的连接文件的创建/复制,而非复制文件本身。
- -r: 递归复制,用于复制目录。
- -s: 复制成"快捷方式"文件。
- -u: 如果目标文件比源文件旧,则更新目标文件。

示例如下:

(1) 复制/root 目录下的 .bashrc 文件到/tmp 目录下,并重命名为 bashrc 文件。

```
[root@locahost ~]# cp ~/.bashrc /tmp/bashrc
```

(2) 复制文件属性以及文件。

```
[root@locahost ~]# cp -a /var/log/tmp_1 tmp_2          <==复制文件及其属性
[root@locahost ~]# ls -l /var/log/tmp_1 tmp_2          <==查看两个文件属性
-rw-rw-r-- 1 root root 96384 Sep 11 12:00 /var/log/tmp_1
-rw-rw-r-- 1 root root 96384 Sep 11 12:00 tmp_2
```

(3) 复制多个文件到目录。

```
[root@locahost ~]# cp -a tmp_1 tmp_2 tmp_3 /tmp   <==复制多个文件到/tmp目录
```

6. mv 命令

mv 命令是 move 的简称,move 字面意思是"移动",所以 mv 命令的作用就是移动文件,使用 mv 命令可以移动文件与目录,或重命名。mv 命令的参数可以不设置,源文件也可以有多个,并以空格隔开。

语法:

```
mv [参数] [源文件] [目标文件]
```

参数：
- -f：f 参数表示强制执行，例如文件已存在，不询问就可以执行覆盖等。
- -i：如目标文件存在，需要询问才可以覆盖（i 为默认参数）。

示例如下：将/tmp 目录下的 Hello 文件移动到/root 目录。

```
[root@locahost ~]# mv /tmp/Hello /root
```

7. rm 命令

rm 命令是 remove 的简称，remove 的字面意思是"删除"，所以 rm 命令的作用是删除文件，可以删除文件或目录。

语法：

```
rm [参数] [文件或目录]
```

参数设置：
- -f：强制执行，忽略不存在、警告等信息。
- -r：递归删除，常用于目录的删除。

示例如下：

（1）删除/tmp 目录下的多个文件。

```
[root@locahost ~]# cd /tmp                <==进入/tmp目录
[root@locahost tmp]# ls                   <==列出当前目录下的所有文件/文件夹
Test1  Test2                              <==列出的文件/文件夹
[root@locahost tmp]# rm Test1 Test2       <==删除Test1、Test2
rm: 是否删除普通文件 "Test1"？y            <==y表示"是"；n表示"否"
rm: 是否删除普通文件 "Test2"？y            <==y表示"是"；n表示"否"
```

（2）删除/tmp 目录下的 etc 文件夹。

```
[root@locahost tmp]# rm ./etc             <==删除当前目录下的etc，会出现如下提示
rm: 无法删除"./etc": 是一个目录
[root@locahost tmp]# rm -r ./etc          <==在rm命令后，加-r参数即可
rm: 是否删除目录 "./etc"？ y               <==输入y即可
```

8. cat 命令

cat 命令是 concatenate 的简称，用于显示指定文件内容的命令，同时查看文件内容的命令还有 tac、nl 等，但 tac 命令与 cat 命令是相反的，因为 cat 命令是从文件的第一行开始显示，而 tac 命令是从文件的最后一行开始显示。

参数：
- -b：列出行号，但空白行不显示当前的行号。
- -A：将结尾的断行符$显示出来。
- -n：列出行号，且空白行的行号也会显示出来。

示例如下：

（1）查看/tmp/Hello 文件内容。

```
[root@locahost ~]# cat /tmp/Hello      <==查看/tmp目录中的Hello文件内容
hello bunfly                           <== "hello bunfly"为Hello文件内容
```

（2）查看/tmp/Hello 文件内容以-n 参数显示。

```
[root@locahost ~]# cat -n /tmp/Hello   <==查看/tmp目录中的Hello文件内容
1    hello bunfly                      <==显示文件内容并将行号显示出来
```

（3）查看/tmp/Hello 文件内容以-A 参数显示。

```
[root@locahost ~]# cat -A /tmp/Hello   <==将文件的特殊字符显示出来
hello bunfly$                          <==将结尾断行符$显示出来
```

2.3 权限与目录

关于权限与目录命令的讲解视频可扫描二维码观看。

2.3.1 权限

Linux 中一切设备皆文件，而所有文件都是有权限的，查看文件权限等详细信息可以使用 ls -l 命令。

示例如下：查看/tmp 目录下所有文件/文件夹的详细信息。

```
[root@locahost tmp]# ls -l          <==查看当前目录下所有文件的详细信息
总用量 12
-rw-r--r--. 1 root root     0 12月 24 19:57 2
-rw-r--r--. 1 root root 10240 12月 24 22:24 2.tar
```

如上例所示，我们通过 2.2.3 节了解到 ls 命令返回的结果中列出的第一列信息尤其重要，类似于"-rw-------"，这里的"-rw-------"表示用户对文件可操作的权限。权限分为 4 组，如图 2.34 所示。

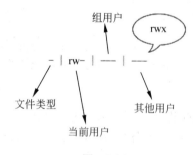

图 2.34

第一组"-"代表这个文件是一个普通文件。"d"代表这个文件是一个目录。"ln"代表这个文件是一个软链接文件。

第二组"rw-"代表当前用户对这个文件只有"读"和"写"的操作权限。

第三组"---"代表组用户对这个文件没有任何权限。

第四组"---"代表其他用户对这个文件没有任何权限。

第一组、第二组和第三组以三个字母为一组出现，这三个字母分别是 r（read）、w（write）、

x（execute），因此 rwx 三个字母顺序是固定的，r 代表这个文件可读，w 代表这个文件可写，x 代表这个文件可以执行。如果不给这个文件赋权限，只需要在对应位置用"-"代替即可。

2.3.2 目录

Linux 的文件路径都带有一个 /，这条斜杠单独出现时称为根目录，所有文件和目录都存放在根目录之下，可以用 ls /命令进行查看：

```
[root@locahost ~]# ls /                    <==查看根目录下所有文件/文件夹
bin    dev   home    lib64        media    opt   root    selinuxsys   usr
boot   etc   lib     lost+found   mnt      proc  sbin    srv    tmp   var
[root@locahost ~]#
```

ls /usr/src 里面的 / 是分隔分层的意思，即显示 usr 目录中的 src 目录中的所有文件及目录。只有 / 单独出现时才代表根目录，如下：

```
[root@locahost ~]# ls /usr/src             <==查看/usr/src目录下所有文件/文件夹
debug    kernels                           <==查看/usr/src目录下所有文件/文件夹结果
[root@locahost ~]#
```

Linux 的目录结构比较简单，一般在 etc 目录下的文件是配置文件，在 bin 下的文件是二进制可执行文件，在 lib 下的文件是一些应用库文件。

每一个登录系统的使用者都会有一个家目录，默认是在/home 文件夹下，并以使用者用户名命名的文件夹。这个目录属于使用者的家目录，可以在里面任意操作，并不会对整个系统产成破坏性影响。但如果是 root 用户，家目录默认是/root，操作时就要谨慎。因为 root 的权限很大，它可以忽略任何限制，如果操作不当可能会对系统造成破坏。

表 2.1 是根目录内的主要目录说明。

表 2.1

目录	应放置档案内容
/bin	在/bin 下的指令可以被 root 或普通用户使用，主要有 cat、chmod、chown、date、mv、mkdir、cp、bash 等常用的指令
/boot	主要放置引导加载程序相关的文件，包括 Linux 核心文件以及开机选单与开机所需设定档案等
/dev	在 Linux 系统上，任何装置与设备都是以文件的形态存在于这个目录当中。只要通过存取这个目录下的某个文件，就等于存取某个装置。这些包括终端设备、USB 或连接到系统的任何设备。例如/dev/tty1、/dev/usbmon0
/etc	包含所有程序所需的配置文件，也包含了用于启动/停止单个程序的启动和关闭 shell 脚本。例如/etc/resolv.conf、/etc/logrotate.conf
/home	所有用户用 home 目录来存储他们的个人档案。例如/home/hadoopuser、/home/otheruser
/lib	/lib 是用来放置在开机时会用到的函数库，以及在/bin 或/sbin 下的指令会呼叫的函数库
/media	用于挂载可移动设备的临时目录。例如挂载 CD-ROM 的/media/cdrom，挂载软盘驱动器的/media/floppy
/mnt	临时安装目录，系统管理员可以挂载文件系统
/root	系统管理员的家目录
/sbin	Linux 有非常多指令是用来设定系统环境的，这些指令只有 root 用户才能用来设定系统，其他用户最多只能用来查询而已。放在/sbin 下的为开机过程所需要的，包括开机、修复、还原系统所需要的指令
/tmp	包含系统和用户创建的临时文件。这个目录是任何人都能够存取的，所以需要定期清理
/usr	/usr 是 Linux 操作系统软件资源所放置的目录

2.4 文件操作

关于文件操作命令的讲解视频可扫描二维码观看。

2.4.1 文件与目录管理

在 Linux 系统下，一切皆文件。因此，使用光盘首先要建立一个目录文件，然后通过操作这个目录来操作光盘，连同鼠标、键盘都被看作文件。文件的类型主要分为 5 类：普通文件、目录文件、设备文件、链接文件、管道文件。其中，前三个都是基本的文件。

普通文件：文本文件、二进制文件。文本文件以 ASCII 码形式存储，人类能够读懂，可以编辑修改；二进制文件是以二进制存储的，要借助一定的软件工具才可以看懂，一般是声音、图像、可执行程序。

设备文件：把每一个 I/O 看作一个设备文件，即把 I/O 看作普通的文件进行写入和读取操作。用户不必了解设备的细节，对设备的使用就像操作文件一样。设备文件又分为块设备和点设备，块设备指硬盘光驱之类的以字符串为单位存取的。点设备指键盘鼠标之类的以单个字符为单位存取的。每一个设备对应一个设备文件，存放在/dev 目录中。

链接文件：分为软链接和硬链接。软链接其实是给文件指定的别名，可以理解为 Windows 系统中的快捷方式。硬链接与软链接基本相同，都是为了解决文件的共享使用，同时带来隐藏文件路径、增加权限安全及节省存储等好处。

那么要怎么区分这些文件呢？这时可以用 ll 命令来查看打印文件的详细信息，每一行的第一个字段中的第一个字符代表的就是这个文件的类型。如下：

- -：普通文件。
- d：目录。
- l：连接文件。
- b：块设备文件。
- c：字符设备文件。
- s：套接口文件。
- p：管道。

1. 绝对路径与相对路径

在 Linux 中什么是一个文件的路径呢？所谓文件的路径，就是文件存在的地址。如同快递寄送东西需要寄件地址，这个地址在 Linux 文件中就是它的路径。例如/root/mfkddd/file，file 是一个文件，它的路径就是/root/mfkddd。在 Linux 中，路径又分为绝对路径和相对路径两种。

绝对路径：路径的写法一定是由根目录 / 开始。例如 cat /root/mfkddd/file。这条语句的意思是查看 file 文件中的内容，cat 后面跟的是绝对路径。

相对路径：路径的写法不是由根目录开始的。例如，假如进入到 mfkddd 目录，可以用 cat file 这个命令直接查看 file 文件中的内容，这里的 cat 后面跟的 file 文件相对 mfkddd 而言就是相对路径。

2. 操作目录的相关命令

1）linux 系统中常见的特殊目录（见表 2.2）

表 2.2

符号	说明
.	代表当前目录
..	代表上一层目录
-	代表前一个工作目录
~	代表当前用户的家目录（家目录又称为 home 目录）
~account	代表 account 这个用户的家目录

每个目录下都有两个目录："."和".."，分别代表当前目录和上层目录。在根目录下使用 ls -a 命令去查询，可以看到根目录下存在"."和".."两个目录，这两个目录的属性和权限完全一致，这说明根目录的"."和".."是同一个目录，如表 2.2 所示。

2）常用操作目录的命令

（1）cd：切换目录。

```
[root@locahost ~]# cd /tmp              <==切换到/tmp目录中
[root@locahost tmp]#                    <==由此可知，此目录为tmp
```

如上操作，cd 命令是切换命令，cd 后面跟的 tmp 是相对路径，cd tmp 命令的意思是从当前目录切换到 tmp 目录。这里使用的是 tmp 的相对路径，使用相对路径的前提是目标目录必须事先存在。

如果只输入 cd 或者 cd ～，代表回到使用者的家目录。输入 cd -代表回到上一个工作目录。输入 cd /root/mfkddd 代表进入 mfkddd 目录，这里 cd 后面跟的是绝对路径，所以在任何目录下输入 cd /root/mfkddd 命令都可以进入 mfkddd 目录。

（2）pwd：显示当前目录的路径。

例如，先在 root 目录下创建一个 mfkddd 目录，并在里面创建一个 file 文件。

```
[root@localhost /]# cd /root                    <==首先进入root目录
[root@localhost root]# mkdir mfkddd             <==创建目录
[root@localhost root]# cd /root/mfkddd          <==再进入mfkddd目录
[root@localhost mfkddd]# vi file                <==创建文件并进入文件
```

创建文件时在文件中输入想输入的内容，如"www.bunfly.com"，按 Esc 键后输入":wq"，保存并退出。

做好所有准备后，可以使用 pwd 命令来查询文件路径了。在 mfkddd 目录下输入 pwd 命令，打印出来的/root/mfkddd 路径是 mfkddd 目录的绝对路径，如下：

```
[root@localhost mfkddd]# pwd
/root/mfkddd
```

（3）mkdir：建立一个新的目录。

mkdir 命令后面可以选择 m 或 p 参数。m 表示可以给创建的目录设置权限，p 表示可以创建多级目录。

如果没有加 p 参数创建多级目录，终端将会打印出"没有这样的文件或目录"的提示，因为当前目录下没有 test1 这个目录，所以找不到它，后面的 test2 和 test3 自然也不能被创建。

```
[root@locahost mfkddd]# mkdir test1/test2/test3          <==创建目录
mkdir: 无法创建目录"test1/test2/test3"：没有那个文件或目录  <==创建失败
root@locahost mfkddd]#
```

为了解决上述问题，在 mkdir 后面加上一个参数"-p"，系统就会默认地先创建 test1，然后再创建 test2，最后创建 test3，如下操作。

```
[root@locahost mfkddd]# mkdir -p test1/test2/test3       <==创建多级目录
[root@locahost mfkddd]#
```

现在执行了"mkdir –p test1/test2/test3"命令，没有看到效果，这就说明创建成功，可以通过切换命令进入 test3 目录，如果能成功进入，表示多级目录创建成功，如下操作：

```
[root@locahost mfkddd]# cd test1/test2/test3            <==切换目录
[root@locahost test3]#                                   <==已切换到此目录
```

（4）rmdir：删除一个空的目录。

rmdir 命令删除目录时需要一层一层地进行，而且被删除的目录必须是空目录。如果要将目录下的所有内容都删除，需要在 rmdir 命令后面加上"-p"参数。

当直接用 rmdir 命令删一个多级目录时，将出现提示错误信息：目录不为空。

```
[root@locahost mfkddd]# rmdir test1/         <==删除test1目录
rmdir: 删除 "test1/" 失败：目录非空            <==删除失败
[root@locahost mfkddd]#
```

为了解决上述问题，在 rmdir 命令后面加上参数"-p"。rmdir 只是针对目录，并且目录内没有其他文件的情况使用。如果既要删除目录又要删除文件，可以用 rm 命令来实现。

```
[root@locahost mfkddd]# rmdir -p test1/test2/   <==删除多层目录，加-p参数
[root@locahost mfkddd]#
```

3）关于执行文件路径的变量$PATH

当执行一个命令时，如 ls，系统会按照 PATH 的设定去每个 PATH 定义的目录下查找文件名为 ls 的可执行文件。如果在 PATH 定义的目录中含有多个文件名为 ls 的可执行文件，那么先查询到的同名命令就先被执行。

使用 echo $PATH 命令可以查看有哪些目录被定义。echo 命令的作用是显示或打印文件内容，而 PATH 前面加的$表示后面接的是变量，所以会显示目前的 PATH。

```
[root@locahost ~]# echo $PATH
/usr/local/sbin:/usr/local/bin:/sbin:/bin:/usr/sbin:/usr/bin:/root/bin
```

PATH 一定要大写，这个变量的内容由一大串目录组成，每个目录使用 : 分隔，每个目录有顺序之分。无论是 root 还是其他用户都有/bin 或/usr/bin 这个目录在 PATH 变量内，所以就能在任何地方执行 ls 命令来找到/bin/ls 执行文件。

（1）在 PATH 当中加入目录的方法如下。例如在任何目录均可执行/root 目录下的命令，那么就将/root 加入到 PATH 当中即可。

```
[root@locahost ~]# PATH="${PATH:/root}"
```

（2）PATH 的特点如下：
- 不同身份用户预设的 PATH 不同，预设能够随意执行的指令也不同。
- PATH 是可以修改的。
- 使用绝对路径或相对路径直接指定某个命令的文件名进行执行，会比查找 PATH 正确性更高。
- 命令应该要放到正确的目录下，执行才会比较方便。
- 本地目录（.）最好不要放到 PATH 当中。

3. 文件与目录管理

1）查看文件与目录 ls

ls 是 list 的简称，是列出列表的意思。ls 命令是列出文件或目录中的所有文件。语法如下。

语法：

```
ls [参数] [文件]
```

参数：
- -a：全部的文件，连同隐藏文件一起列出。
- -f：直接列出结果，而不进行排序。
- -R：连同子目录的内容一起列出来，这样该目录下的所有内容都会显示。
- -S：以文件的大小排序，而不是用文件名排序。
- -t：以文件的更新时间排序，而不是文件名排序。

因为常常使用"-l"这个参数，所以快捷方式 ll 等同于 ls -l。

示例如下。

（1）列出/查看 tmp 目录下的所有文件/文件夹。

```
[root@locahost tmp]# ls /tmp           <==查看tmp目录下的所有文件
bunfly  bunfly.tar
[root@locahost tmp]#
```

（2）直接用 ls 命令，表示查看当前目录下的所有文件/文件夹。

```
[root@locahost tmp]# ls                <==查看当前目录下的所有文件
bunfly  bunfly.tar
[root@locahost tmp]#
```

（3）查看当前目录下所有文件的详细属性/信息。

```
[root@locahost tmp]# ls -l             <==查看当前目录下所有文件的详细信息
总用量 16
-rw-r--r--. 1 root root    52 12月 25 00:09 bunfly
```

```
-rw-r--r--. 1 root root 10240 12月 24 22:24 bunfly.tar
[root@locahost tmp]#
```

2）复制、删除、移动：cp、rm、mv

具体可参考 2.2 节的基本命令的内容。

4. 查看文件内容

1）直接查看文件内容：cat、tac、nl

操作查看命令前先在/tmp 目录下用 vi bunfly 命令新建一个 bunfly 文件，并在里面写入 1~3 行的内容。操作如下：

```
[root@locahost tmp]# vi bunfly            <==使用vi编辑器打开文件
1    Hello bunfly 001
2    Hello bunfly 002
3    www.bunfly.com
```

在 2.2.3 节中介绍了 cat 命令的使用，本节将介绍 tac 命令的使用和与 cat 命令的区别。

（1）tac 也是用来查看文件内容的命令，只不过 tac 是从最后一行开始显示文件的信息，与 cat 命令刚好相反。tac bunfly 命令如下：

```
[root@locahost tmp]# tac bunfly           <==从最后一行开始显示bunfly文件
3    www.bunfly.com
2    Hello bunfly 002
1    Hello bunfly 001
[root@locahost tmp]#
```

（2）nl 命令是查看命令，它与 cat 或 tac 的区别在于 nl 命令默认带行号显示内容。

语法：

```
nl [参数] [文件]
```

参数：
- -b：指定行号的方式，主要有两种。
    - -b a：表示不论是否为空行，都同样列出行号（类似 cat -n）；
    - -b t：如果有空行，空的那一行不列出行号（默认值）。
- -n：列出行号表示的方法，主要有三种。
    - -n ln：行号在屏幕的最左方显示；
    - -n rn：行号在所在栏的最右方显示，且不加 0；
    - -n rz：行号在所在栏的最右方显示，且加 0。
- -w：行号所在栏占用的字符数。

nl –b a bunfly 命令示例如下：

```
[root@locahost tmp]# nl -b a bunfly
1    Hello bunfly 001
2    Hello bunfly 002
3    www.bunfly.com
```

```
4                              <==空行（没有数据的一行）
5                              <==空行（没有数据的一行）
[root@locahost tmp]#
```

nl –b t bunfly 命令示例如下：

```
[root@locahost tmp]# nl -b t bunfly
1    Hello bunfly 001
2    Hello bunfly 002
3    www.bunfly.com
[root@locahost tmp]#
```

2）翻页查看文件内容：more、less

（1）more 命令是一页一页地显示文件信息。在 more 命令运行过程中，可以使用以下按键进行后续操作，如表 2.3 所示。

表 2.3

按键	功能
Enter	代表向下翻一行
/	代表在这个显示内容中，向下查询"字符"这个关键字
:f	立刻显示出文件名以及目录显示的行数
Q	代表立刻离开 more，不再显示该文件内容
B	代表往回翻页，该操作只对文件有用
N	重复搜索同一个字符

（2）less 命令与 more 命令功能类似，区别是 less 命令在运行过程中可以使用如表 2.4 所示的按键进行后续操作。

表 2.4

按键	功能
空格键	向下翻一页
PgDn	向下翻一页
PgUp	向上翻一页
/	向下搜索"字符"的功能
?	向上搜索"字符"的功能
n	重复前一个搜索
N	反向地重复前一个搜索
g	显示到这个文件的第一行去
G	显示到这个文件的最后一行去
q	离开 less 这个程序

3）获取资料命令：head、tail

（1）head 命令是查询文件内容的命令，它可以指定参数从前往后显示指定的行数。
语法：

```
head [参数] [文件名]
```

参数:
-n: n 代表 int 类型数字,假设数字设置为 3,显示的内容从前往后显示前 3 行,如果不指定参数,则默认显示前 10 行。

head bunfly 命令示例如下:

```
[root@locahost tmp]# head bunfly
1    Hello bunfly 001
2    Hello bunfly 002
3    www.bunfly.com
4
5
[root@locahost tmp]#
```

(2) tail 命令也是查询命令,它是从后往前显示指定的行数,并且空格也被显示出来。
语法:

```
tail [参数] file
```

选项与参数:
- -n: n 代表 int 类型数字,假设数字设置为 3,显示的内容从前往后显示前 3 行,如果不指定参数,则默认显示后 10 行。
- -f: 代表实时显示。

tail -10 bunfly 命令示例如下:

```
[root@locahost tmp]# tail -3 bunfly
3    www.bunfly.com
4
5
[root@locahost tmp]#
```

4) od 命令是查询非纯文字文件命令
语法:

```
od [参数] [文件名]
```

参数:
- -t: 后面可以接类型(type)的输出。
- a: 利用默认的字符进行输出。
- c: 使用 ASCII 字符进行输出。

od –t c bunfly 命令示例如下:

```
[root@locahost tmp]# od -t c bunfly
0000000   H   e   l   l   o       b   u   n   f   l   y       0   0   1
0000020  \n   H   e   l   l   o       b   u   n   f   l   y       0   0
0000040   2  \n   w   w   w   .   b   u   n   f   l   y   .   c   o   m
0000060  \n      \n  \n
```

```
0000064
[root@locahost tmp]#
```

最左边第一列是以八进制表示的 bytes 数。

5. 文件与目录的默认权限与隐藏权限

1) umask 命令用于设置默认文件权限命令

umask 命令是用于设置用户在创建文件时的默认权限，在系统中创建目录或文件时，目录或文件所具有的默认权限就是由 umask 值决定的。

```
[root@locahost ~]# umask
0022
[root@locahost ~]#
[root@locahost ~]# umask -S
u=rwx,g=rx,o=rx
[root@locahost ~]#
```

若使用者创建文件，则默认没有可执行（x）权限，只有 r（读）和 w（写）两个权限，也就是最大限为 666，默认权限如下：-rw-rw-rw-。

若使用者创建目录，则由于 x 与是否可以进入此目录有关，因此默认开放所有权限，即 777，默认权限如下：drwxrwxrwx。

在默认情况下，r、w、x 的分值分别是 4、2、1 分，umask 的分数指"该默认值需要减去的权限"，即当需要拿掉"写"的权限，则为 2 分；而如果要拿掉读的权限，则为 4 分。上述 umask 为 002，表示 user、group 并没有被拿掉任何权限，不过 others 的权限被拿掉了 2 分，也就是说 others 被拿掉了写权限。

2) 文件隐藏属性命令：chattr、lsattr

（1）chattr 命令是设置文件隐藏属性的命令。

语法：

```
chattr [+ -][参数] [文件]
```

参数：

- +：增加某一个特殊参数，其他原本存在参数则不动。
- -：移除某一个特殊参数，其他原本存在参数则不动。
- a：当设置 a 后，这个文件将只能增加信息，不能修改、删除信息，只有 root 才能设定这个参数。
- i：当设置 i 后，则该文件"不能被删除、改名、设置链接也无法定稿或新增信息"，可增加系统安全性，只有 root 能设置该属性。

常见的属性是 a 和 i，且很多属性只有 root 才能设置。

```
[root@locahost tmp]# touch attrtest              <==创建attrtest文件
[root@locahost tmp]# chattr +i attrtest          <==给attrtest文件赋予i的权限
[root@locahost tmp]# rm attrtest                 <==删除 attrtest
rm: 是否删除普通空文件 "attrtest"？Y             <==是否删除，设置为Y（是）
rm: 无法删除"attrtest": 不允许的操作              <==删除失败
```

```
[root@locahost tmp]# chattr -i attrtest      <==给attrtest文件去掉i的权限
[root@locahost tmp]# rm attrtest             <==删除attrtest
rm：是否删除普通空文件 "attrtest"？Y         <==是否删除，设置为Y（是）
[root@locahost tmp]#                         <==没有任何错误，表示删除成功
```

在上述例子中用 touch attrtest 命令创建了一个 attrtest 文件，并用 chattr +i attrtest 命令给 attrtest 文件赋予 i 的权限，然后用 rm attrtest 命令删除 attrtest 文件。此时 rm 发出提示"rm: cannot remove 'attrtest' : Operation not permitted"，不允许执行删除操作，只有通过 chattr –I attrtest 命令减去 attrtest 文件的 i 权限后才能删除 attrtest 文件。

（2）lsattr 命令是显示文件隐藏属性的命令。

语法：

```
lsattr [参数] [文件]
```

参数：
- -a：显示隐藏属性。
- -d：如果接的是目录，则仅显示目录本身的属性而不是目录内的文件名。
- -R：连同子目录的文件一起显示。

```
[root@locahost ~]# cd /tmp                   <==切换目录到/tmp
[root@locahost tmp]# touch bunfly            <==创建文件名为bunfly
[root@locahost tmp]# chattr +aiS bunfly      <==为bunfly文件设置属性
[root@locahost tmp]# lsattr bunfly           <==显示bunfly文件隐藏的属性
--S-ia-------e- bunfly                       <==查看bunfly文件属性结果
[root@locahost tmp]#
```

3）查看文件类型：file

查看文件属于哪类文件的基本信息，如文件属于 ASCII、data 或者 binary 等。显示文件的类型是 ASCII 的纯文字文件。

```
[root@locahost ~]# file ~/.bashrc
/root/.bashrc: ASCII text
```

显示 passwd 的文件信息，如文件的 suid 权限、兼容 Intel x86-64 的硬件平台、使用 Linux 核心 2.6.18 的动态方法库连接等。

```
[root@locahost ~]# file /usr/bin/passwd
/usr/bin/passwd: setuid ELF 64-bit LSB shared object, x86-64, version 1
(SYSV), dynamically linked (uses shared libs), for GNU/Linux 2.6.18, stripped
```

6. 查找命令和文件

1）which 命令是查找指定命令所在路径的命令。

which 根据环境变量 PATH 所规范的路径查询执行文件的文件名。

语法：

```
which [参数] [命令]
```

参数：
-a：将所有由 PATH 目录中可以找到的命令均列出，而不止第一个找到的命令名称。
示例如下：
（1）查询 ifconfig 命令的完整文件名。

```
[root@locahost ~]# which ifconfig
/sbin/ifconfig
```

（2）用 which 找出 which 的文件名。显示两个 which，其中一个是 alias 命令别名，输入 which 等于后面接的那串命令。

```
[root@locahost ~]# which which
alias which='alias | /usr/bin/which --tty-only --read-alias --show-dot --show-tilde'
/usr/bin/which
```

（3）查询 history 命令的完整文件名。

```
[root@locahost ~]# which history
/usr/bin/which: no history in (/usr/local/sbin:/usr/local/bin:/sbin:/bin:/usr/sbin:/usr/bin:/root/bin)
```

在最后一个应用中，history 这个常用命令找到，是因为 history 是 bash 的内建命令，而 which 默认查找 PATH 内所规范的目录。
2）查找文件的文件名：whereis、find
（1）whereis 命令。用于在一些特定的目录中查询指定命令路径。
语法：

```
whereis [参数] [文件]
```

参数：
- -b：只查找 binary 格式的文件。
- -m：只查找在说明 manual 路径下的文件。

示例如下：
查找 ifconfig 的文件名。

```
[root@locahost ~]# whereis ifconfig
ifconfig: /sbin/ifconfig /usr/share/man/man8/ifconfig.8.gz
```

只查询在 man 里面的 passwd 文件。

```
[root@locahost ~]# whereis -m passwd
passwd: /usr/share/man/man1/passwd.1.gz
```

whereis 命令主要是针对/bin/sbin 目录下的执行文件，以及/usr/share/man 目录下的 man 文件，或者几个特定的目录进行查找，所以速度较快。可以使用"whrereis –l"查看 whereis

查找的目录。

```
[root@locahost ~]# whereis -l
whereis [ -sbmu ] [ -SBM dir ... -f ] name...
[root@locahost ~]#
```

（2）find 命令。

Linux 下 find 命令的作用是在目录结构中搜索文件，并执行指定的操作。Linux 下 find 命令提供了相当多的查找条件，功能非常强大，所以 find 的选项也非常多。本节介绍 find 的选项的功能和简单的 find 使用。

语法：

```
find [查询路径] [选项] [参数] [查询内容]
```

选项：
- -name：按照文件名称查找文件。
- -path：指定目录下文件匹配的路径。
- -type：查找某一类型的文件。

参数：
- b：块设备文件。
- d：目录。
- c：字符设备文件。
- p：管道文件。
- l：符号链接文件。
- f：普通文件。

示例如下：

查找指定时间内修改过的文件。例如，通过 find –atime -1 命令可以查询到一天内修改过的文件。

```
[root@locahost ~]# find -atime -1
.
./.ssh
./.ssh/known_hosts
./.bash_history
./.vimrc
[root@locahost ~]#
```

根据关键字查找。例如，通过 find / -name mysql 命令可以在根目录下全盘搜索 mysql 关键字。

```
[root@locahost ~]# find / -name mysql
/usr/share/mysql
/usr/lib64/mysql
[root@locahost ~]#
```

按类型查找。例如，通过 find / -type f -name *.log 命令可在根目录下查询所有以.log 接尾的普通文件。

```
[root@locahost ~]# find / -type f -name *.log
/root/install.log
[root@locahost ~]#
```

查找当前所有目录并排序。例如，通过 find /tmp –type d | sort 命令可以查询到 tmp 下的所有目录，并排序显示出来。

```
[root@locahost ~]# find /tmp -type d |sort
/tmp
/tmp/.ICE-unix
/tmp/Jetty_0_0_0_0_50070_hdfs____w2cu08
[root@locahost ~]#
```

按文件大小查找文件。

```
[root@locahost ~]# find /tmp -size +1000c -print
/tmp
/tmp/Jetty_0_0_0_0_50070_hdfs____w2cu08
[root@locahost ~]#
```

2.4.2 用户和用户组管理

Linux 系统是一个多用户多任务的操作系统。任何一个要使用系统资源的用户，都必须先向系统管理员申请一个账号，然后使用所申请账号登录系统。

系统管理员用户可以对所有申请的账号的普通用户进行跟踪，并控制他们对系统资源的访问；系统管理员用户也可以帮助用户组织文件，并为用户提供安全性保护。每个用户账号都拥有一个唯一的用户名和口令，用户在登录时输入正确的用户名和口令后，才能够进入系统。

1. Linux 系统用户账号的管理

用户账号的管理工作主要涉及用户账号的添加、修改和删除。

1）添加账号

关于 useradd 命令的讲解视频可扫描二维码观看。

添加新的用户账号使用 useradd 命令。

语法：

useradd [参数] [用户名]

参数：
- -d：指定用户主目录。
- -u：指定用户的用户号。

示例如下：

（1）创建一个用户。通过 useradd hadoop 命令在 Linux 系统中创建一个名字为 hadoop

的用户，系统将在 home 目录下为 hadoop 用户创建一个以自己命名的文件夹。

```
[root@locahost /]# useradd hadoop         <==创建名为hadoop的用户
[root@locahost /]#                         <==未提示任何信息，一般表示创建成功
```

（2）创建一个用户并给它指定家目录地址。useradd –d /home/spark hadoop3 命令在 Linux 系统中创建了一个名字为 hadoop3 的用户，并重新指定 hadoop3 用户的家目录在 /home/spark 目录下。/home/spark 目录不能存在，当执行 useradd –d /home/spark hadoop3 命令时会自动生成该目录。

```
[root@locahost /]# useradd -d /home/spark hadoop3
```

（3）指定用户家目录且创建 hadoop3 用户，可以通过 cat /etc/passwd 命令查看所有用户及用户信息。

```
[root@locahost /] cat /etc/passwd
hadoop:x:500:0::/home/hadoop:/bin/bash
hadoop3:x:501:502::/home/spark:/bin/bash
[root@locahost /]#
```

2）删除账号

关于 userdel 命令的讲解视频可扫描上一页的二维码观看。

删除已有的用户使用 userdel 命令。

如果一个用户的账号不再使用，则可以利用 userdel 命令从系统中删除。删除用户账号相当于在/etc/passwd 文件和相关文件中将指定的用户记录删除，必要时还需要删除用户的主目录。

语法：

```
userdel [参数] [用户名]
```

参数：

-r：把用户的主目录一起删除。

示例如下：

（1）删除用户。通过 userdel hadoop2 命令仅删除 hadoop1 在/etc/passwd 文件中的记录，但主目录并没有删除。

```
[root@locahost home]# userdel hadoop2    <==删除hadoop2用户，不删除其主目录
[root@locahost home]# ls                  <==查看各用户的主目录
hadoop  hadoop2  Master0  spark          <==可知hadoop2用户的主目录未被删除
```

（2）删除用户所有信息。通过 userdel –r hadoop2 命令删除 hadoop2 在/etc/passwd 文件中的记录，并把 hadoop2 用户的主目录一并删除。

```
[root@locahost home]# useradd hadoop2     <==重新创建hadoop2用户
[root@locahost home]# userdel -r hadoop2  <==删除hadoop2及其主目录
[root@locahost home]# ls
```

```
hadoop   Master0   spark                          <==可知hadoop2主目录已删除
```

3）修改账号

修改用户账号使用 usermod 命令。修改用户账号就是根据实际情况更改用户的有关属性，如用户号、主目录、用户组、登录 shell 等。

语法：

```
usermod [参数] [用户名]
```

参数：
- -e：修改账号的有效期限。
- -l：修改用户账号名称。
- -L：锁定用户密码，使用密码无效。
- -U：解除密码锁定。

4）用户口令的管理

passwd 命令用于对用户密码进行管理作用。用户管理主要内容是用户口令的管理，用户账号创建时没有口令，且被系统锁定，无法使用，必须为其指定口令后才可使用（即使口令指定为空）。

指定和修改用户口令的 shell 命令是 passwd，管理用户可以为自己和其他的普通用户指定口令，普通用户只能修改自己的口令。

语法：

```
passwd [参数] [用户名]
```

参数：
- -l：锁定口令、禁用账号。
- -u：口令解锁。
- -d：使账号无口令。
- -f：强迫用户下次登录时修改口令。

如果 passwd 命令没有输入用户，则默认修改当前用户口令。仅管理员用户可指定任何用户的口令，普通用户仅能修改自己的口令。普通用户修改自己口令时，passwd 命令会先询问原口令，验证后再要求用户输入两遍新口令，如果两次输入的口令一致，则将这个新口令指定给用户。而管理员用户为普通用户指定口令时，则不需要知道原口令。为了系统安全起见，用户应该选择比较复杂的口令，如"大写+小写+符号+数字"等组合命令。

passwd –d hadoop2 命令是为 hadoop2 用户指定一个空命令，可以通过 exit 命令退出当前用户，再用之前创建的 hadoop2 用户登录，用 hadoop2 用户登录到 Linux 系统是不需要输入密码的。

```
[root@localhost home]# passwd -d hadoop2        <==删除hadoop2用户的密码
清除用户的密码 hadoop2。
passwd: 操作成功
[root@localhost home]# su hadoop2               <==切换用户hadoop2，不需要密码
```

2. Linux 系统用户组的管理

每个用户都有一个用户组，系统可以对一个用户组中的所有用户进行集中管理。用户组的管理涉及用户组的添加、删除和修改。用户组的添加、删除和修改实际上就是对 /etc/group 文件的更新。

1）groupadd 命令

使用 groupadd 命令可以增加一个新的用户组。

语法：

```
groupadd [参数] [新用户组]
```

参数：

- -g：指定新用户组的组标识（GID）。
- -o：表示新用户组的 GID 可以与系统已有用户组的 GID 相同。

示例如下：

（1）向系统中增加 group1 组，新组的组标识号是在当前已有的最大组标识号的基础上加 1。

```
[hadoop@locahost ~]# groupadd group1      <==添加用户组
[hadoop@locahost ~]#
```

（2）向系统中增加 group2 组，同时指定新组的组标识号是 101。

```
[hadoop@locahost ~]# groupadd -g 101 group2
[hadoop@locahost ~]#
```

2）groupdel 命令

如果要删除一个已有的用户组，可以使用 groupdel 命令。

语法：

```
groupdel [用户组]
```

示例如下：将系统中的 group1 组删除。

```
[hadoop@locahost ~]# groupdel group1
[hadoop@locahost ~]#
```

3）groupmod 命令

如果要修改用户组的属性，可以使用 groupmod 命令。

语法：

```
groupmod [参数] [用户组]
```

参数：

- -g：GID 为用户组指定新的组标识号。
- -n：将用户组的名字改为新名字。

示例如下：
（1）将 group2 组的组标识修改为 102。

```
[hadoop@locahost ~]# exit                          <==退出当前用户，进入root用户
[root@locahost ~]# groupmod -g 102 group2   <==将group2的组标识修改为102
```

（2）将 group2 组的组标识号修改为 10000，并且将 group2 组的组名修改为 group3。

```
[root@locahost ~]#groupmod -g 10000 -n group3 group2
```

4）多组用户

如果一个用户同时属于多个用户组，那么用户可以在多个用户组之间切换，方便具有其他用户组的权限。用户可以在登录后，使用 newgrp 命令切换到其他用户组，这个命令的参数就是目标用户组。

示例如下：

```
[hadoop@localhost ~]$ newgrp root
```

上述命令将当前用户切到 root 用户组，前提条件是 hadoop 用户确实属于 root 组或附加组，类似于用户账号的管理，用户组的管理也可以通过集成的系统管理工具来完成。

3. 与用户账号有关的系统文件

完成用户管理的工作有许多种方法，但每一种方法实际上都是对有关的系统文件进行修改。把用户和用户组相关的信息都存放在一些系统文件中，这些文件包括/etc/passwd、/etc/shadow、/etc/group 等，下面分别介绍这些文件的内容。

1）/etc/passwd

/etc/passwd 是用户数据库，其中的域给出了用户名、加密口令和用户的其他信息。
格式：

```
name: password: uid: gid: comment: home: shell
```

解释：
- name：用户登录名。
- password：用户口令。此域中的口令是加密的，常用 x 表示。当用户登录系统时，系统对输入的口令采取相同的算法，与此域中的内容进行比较。如果此域为空，表明该用户登录时不需要口令。
- uid：指定用户的 UID。用户登录进系统后，系统通过该值，而不是用户名来识别用户。
- gid：如果系统要对相同的一群人赋予相同的权利，则使用该值。
- comment：用来保存用户的真实姓名和个人细节。
- home：指定用户的主目录的绝对路径。
- shell：如果用户登录成功，则要执行的命令的绝对路径放在这一区域中，它可以是任何命令。

示例如下：

```
root: x: 0: 0: root: /root: /bin/bash
```

root 用户记录信息被 6 个 ":" 符号分隔开，一共有 7 个区域，下面将解释这 7 个区域的每个信息。

第一段：用户名。
第二段：加密后的密码。
第三段：UID 用户标识。
第四段：GID 组标识。
第五段：用户命名。
第六段：开始目录。
第七段：对登录命令进行解析的工具。

2）/etc/shadow

/etc/shadow 文件中的记录行与/etc/passwd 中的一一对应，它由 pwconv 命令根据/etc/passwd 中的数据自动产生，它的文件格式与/etc/passwd 类似，由若干个字段组成，字段之间用 ":" 隔开。

格式：

```
name: passwd: 13675: 0: 99999: 7 : : :
```

每一行给一个特殊账户定义密码信息，每个字段用 ":" 隔开。

字段 1：定义与这个 shadow 条目相关联的特殊用户账户。
字段 2：包含一个加密的密码。
字段 3：自 1/1/1970 起，密码被修改的天数。
字段 4：密码将被允许修改之前的天数（0 表示"可在任何时间修改"）。
字段 5：系统将强制用户修改为新密码之前的天数（1 表示"永远都不能修改"）。
字段 6：密码过期之前，用户将被警告过期的天数（-1 表示"没有警告"）。
字段 7：密码过期之后，系统自动禁用账户的天数（-1 表示"永远不会禁用"）。
字段 8：该账户被禁用的天数（-1 表示"该账户被启用"）。
字段 9：保留供将来使用。

示例如下：

```
hadoop: : 17480: 0: 99999: 7 : : :
```

第一段：hadoop 用户名。
第二段：密码为空。
第三段：上次修改密码的时间。
第四段：密码不可被变更的天数。
第五段：密码需要被重新变更的天数，99999 表示不需要变更。
第六段：密码变更前提前几天提醒。
第七段：账号失效时间。
第八段：账号取消时间。
第九段：保留字段。

3）/etc/group

将用户分组是 Linux 系统中对用户进行管理及控制访问权限的一种方式。每个用户都属于某个用户组，一个组中可以有多个用户，一个用户也可以属于多个不同的组。当一个用户同时是多个组中的成员时，/etc/passwd 文件中记录的是用户所属的主组，也就是登录时所属的默认组，而其他组称为附加组。

用户要访问属性附加组的文件时，必须先使用 newgrp 命令使自己成为所要访问组中的成员。用户组的所有信息都存放在/etc/group 文件中，此文件的格式也类似于/etc/passwd 文件，由"："符号隔开若干字段。

格式：

```
name: passwd: gid: list
```

每一行给一个特殊账户定义密码信息，每个字段用"："隔开。
- name：用户名。
- passwd：密码。
- gid：组标识。
- list：组内用户列表。

4. 批量添加用户

添加和删除用户是每位 Linux 系统管理员的必备技能。比较棘手的是如果要添加几十个、几百个甚至几千个用户，不太可能使用 useradd 命令逐一添加，必然要找一种创建大量用户的简便方法。

Linux 系统提供了创建大量用户的接口，可以即刻创建大量用户，步骤如下。

1）编辑一个文本用户文件

每一列按照/etc/passwd 密码文件的格式书写，要注意每个用户的用户名、UID、home 目录都不可以相同，其中密码栏可以留做空白或输入 x 号。

范例文件 user.txt 内容如下：

```
user001:: 601: 100: user: /home/user001: /bin/bash
user002:: 602: 100: user: /home/user002: /bin/bash
user003:: 603: 100: user: /home/user003: /bin/bash
user004:: 604: 100: user: /home/user004: /bin/bash
user005:: 605: 100: user: /home/user005: /bin/bash
...
user00n:: 60n: 100: user: /home/user00n: /bin/bash
```

2）将 user.txt 数据导入 passwd

以 root 身份执行命令/usr/sbin/newusers，从刚创建的用户文件 user.txt 中导入数据，创建用户：

```
[root@localhost ~] # newusers < user.txt
```

然后可以用 cat 或 vi 命令检查/etc/passwd 文件中是否出现刚刚导入的用户信息。然后再查询/home 目录下是否出现用户的家目录。

3）执行命令/usr/sbin/pwunconv

将/etc/shadow 产生的 shadow 密码解码，然后回写到 /etc/passwd 中，并将/etc/shadow

的 shadow 密码栏删掉。这是为了方便"下一步"的密码转换工作,即先取消 shadow password 功能。

```
[root@localhost ~]# pwunconv
```

4)编辑每个用户的密码对照文件

范例文件 passwd.txt 内容如下:

```
user001：密码
user002：密码
user003：密码
user004：密码
user005：密码
…
user00n：密码
```

以 root 身份执行命令/usr/sbin/chpasswd。

创建用户密码,chpasswd 会将经过/usr/bin/passwd 命令编码过的密码写入/etc/passwd 的密码栏。

```
[root@localhost ~]# chpasswd < passwd.txt
```

确定密码经编码写入/etc/passwd 的密码栏后,执行/usr/sbin/pwconv 命令将密码编码为 shadow password 格式,并将结果写入/etc/shadow。

```
[root@localhost ~]# pwconv
```

这样就完成了大量用户的创建,之后回到/home 下检查这些用户的主目录权限设置是否都正确,并登录验证用户密码是否正确。

2.5 习题与思考

1. 尝试独立在虚拟平台中安装 Linux 虚拟系统,配置网络设置,使 XShell 工具能够正常连接 Linux 虚拟机。

2. 尝试在 Linux 系统中使用 tar 命令将/etc 目录压缩成 etc.tar.gz 文件,将 etc.tar.gz 文件剪切到/usr/tmp 目录,并对 etc.tar.gz 文件解压。

3. 在/usr/tmp 中创建 tmpfile.txt 文件,对 tmpfile.txt 设置所有用户均有读、写权限。

4. 尝试在 Linux 系统中创建 Hive 用户,并在 Hive 目录家目录下创建 software 文件夹。

第 3 章　任务命令

3.1　脚本配置

3.1.1　Shell 脚本

Shell 是用 C 语言编写的程序，它是用户使用 Linux 内核的桥梁。Shell 既是一种命令语言，又是一种程序设计语言。Shell 应用程序提供了一个界面，用户通过这个界面可以访问操作内核的服务。关于 Shell 的讲解视频可扫描二维码观看。

Shell 脚本（Shell Script）是一种为 Shell 编写的脚本程序。业界所说的 Shell 通常是指 Shell 脚本，但 Shell 和 Shell Script 是两个不同的概念。

Shell 编程跟 Java、PHP 编程一样，只需要一个能编写代码的文本编辑器和一个能解释执行的脚本解释器。

Linux 的 Shell 种类众多，常见的有：

- Bourne Shell（/usr/bin/sh 或/bin/sh）
- Bourne Again Shell（/bin/bash）
- C Shell（/usr/bin/csh）
- K Shell（/usr/bin/ksh）
- Shell for Root（/sbin/sh）

……

3.1.2　Shell 变量

关于 Shell 变量的讲解视频可扫描二维码观看。

Linux 的 Shell 编程是一种非常成熟的编程语言，它支持各种类型的变量。有三种主要的变量类型：环境变量、局部变量和 Shell 变量。

环境变量：所有的程序，包括 Shell 启动程序，都能访问环境变量。有些程序需要环境变量来保证其正常运行，必要的时候 Shell 脚本也可以自定义环境变量。

局部变量：局部变量是在脚本或命令中定义，仅在当前 Shell 实例中有效，其他 Shell 程序不能访问的局部变量。

Shell 变量：Shell 变量是由 Shell 程序设置的特殊变量。Shell 变量中有一部分是环境变量，有一部分是局部变量，这些变量保证了 Shell 的正常运行。

Shell 编程和其他编程语言的主要不同之处是：在 Shell 编程中，变量是非类型性质的，不必指定变量是数字类型还是字符串类型。

1. 局部变量

Shell 编程中，使用局部变量无须事先声明，同时变量名的命名须遵循如下规则：
- 首个字符必须为字母（a～z，A～Z）。
- 中间不能有空格，可以使用下画线（_）。
- 不能使用标点符号。
- 不能使用 bash 中的关键字（可以用 help 命令查看保留关键字）。

2. 局部变量赋值

变量赋值的格式：

```
变量名=值
```

访问变量值：取用一个变量的值，只需在变量名前面加一个$。
示例如下：

```
#!/bin/bash
# 对变量赋值:
a="hello world"    #等号两边均不能有空格存在
# 打印变量a的值:
echo -e "A is: $a\n"
```

备注：bash 中变量赋值，等号两边均不能有空格存在。

可以使用自己喜欢的编辑器，输入上述内容，并保存为文件 test_hello.bsh，然后执行 chmod +x test_hello.bsh 使其具有执行权限，最后输入"./test_hello"或"bash test_hello.bsh"执行该脚本。

程序运行结果：

```
A is: hello world
```

有时候变量名可能会和其他文字混淆，例如：

```
num=1
echo "this is the $numst"
上述脚本并不会输出"this is the 1st"而是"this is the "，这是由于Shell会去搜索变量
numst的值，而实际上这个变量并未赋值，可以用大括号来告诉 Shell 把 num 变量跟其他部分
分开。num=1
echo "this is the ${num}st"
```

程序运行结果：

```
this is the 1st
```

3.1.3 Shell 传递参数

关于 Shell 传递参数的讲解视频可扫描二维码观看。

1. 普通字符

可以在执行 Shell 脚本时，向脚本传递参数，脚本内获取参数的格式为$n。n 代表一个

数据，n=1 为执行脚本的第一个参数，n=2 为执行脚本的第二个参数，以此类推。

示例如下：以下代码向脚本传递三个参数，并分别输出。

```
#!/bin/bash
echo "Shell传递参数实例!";
echo "第一个参数为$1";
echo "第二个参数为$2";
echo "第三个参数为$3";
```

为脚本设置可执行权限后，并执行脚本，输出结果如下所示：

```
[root@localhost ~]# chmod u+x test.sh
[root@localhost ~]# ./test.sh 1 8 89
```

打印结果：

```
Shell传递参数实例!
第一个参数为：1
第二个参数为：8
第三个参数为：89
```

2. 字符

除普通字符外，还有一些特殊字符可以用来处理参数，如表 3.1 所示。

表 3.1

参数	说明
$#	传递到脚本的参数个数
$*	以一个单字符串显示所有向脚本传递的参数
$$	脚本运行的当前进程 ID 号
$!	后台运行的最后一个进程的 ID 号
$@	与$*相同，但是使用时加引号，并在引号中返回每个参数
$-	显示 Shell 使用的当前选项，与 set 命令功能相同
$?	显示最后命令的退出状态。0 表示没有错误，其他任何值表明有错误

示例如下：

```
#!/bin/bash
echo "Shell传递参数实例!";
echo "第一个参数为$1";
echo "参数个数为$#";
echo "传递的参数作为一个字符串显示：$*";
```

为脚本设置可执行权限后，并执行脚本，输出结果如下所示：

```
[root@localhost ~]# chmod u+x test.sh
[root@localhost ~]# ./test.sh 1 8 89
```

打印结果：

```
Shell 传递参数实例
第一个参数为：1
参数个数为：3
传递的参数作为一个字符串显示：1 8 89
```

3.1.4 Shell 数组

数组中可以存放多个值，但 BASH Shell 只支持一维数据，初始化时不需要定义数据大小。与大部分编程语言类似，数据元素的下标是由 0 开始的。

（1）Shell 数组用括号来表示，元素用"空格"符号分隔开。创建数组的语法格式如下：

```
array_name=（value1 value2 value…）
```

或

```
array_name[0]=value0
array_name[1]=value1
array_name[2]=value2
```

示例如下：

```
#!/bin/bash
my_array=（A B "C" D）
```

（2）读取数组元素值的一般格式如下：

```
[roo@localhost ~]# {array_name[index]}
```

示例如下：

```
#!/bin/bash
my_array={A B "C" D}
echo "第一个元素为: ${my_array[0]}"
echo "第二个元素为: ${my_array[1]}"
echo "第三个元素为: ${my_array[2]}"
echo "第四个元素为: ${my_array[3]}"
```

为脚本设置可执行权限后，执行脚本，输出结果如下所示：

```
[root@localhost ~]# chomod u+x test.sh
[root@localhost ~]# ./test.sh
```

打印结果：

```
第一个元素为：A
第二个元素为：B
第三个元素为：C
第四个元素为：D
```

（3）获取数组中的所有元素。

使用"@"或"*"可以获取数组中的所有元素。

示例如下：

```bash
#!/bin/bash
my_array[0]=A
my_array[2]=B
my_array[3]=C
my_array[4]=D
echo "数组元素个数为：${my_array[*]}"
echo "数组元素个数为：${my_array[@]}"
```

为脚本设置可执行权限后，执行脚本，输出结果如下所示：

```
[root@localhost ~]# chomod u+x test.sh
[root@localhost ~]# ./test.sh
```

打印结果：

```
数组元素个数为：A B C D
数组元素个数为：A B C D
```

（4）获取数组的长度

获取数组长度的方法与获取字符串长度的方法相同，均可使用"${#变量[*]}"或"${#变量[@]}"获取。

示例如下：

```bash
#!/bin/bash
my_array[0]=A
my_array[2]=B
my_array[3]=C
my_array[4]=D
echo "数组元素个数为：${#my_array[*]}"
echo "数组元素个数为：${#my_array[@]}"
```

为脚本设置可执行权限后，执行脚本，输出结果如下所示：

```
[root@localhost ~]# chomod u+x test.sh
[root@localhost ~]# ./test.sh
```

打印结果：

```
数组元素个数为：4
数组元素个数为：4
```

3.1.5　Shell 运算符

关于 Shell 运算符的讲解视频可扫描二维码观看。

Shell 和其他编程语言一样,支持多种运算符,包括:
- 算术运算符;
- 关系运算符;
- 布尔运算符;
- 字符串运算符;
- 文件测试运算符。

原生 Shell BASH 不支持简单的数学运算,但是可以通过加入辅助命令来实现,如:awk 和 expr,其中 expr 比较常用。

expr 是一款表达式计算工具,用来完成表达式的求值操作。例如"val='expr 2 + 2'",执行脚本后输出结果为 4。注意,表达式和运算符之间要有空格,例如"2+2"是错误的书写格式,必须写成"2 + 2"。另外,完整的表达式要被"' '"包含,这个符号不是常用的单引号,而是位于 Esc 键之下的 `~` 符号。

1. 算术运算符

表 3.2 列出了常用的算术运算符,假定变量 a 为 5,变量 b 为 10。

表 3.2

运算符	说明	举例
+	加法	'expr $a + $b'结果为 15
-	减法	'expr $b - $a'结果为 5
*	乘法	'expr $b * $a'结果为 50(注意:乘号前边必须加反斜杠,才能实现乘法运算)
/	除法	'expr $b / $a'结果为 2
%	取余	'expr $b % $a'结果为 0
=	赋值	a=$b 将变量 b 的值赋给 a
==	相等:相同则返回 true	[$a == $b]返回 false
!=	不相等:不相同则返回 true	[$a == $b]返回 true
注意:条件表达式要放在方括号之间,并且要有空格,例如,[$a==$b]是错误的,必须写成[$a == $b]			

示例如下:

```
#!/bin/bash
a=5
b=10
val=`expr $a + $b`
echo "a +b:$val"
val=`expr $b - $a`
echo "b - a:$val"
val=`expr $a \* $b`
echo "a*b:$val"
val=`expr $b \ $a`
echo "a \ b:$val"
```

执行脚本后,打印结果如下:

```
a+b:15
b-a:5
a*b:50
b\a:2
```

2. 关系运算符

关系运算符只支持数字，不支持字符串，除非字符串的值是数字。表 3.3 列出了常用的关系运算符，假定变量 a 为 5，变量 b 为 10。

表 3.3

运算符	说明	举例
-eq	检测两个数是否相等，相等返回 true	[$a -eq $b] 返回 false
-ne	检测两个数是否相等，不相等返回 true	[$a -ne $b] 返回 true
-gt	检测左边的数是否大于右边的，如果是，则返回 true	[$a -gt $b] 返回 false
-lt	检测左边的数是否小于右边的，如果是，则返回 true	[$a -lt $b] 返回 true
-ge	检测左边的数是否大于或等于右边的，如果是，则返回 true	[$a -ge $b] 返回 false
-le	检测左边的数是否小于或等于右边的，如果是，则返回 true	[$a -le $b] 返回 true

示例如下：

```
#!/bin/bash
a=5
b=10
if[$a -eq $b]
then
    echo "a不等于b"
else
    echo "a等于b"
fi
if[$a -le $b]
then
    echo "a小于或等于b"
else
    echo "a不小于或等于b"
fi
```

执行脚本后，打印结果如下：

```
a不等于b
a小于或等于b
```

3. 布尔运算符

表 3.4 列出了常用的布尔运算符，假定变量 a 为 5，变量 b 为 10。

表 3.4

运算符	说明	举例
!	非运算，表达式为 true 则返回 false，否则返回 true	[! false]返回 true
-o	或运算，有一个表达式为 true，则返回 true	[$a -lt 20 -o $b -gt 100] 返回 true
-a	与运算，两个表达式都为 true，才返回 true	[$a -lt 20 -a $b -gt 100] 返回 false

示例如下：

```
#!/bin/bash
a=5
b=10
if[$a -lt 20 -o $b -gt 100]
then
    echo "执行了或操作"
else
    echo "或，操作失败"
fi
if[$a -lt 20 -a $b -gt 100]
then
    echo "执行了并且操作"
else
    echo "并且，操作失败"
fi
```

执行脚本后，打印结果如下：

```
执行了或操作
并且，操作失败
```

4. 字符串运算符

表 3.5 列出了常用的字符串运算符，假定变量 a 为 "abc"，变量 b 为 "efg"。

表 3.5

运算符	说明	举例
=	检测两个字符串是否相等，相等返回 true	[$a = $b] 返回 false
!=	检测两个字符串是否相等，不相等返回 true	[$a != $b] 返回 true
-z	检测字符串长度是否为 0，为 0 返回 true	[-z $a] 返回 false
-n	检测字符串长度是否为 0，不为 0 返回 true	[-n $a] 返回 true
str	检测字符中是否为空，不为空返回 true	[str $a] 返回 true

示例如下：

```
#!/bin/bash
a="abc"
b="efg"
if[$a = $b]
then
    echo "两个字符串相等"
else
    echo "两个字符串不相等"
if[str $a]
then
    echo "字符串不为空"
else
    echo "字符串是空的"
```

```
fi
```

执行脚本后，打印结果如下：

```
两个字符串不相等
字符串不为空
```

5. 文件测试运算符

文件测试运算符如表 3.6 所示。

表 3.6

字符串	说明	举例
-b	检测文件是否是块设备文件，如果是，则返回 true	[-b $file]
-c	检测文件是否是字符设备文件，如果是，则返回 true	[-c $file]
-d	检测文件是否是目录，如果是，则返回 true	[-d $file]
-f	检测文件是否是普通文件，如果是，则返回 true	[-f $file]
-g	检测文件是否设置了 SGID 位，如果是，则返回 true	[-g $file]
-k	检测文件是否设置了 sticky bit，如果是，则返回 true	[-k $file]
-p	检测文件是否是管理，如果是，则返回 true	[-p $file]
-u	检测文件是否设置了 SUID 位，如果是，则返回 true	[-u $file]
-r	检测文件是否可读，如果是，则返回 true	[-r $file]
-w	检测文件是否可写，如果是，则返回 true	[-w $file]
-x	检测文件是否可执行，如果是，则返回 true	[-x $file]
-s	检测文件是否为空，不为空返回 true	[-s $file]
-e	检测文件是否存在，如果是，则返回 true	[-e $file]

3.1.6　Shell echo 命令

Shell 的 echo 指令与 PHP 的 echo 指令类似，都是用于字符串的输出。
语法格式：

```
echo "string"
```

echo 的使用如下。

1. 显示普通字符串

```
echo "www.bunfly.com"
```

这里的双引号可以省略，以下命令与上面实例效果一致：

```
echo www.bunfly.com
```

执行后打印结果如下：

```
www.bunfly.com
```

2. 显示转义字符

通过 \\ 实现转义，示例如下：

```
echo "\"www.bunfly.com"\"
```

执行后打印结果如下:

```
"www.bunfly.com"
```

3. 显示变量

read 命令从标准输入中读取一行,并把输入行的每个字段的值指定给 Shell 变量。

```
#!/bin/sh
read name
eche "$name www.bunfly.com"
```

以上代码保存为 test.sh,name 接收标准输入的变量,结果如下:

```
[root@localhost ~]# sh test.sh
[root@localhost ~]# we http is
[root@localhost ~]# we http is www.bunfly.com
```

4. 显示换行

```
echo -e "www.bunfly.com! \n"
echo "It is a test"
```

执行后打印结果如下:

```
www.bunfly.com!
It is a test
```

其中"-e"的作用是开启转义,并换行。

5. 显示不换行

```
echo -e "www.bunfly.com! \c"
echo "It is a test"
```

执行后打印结果如下:

```
www.bunfly.com!It is a test
```

其中"-e"的作用是开启转义,但不换行。

6. 显示结果重定向至文件

```
echo "www.bunfly.com!" > myfile
```

">"是把 echo 的内容通过覆盖的方式传到 myfile 文件中,而">>"是通过追加的方式把 echo 的内容传到 myfile 文件中。

7. 原样输出字符串

如果不进行转义或取变量,则可以使用单引号。示例如下:

```
echo '$name\" '
```

输出结果：

```
[root@localhost ~]# $name\"
```

8. 显示命令执行结果

显示当前时间，可以用反引号"""。示例如下：

```
echo 'date'
```

输出结果：

```
[root@localhost ~]# Thu Jul 24 08:03:23 CST 2017
```

3.1.7 Shell printf 命令

printf 是 Shell 中的输出命令，类似 Java 中的 print()。printf 使用可以引用文本或空格分隔的参数，外面可以在 printf 中使用格式化字符串，还可以指定字符串的宽度、左右对齐方式等。默认 printf 不会像 echo 自动添加换行符，可以手动添加"\n"符号达到换行目的。

语法格式：

```
printf format-string [arguments]
```

选项或参数：
- format-string：为格式控制字符串。
- arguments：为参数列表。

示例如下：

```
[root@localhost ~]# echo "Hello,Shell"
Hello,Shell
[root@localhost ~]#printf "Hello,Shell"
Hell,Shell[root@localhost ~]#
```

printf 还可以通过格式替代符进行赋值。

示例如下：

```
#!/bin/bash
printf "%-10s %-8s %-4s \n" 姓名 性别 体重
printf "%-10s %-8s %-4s \n" 张三 男 70
printf "%-10s %-8s %-4s \n" 李四 女 50
printf "%-10s %-8s %-4s \n" 王二 男 80
```

执行脚本，输出结果如下：

```
姓名     性别   体重
张三     男     70
李四     女     50
王二     男     80
```

上述示例中的"%-10s""%-8s""%-4s"都是格式替代符。其中"%-10s"表示字符宽度为 10 个字符，任何字符都会被显示在 10 个字符宽的字符内，如果不足则自动以空格填充，超过也会将内容全部显示出来。"\n"表示换行符，将逐行显示数据。

printf 的转义符如表 3.7 所示。

表 3.7

转义符	说明
\a	警告字符
\b	后退
\c	不显示输出结果中任何结尾的换行字符，而且，任何留在参数中的字符、任何接下来的参数以及任何留在格式字符串中的字符都被忽略
\f	换页
\n	换行
\r	回车
\t	空格
\v	垂直制表符
\\	一个字面上的反斜杠字符
\ddd	表示 1～3 位数八进制值的字符，只在格式字符串中有效
\0ddd	表示 1～3 位的八进制值字符

3.1.8 Shell test 命令

Shell 中的 test 命令是用于检查某个条件是否成立的，它可以进行数值、字符和文件三个方面的测试。

1. 数值测试

表 3.8 给出了数值测试的参数。

表 3.8

参数	说明
-eq	等于则为真
-ne	不等于则为真
-gt	大于则为真
-ge	大于或等于则为真
-lt	小于则为真
-le	小于或等于则为真

示例如下：

```
num1=100
num2=100
if test $[num1] -eq $[num2]
then
    echo '两个数相等！'
else
    echo '两个数不相等！'
```

```
fi
```

输出结果:

两个数相等!

代码中可以在"$[ ]"中执行基本的算术运算,如:

```
#!/bin/bash
a=5
b=6
result=$[a+b]  #注意等号两边不能有空格
echo "result为: $result"
```

输出结果:

```
result为: 11
```

2. 字符测试

表 3.9 给出了字符测试的参数。

表 3.9

参数	说明
=	等于则为真
!=	不相等则为真
-z	字符串的长度为零则为真
-n	字符串的长度不为零则为真

示例如下:

```
num1="rulnoob"
num2="runoob"
if test $num1 = $num2
then
    echo '两个字符串相等!'
else
    echo '两个字符串不相等!'
fi
```

输出结果:

两个字符串不相等!

3. 文件测试

表 3.10 给出了文件测试的参数。

表 3.10

参数	说明
-e	如果文件存在,则为真

续表

参数	说明
-r	如果文件存在且可读，则为真
-w	如果文件存在且可写，则为真
-x	如果文件存在且可执行，则为真
-s	如果文件存在且至少有一个字符，则为真
-d	如果文件存在且为目录，则为真
-f	如果文件存在且为普通文件，则为真
-c	如果文件存在且为字符型特殊文件，则为真
-b	如果文件存在且为块特殊文件，则为真

示例如下：

```
cd /bin
if test -e ./bash
then
    echo '文件已存在！'
else
    echo '文件不存在！'
fi
```

输出结果：

文件已存在！

另外，Shell 还提供了"-a""-o""!"三个逻辑操作符用于将测试条件连接起来，其优先级为："!"→"-a"→"-o"。

示例如下：

```
cd /bin
if test -e ./notFile -o -e ./bash
then
    echo '有一个文件存在！'
else
    echo '两个文件都不存在！'
fi
```

输出结果：

有一个文件存在！

3.1.9　Shell 流程控制

关于 Shell 流程控制的讲解视频可扫描二维码观看。

Shell 的流程控制和 Java、PHP 等语言的流程控制有一定区别，Shell 的流程控制不可为空，而 Java 或 PHP 语言的流程控制中可以为空，以 Java 语言为例，其中

Java 语言的流程控制可以用如下写法：

```
if (true) {
    system.out.print("可以正常输出");
} else {
    //不做任何事情
}
```

但在 Shell 中不能这么写，如果 else 分支没有语句执行，就不用写这个 else。

1. if else

if 语法格式如下：

```
    if coundition
then
        command1
        command2
        ...
        commandN
fi
```

如果写成一行，则需要加";"符，用于区分，并且以 if 开始，以 fi 结尾。Shell 语法中大多数据控制语句都是以特定单词开始，以特定单词倒序收尾，例如，if…fi、case…esac 等。

```
if [ $(ps -ef | grep -c "ssh") -gt 1 ]; then echo "true"; fi
```

if else 语法格式如下：

```
if condition
then
        command1
        command2
        ...
        commandN
else
        command
fi
```

if else…if else 语法格式如下：

```
if coundition
then
        command1
elif
        command2
else
        command
fi
```

以下示例判断两个变量是否相等：

```
a=10
b=20
if [ $a -gt $b ]
then
    echo "a大于b"
elif [ $a -lt $b ]
then
    echo "a小于b"
else
    echo "没有符合的条件"
fi
```

输出结果：

```
a小于b
```

2. for 循环

for 循环是 Shell 程序设计中最有用的循环语句之一，一个 for 循环可以用来重复执行某条语句，直到某个条件得到满足。for 语法格式如下：

```
for var in item1 item2 … itemN
do
    command1
    command2
    …
    commandN
done
```

写成一行：

```
for var in item1 intem2 … itemN; do command1; command2; …; commandN; done
```

当变量值在列表里，for 循环即执行一次所有命令，使用变量名获取列表中的当前取值。命令可为任何有效的 Shell 命令和语句。in 列表可以包含替换、字符串和文件名。

in 列表是可选的，如果不用它，则 for 循环使用命令行的位置参数。例如，顺序输出当前列表中的数字：

```
for loop in 1 2 3 4 5
do
    echo "The value is:$loop"
done
```

输出结果：

```
The value is:1
The value is:2
```

```
The value is:3
The value is:4
The value is:5
```

顺序输出字符串中的字符：

```
for str in "This is a string"
do
    echo $str
done
```

输出结果：

```
This is a string
```

3. while 语句

while 语句也称条件判断语句，它的循环方式为利用一个条件来控制是否要继续反复执行这个语句。语法格式如下：

```
while condition
do
    command
done
```

以下是一个基本的 while 循环，测试条件是：如果 int 小于等于 5，那么条件返回真；int 从 0 开始，每次循环处理时，int 加 1；运行以上脚本，返回数字 1 到 5，然后终止。

```
#!/bin/sh
int=1
while(( $int<=5 ))
do
    echo $int
    let "int++"
done
```

运行脚本，输出：

```
1
2
3
4
5
```

代码中使用了 Shell 的 let 命令，它用于执行一个或多个表达式，变量计算中不需要加上$来表示变量。

while 循环可用于读取键盘信息。下面例子中，输入的信息被设置为变量 FILM，按 Ctrl+D 组合键结束循环。

```
echo '按下Ctrl+D键退出循环！'
```

```
echo -n '请输入你喜欢的一个网站： '
while read FILM
do
    echo "是的！$FILM是一个好网站！"
done
```

运行脚本，输出如下结果：

```
按下Ctrl+D键退出循环！
请输入你喜欢的一个网站： www.bunfly.com
是的！www.bunfly.com是一个好网站！
```

4. 无限循环

while…do 循环语句与 while 语句类似，while 是先判断是否为真，如果为真，则执行语句。语法格式如下：

```
while :
do
    command
done
```

或者

```
while true
do
    command
done
```

或者

```
for (( ; ; ))
```

5. until 循环

until 循环执行一系列命令直至条件为真时停止。until 循环与 while 循环在处理方式上相反，一般 while 循环优于 until 循环，但在某些极少数情况下，until 循环更加有用。

until 语法格式如下：

```
until condition
do
    command
done
```

条件可为任意测试条件，测试发生在循环末尾，因此循环至少执行一次。

6. case 语句

Shell case 语句为多分支语句，表达式的值必须是整型、字符型或字符串类型，case 语句首先计算表达式的值，如果表达式的值和某个 ")" 符号前的值一致，则执行后面语句。例如，下面代码中 value 为 1，则执行 D 后面的语句。

case 语法格式如下：

```
case value in
1)
```

```
        command1
        command2
        …
        commandN
        ;;
2)
        command1
        command2
        …
        commandN
        ;;
esac
```

case 工作方式如上例所示，取值后边必须为单词 in，每一选项必须以 ")" 括号结束，取值可以为变量或常数，匹配发现取值符合某一选项后，其间所有命令开始执行到 ";;"。

取值将会检测匹配的每一个选项，一旦选项匹配，则执行完匹配选项相应命令后不再继续其他选项，如果无一匹配选项，则使用 "*" 捕获该值，再执行后面的命令。

下面的脚本提示输入 1 到 4，与每种选项进行匹配，如果匹配，则执行后面的语句。

```
echo '输入1 到4之间的数字：'
ehco '你输入的数字为：'
read aNum
case $aNum in
    1) echo '你选择了1'
    ;;
    2) echo '你选择了2'
    ;;
    3) echo '你选择了3'
    ;;
    4) echo '你选择了4'
    ;;
    *) echo '你没有输入1到4之间的数字'
    ;;
esac
```

输入不同的内容，会有不同的结果。例如：

```
输入1到4之间的数字：
你输入的数字为：
2
你选择了2
```

7. 跳出循环

在循环过程中，有时候需要在未达到循环结束条件时强制跳出循环，Shell 使用两个命令来实现该功能：break 和 continue。

1）break 命令

break 命令允许跳出所有循环，并会终止执行后面的所有循环。

下面的例子中，脚本进入死循环直至用户输入数字大于 5，要跳出这个循环，需要返回到 Shell 提示符下，使用 break 命令。

示例如下：

```bash
#!/bin/bash
while
do
    echo -n "输入1到5之间的数字："
    read aNum
    case $aNum in
        1|2|3|4|5) echo "你输入的数字为$aNum!"
        ;;
        *) echo "你输入的数字不是1到5之间的！游戏结束"
            break
        ;;
    esac
done
```

执行以下代码，输出结果为：

```
输入1到5之间的数字： 3
你输入的数字为3！
输入1到5之间的数字： 38
你输入的数字不是1到5之间的！游戏结束"
```

2）continue 命令

continue 命令与 break 命令类似，但有一点区别，当执行 continue 命令时，结束本次循环，开始下一次循环。

示例如下：

```bash
#!/bin/bash
while
do
    echo -n "输入1到5之间的数字："
    read aNum
    case $aNum in
        1|2|3|4|5) echo "你输入的数字为$aNum!"
        ;;
        *) echo "你输入的数字不是1到5之间的！"
            continue
            echo "游戏结束"
        ;;
    esac
done
```

运行代码会发现，当输入大于 5 的数字时，上例中的循环不会结束，语句"echo"游戏结束""永远不会被执行。

3.2 网络配置

关于网络配置的讲解视频可扫描二维码观看。

在 Linux 中设置网络的相关配置均需要管理员权限，所以在设置网络配置时，需要先把用户切换到 root 用户，输入"su –l root"并输入 root 密码即可切换到 root 用户。

1. 修改 ifcfg-eth0 文件

ifcfg-eth0 文件在/etc/sysconfig/network-scripts/目录中，该文件存放的是网络接口的脚本文件，ifcfg-eth0 是默认的第一个网络接口，如果机器中有多网络接口，那么名字就将以此类推：ifcfg-eth1、ifcfg-eth2、ifcfg-eth3。

ifcfg-eth0 中的文件是相当重要的，关系到网络能否正常工作。ifcfg-eth0 中的设定参数如表 3.11 所示。

表 3.11

项目	设定值	说明
DEVICE		接口名（设备，网卡）
USERCTL	[yes | no]	非 root 用户是否可控制该设备
BOOTPROTO	[none | static | bootp | dhcp]	[引导时不使用协议 | 静态分配 IP | bootp 协议 | 动态协议]
HWADDR		MAC 地址
ONBOOT	[yes | no]	系统启动的时候网络接口是否有效
TYPE	Ethemet	网络类型，通常是 Ethemet
NETMASK		网络掩码
IPADDR		IP 地址
IPV6INIT	[yes | no]	IPv6 是否有效
GATEWAY		默认网关 IP 地址
BROADCAST		广播地址
NETWORK		网络地址

配置静态 IP 地址，示例如下：

```
DEVICE=eth0
HWADDR=00:0C:29:70:75:0B
TYPE=Ethernet
UUID=ba418df8-78dc-496c-9240-907f3851ac5e
ONBOOT=yes
NM_CONTROLLED=yes
BOOTPROTO=static
IPADDR=192.168.2.100
GATEWAY=192.168.2.1
NETMASK=255.255.255.0
```

其中 ONBOOT 和 BOOTPROTO 参数最重要，ONBOOT 设置是否开启网络连接，BOOTPROTO 设置获取 IP 的方式，本文是将虚拟机的 IP 地址设置为静态地址（static）。

```
ONBOOT=yes
BOOTPROTO=static
```

插入 IP 地址、掩码和网关。如果是在 VMware 虚拟平台上配置网络，网关地址可以在 VMware 平台的菜单"编辑"→"虚拟网络编辑器"→VMnet8→"NAT 设置"中查询。

```
IPADDR=192.168.1.100
NETMASK=255.255.255.0
GATEWAY=192.168.1.2
```

配置动态 IP 地址，示例如下：

```
DEVICE=eth0
HWADDR=00:0C:29:70:75:0B
TYPE=Ethernet
UUID=ba418df8-78dc-496c-9240-907f3851ac5e
ONBOOT=yes
NM_CONTROLLED=yes
BOOTPROTO=dhcp
```

设置动态 IP 要比设置静态 IP 简单得多，只需修改 ONBOOT 为 yes，并把 BOOTPROTO 类型改为 dhcp 即可。

需要注意，无论是设置动态 IP 还是设置静态 IP，ifcfg-eth0 中的 DEVICE 和 HWADDR 必须与/etc/udev/rules.d/70-persistent-net.rules 最后一条一致。

例如，/etc/udev/rules.d/70-persistent-net.rules 中的内容如图 3.1 所示。

```
[root@Master001 ~]# cat /etc/udev/rules.d/70-persistent-net.rules
# This file was automatically generated by the /lib/udev/write_net_rules
# program, run by the persistent-net-generator.rules rules file.
#
# You can modify it, as long as you keep each rule on a single
# line, and change only the value of the NAME= key.

# PCI device 0x8086:0x100f (e1000)
SUBSYSTEM=="net", ACTION=="add", DRIVERS=="?*", ATTR{address}=="00:0c:29:89:ba:e8", ATT
R{type}=="1", KERNEL=="eth*", NAME="eth0"
[root@Master001 ~]#
```

图 3.1

ifcfg-eth0 中的 DEVICE 应该为"eth0"，HWADDR 应该为"00:0c:29:89:ba:e8"。

2. 修改 resolv.conf 文件

resolv.conf 文件是 DNS 域名解析的配置文件。它的格式很简单，每行以一个关键字开头，后面接配置参数。

resolv.conf 文件的关键字主要有 4 个，分别如下。

- nameserver：定义 DNS 服务器的 IP 地址。
- domain：定义本地域名。
- search：定义域名的搜索列表。

- sortlist：对返回的域名进行排序。

设置网络配置主要是对 nameserver 关键字进行设置，如果没有指定 nameserver 就找不到 DNS 服务器，也就不能连接外网，但其他关键字是可选的。

配置示例如下：

```
[root@localhost ~]# vi /etc/resolv.conf
```

插入如下：

```
nameserver 192.168.1.2          #自己的网关地址
```

3. 重启网络服务

修改了 IP 地址必须要重启网络服务或者重启计算机才会生效。重启计算机命令可以使用 reboot 也可使用 init 6 等其他命令，重启网络服务同样也有多种命令。

方式一：通过 restart 命令重启。

```
service network restart
```

方式二：先停止再启动。

```
service network stop
service network start
```

如果出现如图 3.2 所示的界面，则表示重启成功。

```
[root@Master001 ~]# service network restart
正在关闭接口 eth0：                                    [确定]
关闭环回接口：                                         [确定]
弹出环回接口：                                         [确定]
弹出界面 eth0：                                        [确定]
[root@Master001 ~]#
```

图　3.2

4. 检查 IP 地址是否修改成功

启动网络服务过后，可以通过 ifconfig 命令查看 IP 地址，如果 IP 地址能查到，并且能正常显示，表示设置成功，如图 3.3 所示。

```
[root@Master001 ~]#
[root@Master001 ~]# ifconfig
eth0      Link encap:Ethernet  HWaddr 00:0C:29:89:BA:E8
          inet addr:192.168.233.101  Bcast:192.168.233.255  Mask:255.255.255.0
          inet6 addr: fe80::20c:29ff:fe89:bae8/64 Scope:Link
          UP BROADCAST RUNNING MULTICAST  MTU:1500  Metric:1
          RX packets:940 errors:0 dropped:0 overruns:0 frame:0
          TX packets:310 errors:0 dropped:0 overruns:0 carrier:0
          collisions:0 txqueuelen:1000
          RX bytes:86499 (84.4 KiB)  TX bytes:34825 (34.0 KiB)

lo        Link encap:Local Loopback
          inet addr:127.0.0.1  Mask:255.0.0.0
          inet6 addr: ::1/128 Scope:Host
          UP LOOPBACK RUNNING  MTU:16436  Metric:1
          RX packets:0 errors:0 dropped:0 overruns:0 frame:0
          TX packets:0 errors:0 dropped:0 overruns:0 carrier:0
          collisions:0 txqueuelen:0
          RX bytes:0 (0.0 b)  TX bytes:0 (0.0 b)
```

图　3.3

5. 验证网络

ping 命令是用于验证网络配置是否成功的最好方法，可以用 ping www.baidu.com 去验证外网是否畅通，也可以用 ping 命令去 ping 本地 VMnet8 IP 检测内外是否连通。需要注意的是，ping 外网时，宿主机一定要联网，因为虚拟机使用的是与宿主机共享的网络地址。

用 ping 命令测试内网。如果出现下面情况，则说明连接成功，可以使用 Ctrl+C 组合键退出测试，如图 3.4 所示。

```
[root@Master001 ~]# ping 192.168.233.1
PING 192.168.233.1 (192.168.233.1) 56(84) bytes of data.
64 bytes from 192.168.233.1: icmp_seq=1 ttl=128 time=0.541 ms
64 bytes from 192.168.233.1: icmp_seq=2 ttl=128 time=0.467 ms
64 bytes from 192.168.233.1: icmp_seq=3 ttl=128 time=0.449 ms
64 bytes from 192.168.233.1: icmp_seq=4 ttl=128 time=0.492 ms
^C
--- 192.168.233.1 ping statistics ---
```

图 3.4

用 ping 命令测试外网。如果出现如图 3.5 所示的情况，则说明连接成功，可以使用 Ctrl+C 组合键退出测试。

```
[root@Master001 ~]#
[root@Master001 ~]# ping www.baidu.com
PING www.a.shifen.com (180.97.33.107) 56(84) bytes of data.
64 bytes from 180.97.33.107: icmp_seq=1 ttl=128 time=35.9 ms
64 bytes from 180.97.33.107: icmp_seq=2 ttl=128 time=44.5 ms
64 bytes from 180.97.33.107: icmp_seq=3 ttl=128 time=34.2 ms
64 bytes from 180.97.33.107: icmp_seq=4 ttl=128 time=39.4 ms
64 bytes from 180.97.33.107: icmp_seq=5 ttl=128 time=42.0 ms
```

图 3.5

3.3 习题与思考

1. 尝试在 Linux 中编写 Shell 脚本，打印出"Hellow WWW.bunfly.com"。
2. 尝试在 Linux 中编写 Shell 脚本，使用脚本程序分别输出两个整数的加、减、乘、除运算的结果。

第 4 章　　数据库操作

4.1　数据库简介

4.1.1　MySQL 数据库简介

关于 MySQL 数据库入门知识的讲解视频可扫描二维码观看。

MySQL 是一种关系型数据库管理系统，由瑞典 MySQL AB 公司开发，目前属于 Oracle 旗下产品。MySQL 是现在最流行的关系型数据库管理系统，在 Web 应用方面，MySQL 是最好的 RDBMS（Relational DataBase Management System，关系数据库管理系统）应用软件之一。

MySQL 是一种关联数据库管理系统。关联数据库将数据保存在不同的表中，而不是将所有数据放在一个"大仓库"内，这种方式增加了速度并提高了灵活性。

MySQL 所使用的 SQL 语言是用于访问数据库的最常用的标准化语言。MySQL 软件采用双授权政策，分为社区版和商业版。由于其体积小、速度快、成本低，尤其是具有开放源码这一优势，因此一般中小型网站的开发都选择 MySQL 作为网站数据库。

MySQL 为关系型数据库，这种所谓的"关系型"可以理解为表格的概念，一个关系数据库由一个或多个表格组成，如图 4.1 所示。

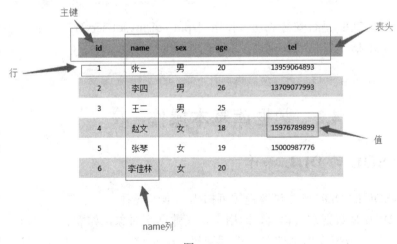

图　4.1

- 表头（header）：每一列的名称。
- 列（column）：具有相同数据类型的数据的集合。

- 行（row）：每一行用来描述某个人或物的具体信息。
- 值（value）：行的具体信息，每个值必须与该列的数据类型相同。
- 键（key）：表中用来识别某个特定的人或物的方法，键的值在当前列中具有唯一性。

本章将介绍 MySQL 在 Linux 环境下的安装和配置，通过介绍 MySQL 的 DDL、DML、DCL、DQL 的常用操作来了解 MySQL 的使用。

4.1.2 安装 MySQL

关于安装 MySQL 的讲解视频可扫描二维码观看。

（1）查看 Linux 系统中是否存在自带数据库。

建议卸载 Linux 系统自带 MySQL 插件，再通过 yum 库安装最稳定的 MySQL 版本。

```
[root@locahost ~]# rpm -qa | grep mysql        <==rpm命令查看软件是否安装
```

（2）卸载 Linux 系统集成的 MySQL 数据库。

卸载分为普通模式和强力模式，强力模式是针对提示有依赖的其他文件时使用。

```
[root@locahost ~]# rpm -e mysql.xx.xx          <==普通删除模式
[root@locahost ~]# rpm -e --nodeps mysql.xx.xx <==强力删除模式
```

（3）通过 yum 来安装 MySQL。

安装前，可以通过"yum list | grep mysql"命令查看 yum 上提供与 MySQL 数据库相关的软件。找到 mysql-server、mysql、mysql-devel 安装包进行安装。

```
[root@locahost ~]#yum list | grep mysql              <==查看yum上的程序
[root@locahost ~]#yum install -y mysql-server        <==安装MySQL-server
[root@locahost ~]#yum install -y mysql               <==安装MySQL
[root@locahost ~]#yum install -y mysql-devel         <==安装MySQL-devel
```

（4）验证 MySQL 是否安装成功。使用"rpm –qi mysql-server"命令，如果出现 MySQL 版本等信息，则表示 MySQL 安装成功。

```
[root@locahost ~]# rpm -qi mysql-server
```

4.2 数据库基本操作

4.2.1 MySQL 的 DDL 操作

关于 MySQL 的 DDL 操作讲解视频可扫描二维码观看。

MySQL DDL 是数据定义语言，即用来定义数据库对象，如库、表、列等，简单说就是对数据库内部的对象进行创建（create）、修改（alter）、删除（drop）的操作语言。

1. 操作数据库

1) create 关键字

create 关键字是 MySQL 中最常用的关键字之一，它可以在 MySQL 中创建数据库，也可以在数据库中创建表。

语法：

```
create database databasename
```

示例如下：

（1）创建一个数据库，数据库名字为 mydb1。

```
mysql> create database mydb1;
```

（2）创建一个数据库，数据库名字为 mydb2，并且指定数据库中的字符集为 utf8。

```
mysql> create database mydb2 character set utf8;
```

2) 查询

（1）查看当前数据库服务器中的所有数据库。

```
mysql> show databases;
```

通过 show databases 命令将查询出 MySQL 下的所有数据库，包括之前创建的 mydb1 和 mydb2 数据库，查询结果如图 4.2 所示。

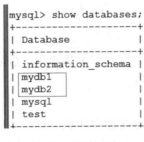

图 4.2

（2）查看之前创建的 mydb2 数据库的定义信息。

```
mysql> show create database mydb2;
```

通过 show create database mydb2 命令可以查询出指定的 mydb2 数据库的定义信息，查询结果如图 4.3 所示。

```
mysql> show create database mydb2;
+----------+-----------------------------------------------------------------+
| Database | Create Database                                                 |
+----------+-----------------------------------------------------------------+
| mydb2    | CREATE DATABASE `mydb2` /*!40100 DEFAULT CHARACTER SET utf8 */  |
+----------+-----------------------------------------------------------------+
1 row in set (0.00 sec)
```

图 4.3

3）修改

通过 alter 关键字可以修改数据库的定义信息。以 mydb2 数据库为例，之前在 MySQL 中创建了 mydb2 数据库，并且给它指定字符集为 utf8，可以通过 alter 将 mydb2 数据库的字符集修改为 gbk。

```
mysql> alter database mydb2 character set gbk;
```

执行 alter database mydb2 character set gbk 代码后，mydb2 的字符集已经从之前的 utf8 改成了 gbk，如图 4.4 所示。

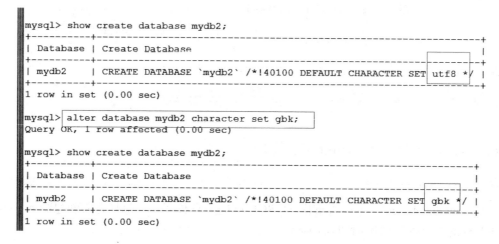

图 4.4

4）删除

可以通过 drop 关键字删除指定数据库。如果出现"Query OK, 0 rows affected (0.00 sec)"，则表示删除成功，如图 4.5 所示。

5）其他

select 关键字可以用于查看数据库的使用状态，如图 4.6 所示。

如果需要切换到其他数据库，则可以使用 use 关键字来实现，如图 4.7 所示。

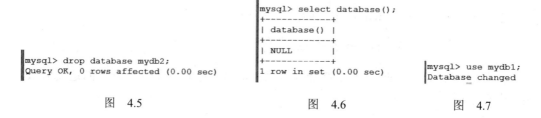

图 4.5　　　　　　　　　　图 4.6　　　　　　　　图 4.7

2. 操作数据表

（1）在 create 关键字后面加上 table 参数，表示创建表语句。

语法：

```
create table tablename (
    column1 type,
```

```
    column2 type,
    …
    columnN type
);
```

MySQL 创建表时可以创建多个列，每个列都有一个数据类型，这些数据类型用来描述列中值的内容。常用的数据类型如表 4.1 所示。

表 4.1

类型	说明
int	整型
double	浮点型
char	固定长度字符串类型
varchar	可变长度字符串类型
text	字符串类型
blob	字节类型
date	日期类型 yyyy-MM-dd
time	时间类型 hh:mm:ss
timestamp	时间戳类型 yyyy-MM-dd hh:mm:ss 会自动赋值
datetime	日期时间类型 yyyy-MM-dd hh:mm:ss

示例如下，在 mydb1 数据库中创建一张 student 表，其中有 id、name、chinese、english、math 五个字段，每个字段类型分别是整数类型、字符串类型和双浮点数类型。

```
create table mydb1.student (
    id int,
    name varchar(20),
    chinese double,
    english double,
    math double
);
```

（2）插入数据可以使用 insert into 关键字来实现，如果需要插入多条数据，则需要在每条语句之间用";"符号分隔。

语法：

```
insert into talbename(column1,column2,…) values(value1,value2,…)
```

列名的名称与插入值需要一一对应（colum1=value1），如果需要全表插入，字段名可以省略，值必须按照建表时的顺序插入。

示例如下，在这里插入了 7 条语句，每条语句用";"隔开。

```
insert into mydb1.student(id,name,chinese,english,math) values(1,'张三',
78.8,98,66);
insert into mydb1.student(id,name,chinese,english,math) values(2,'李四',
88.5,68,96);
```

```
insert into mydb1.student(id,name,chinese,english,math) values(3,'王珊珊',
98,96,96);
insert into mydb1.student(id,name,chinese,english,math) values(4,'张琴',
78,68,66);
insert into mydb1.student(id,name,chinese,english,math) values(5,'李强强',
76,88,86);
insert into mydb1.student(id,name,chinese,english,math) values(6,'陈桌',
58,48,36);
insert into mydb1.student(id,name,chinese,english,math) values(7,'刘星',
88,85,86);
```

或者

```
insert into mydb1.student(id,name,chinese,english,math)
    values(8,'王顾',98,85,96),
    (9,'赵宇',88,75,76),
    (10,'王兴之',68,95,76),
    (11,'张健任',78,95,86);
```

通过 select * from mydb1.student 语句可以查询上述例子中插入的数据内容，从结果可以看出上面执行的两种方式都可以插入数据，如图 4.8 所示。

（3）查询当前数据库中的所有表。

前面介绍了查询数据库的语句，这里将对查询数据库中的表进行阐述。

可以通过 "mysql> use database" 语句进入要查询的数据库，然后再执行 "mysql> show tables" 语句显示该数据库中的所有表，如图 4.9 所示。

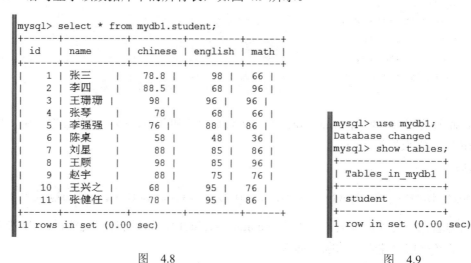

图 4.8　　　　　　　　　　　图 4.9

（4）查看表的字段信息。

desc tablename 语句可以用来查询表字段的详细信息，如图 4.10 所示。

（5）给表添加一列。

假如在使用的过程中发现 student 表中缺少 age 列，则可以使用 alter…add 关键字，给

student 表添加一列，列名为 age，类型为字符串类型 varchar，如图 4.11 所示。

```
mysql> desc mydb1.student;
+---------+-------------+------+-----+---------+-------+
| Field   | Type        | Null | Key | Default | Extra |
+---------+-------------+------+-----+---------+-------+
| id      | int(11)     | YES  |     | NULL    |       |
| name    | varchar(20) | YES  |     | NULL    |       |
| chinese | double      | YES  |     | NULL    |       |
| english | double      | YES  |     | NULL    |       |
| math    | double      | YES  |     | NULL    |       |
+---------+-------------+------+-----+---------+-------+
5 rows in set (0.00 sec)
```

图 4.10

```
mysql> alter table mydb1.student add age varchar(5);
Query OK, 11 rows affected (0.01 sec)
Records: 11  Duplicates: 0  Warnings: 0

mysql> desc mydb1.student;
+---------+-------------+------+-----+---------+-------+
| Field   | Type        | Null | Key | Default | Extra |
+---------+-------------+------+-----+---------+-------+
| id      | int(11)     | YES  |     | NULL    |       |
| name    | varchar(20) | YES  |     | NULL    |       |
| chinese | double      | YES  |     | NULL    |       |
| english | double      | YES  |     | NULL    |       |
| math    | double      | YES  |     | NULL    |       |
| age     | varchar(5)  | YES  |     | NULL    |       |
+---------+-------------+------+-----+---------+-------+
6 rows in set (0.00 sec)
```

图 4.11

（6）修改列。

通过词意得知 age 是年龄，而年龄是以整数形式出现的，这里的 age 使用的却是字符串类型，不符合类型规范，如果要修改列的属性，可以使用 alter…modify 关键字将 age 字段的 varchar 类型修改为 int 类型。

语法：

```
alter table tablename modify column type
```

示例如图 4.12 所示。

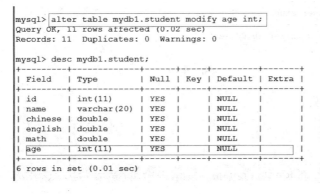

图 4.12

(7)删除列。

如果不需要使用某列,则可以用 alter…drop 关键词删除这一列。

语法:

```
alter table tablename drop cloumn
```

示例如图 4.13 所示。

```
mysql> alter table mydb1.student drop age;
Query OK, 11 rows affected (0.02 sec)
Records: 11  Duplicates: 0  Warnings: 0

mysql> desc mydb1.student;
+---------+-------------+------+-----+---------+-------+
| Field   | Type        | Null | Key | Default | Extra |
+---------+-------------+------+-----+---------+-------+
| id      | int(11)     | YES  |     | NULL    |       |
| name    | varchar(20) | YES  |     | NULL    |       |
| chinese | double      | YES  |     | NULL    |       |
| english | double      | YES  |     | NULL    |       |
| math    | double      | YES  |     | NULL    |       |
+---------+-------------+------+-----+---------+-------+
5 rows in set (0.01 sec)
```

图 4.13

(8)更改表名。

将 student 表的名字重命名为 user_student,可以使用 rename table…to 关键字来实现,示例如图 4.14 所示。

```
mysql> show tables;
+------------------+
| Tables_in_mydb1 |
+------------------+
| student          |
+------------------+
1 row in set (0.00 sec)

mysql> rename table mydb1.student to user_student;
Query OK, 0 rows affected (0.00 sec)

mysql> show tables;
+------------------+
| Tables_in_mydb1 |
+------------------+
| user_student     |
+------------------+
1 row in set (0.00 sec)
```

图 4.14

(9)修改表的字符集。

示例如图 4.15 所示。

(10)更改列名。

```
alter table user_student change name username varchar(20);
```

- user_student:需要修改列的表。

- name：需要修改的列名。
- username varchar(20)：新的列名和类型。

示例如图 4.16 所示。

```
mysql> alter table user_student set gbk;
ERROR 1064 (42000): You have an error in your SQL syntax;
onds to your MySQL server version for the right syntax to
mysql> alter table user_student character set gbk;
Query OK, 11 rows affected (0.01 sec)
Records: 11  Duplicates: 0  Warnings: 0

mysql> show create table mydb1.user_student;
+--------------+------------------------------------------------
----------------------------------------------------------------
----------------------------------------------------------------
--+
| Table        | Create Table
            |
+--------------+------------------------------------------------
----------------------------------------------------------------
----------------------------------------------------------------
--+
| user_student | CREATE TABLE `user_student` (
  `id` int(11) DEFAULT NULL,
  `name` varchar(20) CHARACTER SET latin1 DEFAULT NULL,
  `chinese` double DEFAULT NULL,
  `english` double DEFAULT NULL,
  `math` double DEFAULT NULL
) ENGINE=MyISAM DEFAULT CHARSET=gbk |
+--------------+------------------------------------------------
```

图 4.15

```
mysql> alter table user_student change name user_name varchar(20);
Query OK, 11 rows affected, 11 warnings (0.02 sec)
Records: 11  Duplicates: 0  Warnings: 0

mysql> desc user_student;
+-----------+-------------+------+-----+---------+-------+
| Field     | Type        | Null | Key | Default | Extra |
+-----------+-------------+------+-----+---------+-------+
| id        | int(11)     | YES  |     | NULL    |       |
| user_name | varchar(20) | YES  |     | NULL    |       |
| chinese   | double      | YES  |     | NULL    |       |
| english   | double      | YES  |     | NULL    |       |
| math      | double      | YES  |     | NULL    |       |
+-----------+-------------+------+-----+---------+-------+
5 rows in set (0.00 sec)
```

图 4.16

（11）添加主键约束。

表中的每一行都应该具有可以唯一标识自己的一列，而这个承担标识作用的列称为主键。如果没有主键，则数据的管理会十分混乱，例如，若存在多条一模一样的记录，则删除和修改特定行十分困难。设置主键可以使用 alter table…add constraint primary key(…)语句，如把 id 列设置成主键。

```
mysql> alter table user_student add constraint primary key(id);
```

示例如图 4.17 所示。

图 4.17

（12）删除表。

```
mysql> drop table user_student;
```

示例如图 4.18 所示。

图 4.18

4.2.2 MySQL 的 DML 操作

关于 MySQL 的 DML 操作讲解视频可扫描二维码观看。

DML 是对表中的数据进行插入（insert）、修改（update）、删除（delete）等操作。

（1）在 mydb1 数据库中创建一张 student 表，操作如下：

```
create table mydb1.student (
    id int,
    name varchar(20),
    chinese double,
    english double,
    math double
);
```

（2）数据插入（insert），操作如下：

```
insert into mydb1.student(id,name,chinese,english,math) values(1,'张三',
```

```
78.8,98,66);
insert into mydb1.student(id,name,chinese,english,math) values(2,'李四',
88.5,68,96);
insert into mydb1.student(id,name,chinese,english,math) values(3,'王珊珊',
98,96,96);
insert into mydb1.student(id,name,chinese,english,math) values(4,'张琴',
78,68,66);
insert into mydb1.student(id,name,chinese,english,math) values(5,'李强强',
76,88,86);
insert into mydb1.student(id,name,chinese,english,math) values(6,'陈桌',
58,48,36);
insert into mydb1.student(id,name,chinese,english,math) values(7,'刘星',
88,85,86);
insert into mydb1.student(id,name,chinese,english,math)
    values(8,'王顾',98,85,96),
    (9,'赵宇',88,75,76),
    (10,'王兴之',68,95,76),
    (11,'张健任',78,95,86);
```

（3）查看 student 表中的数据，操作如下：

```
mysql> select * from mydb1.student;
```

示例如图 4.19 所示。

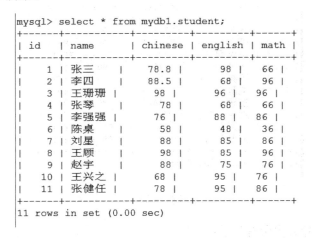

图 4.19

（4）修改（update）。假如发现张健任的语文成绩过低，经核实张健任的语文成绩是 98 分，确实是录入错误，这时可以用 update 语句修改张健任的语文分数，示例如图 4.20 所示。

（5）删除（delete）。

语法：

```
delete from tablename [where cloumn=value]
```

```
mysql> update mydb1.student set chinese=98 where id=11;
Query OK, 1 row affected (0.01 sec)
Rows matched: 1  Changed: 1  Warnings: 0

mysql> select * from mydb1.student;
+------+-----------+---------+---------+------+
| id   | name      | chinese | english | math |
+------+-----------+---------+---------+------+
|    1 | 张三      |    78.8 |      98 |   66 |
|    2 | 李四      |    88.5 |      68 |   96 |
|    3 | 王珊珊    |      98 |      96 |   96 |
|    4 | 张琴      |      78 |      68 |   66 |
|    5 | 李强强    |      76 |      88 |   86 |
|    6 | 陈桌      |      58 |      48 |   36 |
|    7 | 刘星      |      88 |      85 |   86 |
|    8 | 王顾      |      98 |      85 |   96 |
|    9 | 赵宇      |      88 |      75 |   76 |
|   10 | 王兴之    |      68 |      95 |   76 |
|   11 | 张健任    |      98 |      95 |   86 |
+------+-----------+---------+---------+------+
11 rows in set (0.00 sec)
```

图 4.20

① 删除表中序号为"5"的学生记录，如图 4.21 所示。

```
mysql> delete from mydb1.student where id=5;
Query OK, 1 row affected (0.00 sec)

mysql> select * from mydb1.student;
+------+-----------+---------+---------+------+
| id   | name      | chinese | english | math |
+------+-----------+---------+---------+------+
|    1 | 张三      |    78.8 |      98 |   66 |
|    2 | 李四      |    88.5 |      68 |   96 |
|    3 | 王珊珊    |      98 |      96 |   96 |
|    4 | 张琴      |      78 |      68 |   66 |
|    6 | 陈桌      |      58 |      48 |   36 |
|    7 | 刘星      |      88 |      85 |   86 |
|    8 | 王顾      |      98 |      85 |   96 |
|    9 | 赵宇      |      88 |      75 |   76 |
|   10 | 王兴之    |      68 |      95 |   76 |
|   11 | 张健任    |      98 |      95 |   86 |
+------+-----------+---------+---------+------+
10 rows in set (0.00 sec)
```

图 4.21

② 用 delete 删除表中所有记录：

```
mysql> delete from mydb1.student;
```

③ 用 truncate 删除表中所有记录：

```
mysql> truncate table mydb1.student;
```

delete 和 truncate 都可以删除表中的数据。delete 删除表中的数据，表结构还在，删除后的数据还可以找回。truncate 删除是把表全部 drop 掉，然后再创建一个同样的新表，删除的数据不能找回，但它的执行速度比 delete 快。

4.3　数据库用户操作

4.3.1　创建用户

语法：

```
create user username@localhost identified by 'passwd'
```

解释：

- username：用户名。
- localhost：指定远程可以访问数据库的地址。“%”代表所有人都可以访问该数据库。localhost 代表只能本机访问。
- passwd：用户密码。

示例如图 4.22 所示。

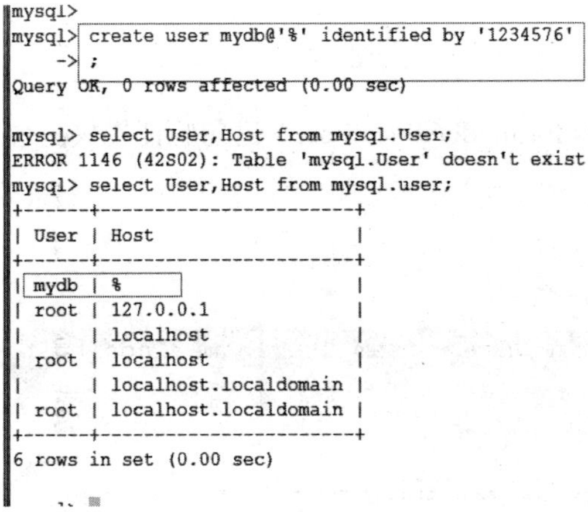

```
mysql>
mysql> create user mydb@'%' identified by '1234576'
    -> ;
Query OK, 0 rows affected (0.00 sec)

mysql> select User,Host from mysql.User;
ERROR 1146 (42S02): Table 'mysql.User' doesn't exist
mysql> select User,Host from mysql.user;
+------+-----------------------+
| User | Host                  |
+------+-----------------------+
| mydb | %                     |
| root | 127.0.0.1             |
|      | localhost             |
| root | localhost             |
|      | localhost.localdomain |
| root | localhost.localdomain |
+------+-----------------------+
6 rows in set (0.00 sec)
```

图　4.22

可以通过“select user,Host from mysql.user”语句验证用户创建是否成功。

4.3.2　给用户授权

语法：

```
grant 权限1，…，权限n on *.* to user_name@localhost
```

解释：

① 权限。

- all：全部权限。
- create：赋予用户创建表的权限。
- alter：赋予用户修改表的权限。
- drop：赋予用户删除表的权限。

- insert：赋予用户插入数据的权限。
- update：赋予用户更新数据的权限。
- 其他权限。

② *.*：第一个"*"代表该用户中的所有数据库，第二个"*"代表该用户中的所有表。

③ user_name：用户名。

④ localhost：指定远程可以访问数据库的地址。"%"代表所有人都可以访问该数据库，localhost 代表只能本机访问。

示例如图 4.23 所示。

图 4.23

可以通过"show grants for mydb@localhost"语句验证授权是否成功。

4.3.3 撤销授权

语法：

revoke 权限1，…，权限n on *.* from user_name@localhost

示例如图 4.24 所示。

图 4.24

从上述示例可以看出，mydb 用户的权限通过 revoke…from 语句撤销后，mydb 用户的 ALL PRIVILEGES 权限变成了 USAGE 权限。

4.3.4 查看用户权限

我们在之前已经使用过查看用户权限语句。

语法：

```
show grants for user_name@localhost
```

解释：
- user_name：用户名。
- localhost：指定远程可以访问数据库的地址。"%"代表所有人都可以访问该数据库。"localhost"代表只能本机访问。

示例如图 4.25 所示。

```
mysql> show grants for mydb@'%';
+----------------------------------------------------------------+
| Grants for mydb@%                                              |
+----------------------------------------------------------------+
| GRANT USAGE ON *.* TO 'mydb'@'%' IDENTIFIED BY
 A46E58C67E71F' |
+----------------------------------------------------------------+
1 row in set (0.00 sec)
```

图 4.25

4.3.5 删除用户

语法：

```
drop user user_name@localhost
```

- user_name：用户名。
- localhost：指定远程可以访问数据库的地址。"%"代表所有人都可以访问该数据库。"localhost"代表只能本机访问。

示例如图 4.26 所示。

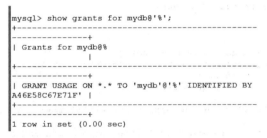

图 4.26

4.3.6 修改用户密码

修改用户密码操作是不常使用的操作,但是如果不小心忘记了密码就需要用到此操作,修改用户密码分为两步:第一步,修改 mysql 数据库下的 user 表中的指定用户的密码;第二步,修改完 user 表后必须执行 flush privileges 重新加载权限表,修改的密码才会生效。

语法:

```
update mysql.user set password=password('passwd') where User='new_username';
flush privileges;
```

示例如图 4.27 所示。

```
mysql> update mysql.user set password=password('bunfly') where user='mydb';
Query OK, 1 row affected (0.00 sec)
Rows matched: 1  Changed: 1  Warnings: 0

mysql> flush privileges;
Query OK, 0 rows affected (0.00 sec)

mysql> exit;
Bye
[root@localhost ~]# mysql -u mydb -pbunfly
Welcome to the MySQL monitor.  Commands end with ; or \g.
Your MySQL connection id is 6
Server version: 5.1.73 Source distribution

Copyright (c) 2000, 2013, Oracle and/or its affiliates. All rights reserved.

Oracle is a registered trademark of Oracle Corporation and/or its
affiliates. Other names may be trademarks of their respective
owners.

Type 'help;' or '\h' for help. Type '\c' to clear the current input statement.

mysql>
```

图 4.27

上述示例使用"update mysql.user set password=password('bunfly') where user='mydb'"语句修改了 mydb 用户的密码,并把密码修改为"bunfly",用掩密方式存储在 user 表中,通过"flush privileges"语句将修改过的 user 表重新加载到权限表,使新密码生效。通过 exit 退出当前登录 mysql 的用户,然后用新用户登录验证密码是否修改成功。

4.4 数据库查询操作

DQL 操作是对数据的操作,执行 DQL 语句不会对数据进行改变,而是让数据库发送结果集给客户端。练习数据参照 4.2.2 节中的 student 表数据。

关于 DQL 的讲解视频可扫描二维码观看。

查询关键字:select

语法:

```
select column1,column2,… from tablename [where | group by | having | order
```

by | limit]

解释：
- column：列名。
- tablename：表名。
- where：条件。
- group by：对结果分组。desc 为降序，asc 为升序（默认）。
- having：分组后的条件。
- order by：对结果排序。
- limit：结果限定。

1. 基础查询

（1）查询所有列可以使用通配符"*"表示，也可以列出所有字段。

```
mysql> select * from mydb1.student;
```

或

```
mysql> select id,name,chinese,english,math from mydb1.student;
```

（2）查询指定列只需要将需要查询的字段列出来即可。

例如，需要查询 id、name 列，代码如下：

```
mysql> select id,name from mydb1.student;
```

2. 条件查询

（1）条件查询就是在查询时给出 where 关键字，在 where 子句中可以使用如表 4.2 所示的运算符及关键字。

表 4.2

运算符和关键字	说明
=	等于
!= 或 <>	不等于
<	小于
<=	小于或等于
>	大于
>=	大于或等于
between…and	在多少到多少数字之间
in	在什么内
is null	是空
and	并且
or	或者
not	不…

（2）查询"语文"成绩小于 80 分的所有学生，如图 4.28 所示。

```
mysql> select name,chinese from mydb1.student where chinese<80;
+-----------+---------+
| name      | chinese |
+-----------+---------+
| 张三      |    78.8 |
| 张琴      |      78 |
| 陈桌      |      58 |
| 王兴之    |      68 |
+-----------+---------+
4 rows in set (0.00 sec)
```

图 4.28

（3）查询"语文"成绩小于 80 分，或者"数学"成绩小于 70 分的所有学生，如图 4.29 所示。

```
mysql> select name,chinese,math from mydb1.student where chinese<80 or math<70;
+-----------+---------+------+
| name      | chinese | math |
+-----------+---------+------+
| 张三      |    78.8 |   66 |
| 张琴      |      78 |   66 |
| 陈桌      |      58 |   36 |
| 王兴之    |      68 |   76 |
+-----------+---------+------+
4 rows in set (0.00 sec)
```

图 4.29

（4）查询"语文"成绩小于 80 分，并且"数学"成绩小于 70 分的所有学生，如图 4.30 所示。

```
mysql> select name,chinese,math from mydb1.student where chinese<80 and math<70;
+--------+---------+------+
| name   | chinese | math |
+--------+---------+------+
| 张三   |    78.8 |   66 |
| 张琴   |      78 |   66 |
| 陈桌   |      58 |   36 |
+--------+---------+------+
3 rows in set (0.00 sec)
```

图 4.30

（5）查询"语文"成绩在 80～100 分的所有学生。
方式一：如图 4.31 所示。

```
mysql> select * from mydb1.student where chinese>=80 and chinese <=100;
+------+----------+---------+---------+------+
| id   | name     | chinese | english | math |
+------+----------+---------+---------+------+
|    2 | 李四     |    88.5 |      68 |   96 |
|    3 | 王珊珊   |      98 |      96 |   96 |
|    7 | 刘星     |      88 |      85 |   86 |
|    8 | 王顾     |      98 |      85 |   96 |
|    9 | 赵宇     |      88 |      75 |   76 |
|   11 | 张健任   |      98 |      95 |   86 |
+------+----------+---------+---------+------+
6 rows in set (0.00 sec)
```

图 4.31

方式二：如图 4.32 所示。

```
mysql> select * from mydb1.student where chinese between 80 and 100;
+------+-----------+---------+---------+------+
| id   | name      | chinese | english | math |
+------+-----------+---------+---------+------+
|    2 | 李四      |    88.5 |      68 |   96 |
|    3 | 王珊珊    |      98 |      96 |   96 |
|    7 | 刘星      |      88 |      85 |   86 |
|    8 | 王顾      |      98 |      85 |   96 |
|    9 | 赵宇      |      88 |      75 |   76 |
|   11 | 张健任    |      98 |      95 |   86 |
+------+-----------+---------+---------+------+
6 rows in set (0.00 sec)
```

图 4.32

（6）查询所有学生的成绩信息，并按"语文"成绩从高到低排序，如图 4.33 所示。

```
mysql> select * from mydb1.student order by chinese desc;
+------+-----------+---------+---------+------+
| id   | name      | chinese | english | math |
+------+-----------+---------+---------+------+
|   11 | 张健任    |      98 |      95 |   86 |
|    3 | 王珊珊    |      98 |      96 |   96 |
|    8 | 王顾      |      98 |      85 |   96 |
|    2 | 李四      |    88.5 |      68 |   96 |
|    9 | 赵宇      |      88 |      75 |   76 |
|    7 | 刘星      |      88 |      85 |   86 |
|    1 | 张三      |    78.8 |      98 |   66 |
|    4 | 张琴      |      78 |      68 |   66 |
|   10 | 王兴之    |      68 |      95 |   76 |
|    6 | 陈桌      |      58 |      48 |   36 |
+------+-----------+---------+---------+------+
10 rows in set (0.00 sec)
```

图 4.33

（7）查询总分排名前三的学生信息，示例如图 4.34 所示。

```
mysql> select * from
    -> (select id,name,(chinese+english+math)score_sum from mydb1.student) t
    -> order by t.score_sum
    -> limit 3;
+------+--------+-----------+
| id   | name   | score_sum |
+------+--------+-----------+
|    6 | 陈桌   |       142 |
|    4 | 张琴   |       212 |
|    9 | 赵宇   |       239 |
+------+--------+-----------+
3 rows in set (0.00 sec)
```

图 4.34

上述示例是通过"select id,name,(chinese+english+math)score_sum from mydb1.student"语句对三门成绩做求和操作，并把结果作为一个新的"临时表"，然后在这个"临时表"的基础上再进行统计。

4.5 习题与思考

1. 在 Linux 系统 MySQL 中创建 bunfly 数据库，并按表 4.3 和表 4.4 要求分别创建 Student 表和 Score 表。

表 4.3

字段名	字段描述	数据类型	主键	外键	非空
Name	姓名	VARCHAR(20)	否	否	是
Sex	性别	VARCHAR(4)	否	否	否
Birth	出生年份	VARCHAR(10)	否	否	否
Department	院系	VARCHAR(20)	否	否	是
Address	家庭住址	VARCHAR(50)	否	否	否
Stu_id	学号	INT（10）	否	否	是

表 4.4

字段名	字段描述	数据类型	主键	外键	非空
Stu_id	学号	INT(10)	否	否	是
C_name	课程名	VARCHAR(20)	否	否	否
Grade	分数	INT(10)	否	否	否

2. 按表 4.5 和表 4.6 内容为 Student 表和 Score 表增加内容。

表 4.5

姓名	性别	出生年份	院系	家庭住址	学号
刘一	男	1994/11/26	数学系	上海	20157259
陈二	男	1993/6/11	数学系	北京	20153174
张三	女	1994/9/21	数学系	北京	20157824
李四	男	1993/1/26	信息工程系	云南	20155367
王五	男	1993/9/18	信息工程系	贵州	20152240
赵六	女	1994/9/26	信息工程系	贵州	20152229
孙七	女	1994/11/5	信息工程系	重庆	20154781
周八	女	1995/1/29	信息工程系	云南	20157395
吴九	男	1994/9/3	信息工程系	浙江	20155985
郑十	女	1994/12/2	机电工程系	山西	20155846

表 4.6

学号	课程名	分数
20157259	语文	90
20157259	数学	58
20157259	外语	39
20152240	语文	91
20152240	数学	95
20152240	外语	75
20152229	语文	60

续表

学号	课程名	分数
20152229	数学	58
20152229	外语	53
20155985	语文	62
20155985	数学	43
20155985	外语	74

3. 尝试查询出数学系数学成绩不及格的学生。
4. 尝试查询出信息工程系所有学生的外语成绩。

第 2 篇　Hadoop 技术

第 5 章　Hadoop 开发环境

5.1　Hadoop 生态圈工具

开发人员通常要使用开发工具，不同的工具有不同的用途。例如，他们可能会用 Eclipse 进行代码编写，用 MySQL 或者 Oracle 进行数据存储，这里的工具都有特定的用处。然而 Hadoop 生态圈就变得有点复杂，那 Hadoop 生态圈是什么呢？

根据官方描述，Hadoop 是一个由 Apache 基金会所开发的分布式系统基础架构，用户可以在不了解分布式底层细节的情况下去开发分布式程序，充分利用集群的威力进行数据存储和数据计算。Hadoop 的框架最核心的设计是 HDFS 和 MapReduce。HDFS 为海量的数据提供了分布式存储，MapReduce 是对数据进行处理。Hadoop 框架是 Hadoop 生态圈的基础，许多工具都是基于 Hadoop 框架中分布式文件存储系统的基础而运行的，包括 Hive、Pig、HBase 等工具，所以通常人们说 Hadoop 就是指 Hadoop 生态圈。

Hadoop 生态圈好比一个厨房，需要各种工具。锅、碗、瓢、盆各有各的用处，相互之间又有重合。例如，可以用汤锅当碗来吃饭喝汤，也可以用小刀削土豆皮，每个工具都有自己的特性。虽然都可以达到最终目的，但是这种工具未必是最佳的选择。Hadoop 生态圈中的这些工具就像是厨房里的各种工具，它们都是基于"厨房"（Hadoop 框架）存在的，它们在这个圈里发挥各自的作用，组成了 Hadoop 生态系统。

Hadoop 生态系统组件如图 5.1 所示。

图　5.1

1. Hadoop

Hadoop 是一个由 Apache 基金会所开发的分布式系统基础架构。用户可以在不了解分布式底层细节的情况下，开发分布式程序，充分利用集群的威力进行高速运算和存储。Hadoop 实现了一个分布式文件系统（Hadoop Distributed File System，HDFS）。HDFS 有高容错性的特点，设计用来部署在低廉的 PC（Personal Computer）上，而且它提供高吞吐量访问应用程序的数据，适合有超大数据集的应用程序。

2. Hive

Hive 是基于 Hadoop 的一个数据仓库工具，可以将结构化的数据文件映射为一张数据库表，并提供类似 SQL 的查询功能，本质是将 SQL 转换为 MapReduce 程序。

Hive 的特点如下。

- 可扩展性：Hive 可以自由地扩展集群的规模，一般情况下不需要重启服务。
- 延展性：Hive 支持用户自定义函数，用户可以根据自己的需求来实现自己的函数。
- 容错性：节点出现问题 SQL 仍可完成执行，容错性良好。

3. Pig

Pig 是一种数据流语言和运行环境，用于检索非常大的数据集。它是一个高级过程语言，适合于使用 Hadoop 和 MapReduce 平台来查询大型半结构化数据集。通过脚本语言方式简化了 Hadoop 的使用，Pig 为复杂的海量数据并行计算提供了一个简单的操作和编程接口。

4. Sqoop

Sqoop 是一款开源的工具，主要用于在 Hadoop 与关系型数据库（MySQL、Oracle、…）间进行数据的传递。Sqoop 项目开始于 2009 年，最早是作为 Hadoop 的一个第三方模块存在的，后来为了实现使用者能够快速部署，以及开发人员能够更快速地迭代开发，Sqoop 独立成为一个 Apache 项目。

5. Flume

Flume 是 Apache 公司提供的一个高可用的、高可靠的、分布式的海量日志采集、聚合和传输的系统，Flume 支持在日志系统中定制各类数据发送方，用于收集数据。同时，Flume 提供对数据进行简单处理，并写到各种数据接收方的能力。

6. Zookeeper

Zookeeper 是一个开放源码的分布式应用程序协调服务，是 Google 公司的 Chubby 的一个开源的实现，是 Hadoop 和 Hbase 的重要组件，是一个为分布式应用提供一致性服务的软件。提供的功能包括配置维护、名字服务、分布式同步、组服务等，目标就是封装好复杂易出错的关键服务，将简单易用的接口和性能高效、功能稳定的系统提供给用户。

7. HBase

HBase 是一个开源的非关系型分布式数据库（NoSQL）。它参考了 Google 公司的 BigTable 建模，实现的编程语言为 Java。它是 Apache 软件基金会 Hadoop 项目的一部分，运行于 HDFS 文件系统之上，为 Hadoop 提供类似于 BigTable 规模的服务。HBase 在列上实现了 BigTable 论文提到的压缩算法、内存操作和布隆过滤器。HBase 的表能够作为 MapReduce 任务的输入和输出，可以通过 Java API 来存取数据，也可以通过 REST、Avro 或者 Thrift 的 API 来访问。HBase 弥补了 Hive 不能随机读写的缺陷。

5.2 环境搭建

0.23 版本之前与之后的 Hadoop 集群都存在单点问题，其中，NameNode 的单点问题尤为严重。由于 NameNode 保存了整个 HDFS 的元数据信息，因此一旦 NameNode 出现故障，整个 HDFS 就无法访问。同时 Hadoop 生态系统中依赖 HDFS 的各个组件，包括 MapReduce、Hive、Pig、HBase 等工具，也都无法正常工作，并且重新启动 NameNode 和进行数据恢复的过程也会比较耗时。这些问题在给 Hadoop 的使用者带来困扰的同时，也极大地限制了 Hadoop 的使用场景。所幸在 Hadoop 2.0 中 NameNode 和 ResourceManager 的单点问题都得到了解决，经过多个版本的迭代和发展，目前已经能用于生产环境。

生产环境中的 Hadoop 大数据集群是由多台服务器组成的集群。为了方便学习和教学，这里采用在 VMware 平台中搭建虚拟机的方式模拟 Hadoop 大数据集群。要想实现 Hadoop 大数据集群高可用环境，最少需要 5 台虚拟机，其中两台 Master 节点（一台作为活动状态（active），另一台作为备用状态（standby））为集群管理节点，三台 Slave 节点做数据存储。Hadoop 高可用大数据集群每台虚拟机所需要的软件如图 5.2 所示。

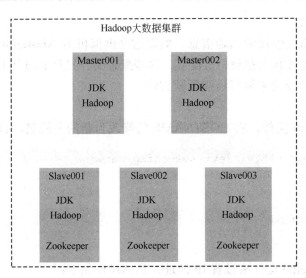

图 5.2

安装高可用大数据集群步骤分为安装虚拟机、安装 JDK、安装 Hadoop、复制虚拟机、设置免密、安装 Zookeeper 和启动 Hadoop 集群等。Linux 系统的虚拟机安装请参考 2.1 节的系统安装。

5.2.1 步骤1——虚拟机安装

虚拟机安装步骤见 2.1.2 节。

5.2.2 步骤2——安装 JDK 和 Hadoop

关于 JDK 和 Hadoop 在 Linux 环境下安装的讲解视频可分别扫描以下三个二维码观看。

为了解决 Hadoop 的高可用问题，Zookeeper 应运而生。高可用分为两种：数据高可用和服务高可用。为了实现高可用必须有两台 Master 节点和多台 Slave 节点（其中 IP 地址第三段"153"需要跟自己的 VMware 工具平台网段一致，可以通过在 VMware 工作平台中选择"编辑"→"虚拟网络编辑器"→VMware8→"子网"来查看），具体配置如表 5.1 所示。

表 5.1

主机名	IP 地址
Master001	192.168.153.101
Master002	192.168.153.102
Slave001	192.168.153.201
Slave002	192.168.153.202
Slave003	192.168.153.203

1. 基础信息配置

首先在一台虚拟机中设置基础信息，假设这台虚拟机为 Master001。在基础信息中需要设置主机名、IP 地址和名称解析等配置，这些配置文件只有 root 用户才有改写权限，所以需要使用 root 用户登录来编写这些配置文件。

1）修改主机名

通过编辑 network 文件，将 HOSTNAME 值修改为新的主机名，具体操作如下：

```
[root@localhost ~]# vi /etc/sysconfig/network
```

改写：

```
HOSTNAME=Master001
```

2）设置静态 IP

通过编辑 ifcfg-eth0 文件来设置 IP 地址，具体操作如下：

```
[root@localhost ~]# vi /etc/sysconfig/network-scripts/ifcfg-eth0
```

改写：

```
ONBOOT=yes
BOOTPROTO=static
```

插入：

```
IPADDR=192.168.153.101
NETMASK=255.255.255.0
GATEWAY=192.168.153.2
```

3）设置 resolv.conf

文件是 DNS 域名解析的配置文件，它的格式很简单，每行以一个关键字开头，后接配置参数。

resolv.conf 有 4 个主要关键字，分别是：

nameserver   #定义 DNS 服务器的 IP 地址
domain   #定义本地域名
search   #定义域名的搜索列表
sortlist   #对返回的域名进行排序

其中最主要的是 nameserver 关键字，如果不指定 nameserver 就找不到 DNS 服务器，其他关键字也是可选的。

具体操作如下：

```
[root@localhost ~]# vi /etc/resolv.conf
```

插入：

```
nameserver 192.168.153.2
```

4）设置 hosts

hosts 文件是 Linux 系统中负责 IP 地址与域名快速解析的文件，需要配置其他几个节点的主机名和 IP 来快速访问集群中的其他节点。

具体操作如下：

```
[root@localhost ~]# vi /etc/hosts
```

插入：

```
192.168.153.101    Master001
192.168.153.102    Master002
192.168.153.201    Slave001
192.168.153.202    Slave002
192.168.153.203    Slave003
```

5）使设置生效

只是修改 IP 地址可以重启网络服务即可以生效，操作如下：

```
[root@localhost ~]# service network restart
```

如果修改了主机名，必须重启虚拟机才能生效，操作如下：

```
[root@localhost ~]# reboot
```

6）验证设置是否成功

启动成功后信息栏从"[root@localhost ~]#"变成"[root@Master001 ~]#"，这时主机名修改成功。

验证 IP 地址设置是否成功可以通过 ifconfig 命令查看 IP 地址，如果出现"eth0"网络名称和 IP 地址，则说明静态 IP 设置成功，这时可以使用 ping 命令进一步验证是否能联通内网，具体操作如下：

```
[root@Master001 ~]# ifconfig
    eth0   Link encap:Ethernet  HWaddr 00:0C:29:45:78:7B
           inet addr:192.168.153.101 Bcast:192.168.153.255 Mask:255.255.255.0
           ...
           RX bytes:8610 (8.4 KiB)  TX bytes:9849 (9.6 KiB)

    lo     Link encap:Local Loopback
           net addr:127.0.0.1  Mask:255.0.0.0
           ...
           RX bytes:0 (0.0 b)  TX bytes:0 (0.0 b)

[root@Master001 ~]# ping 192.168.153.1
    PING 192.168.153.1 (192.168.153.1) 56(84) bytes of data.
    64 bytes from 192.168.153.1: icmp_seq=1 ttl=128 time=0.450 ms
    ...
    64 bytes from 192.168.153.1: icmp_seq=4 ttl=128 time=0.522 ms
    ^C
    --- 192.168.153.1 ping statistics ---
    4 packets transmitted, 4 received, 0% packet loss, time 4019ms
    rtt min/avg/max/mdev = 0.450/0.501/0.572/0.055 ms
```

验证外网是否联通。VMware 平台中的虚拟是通过与虚拟机共享主机的 IP 地址来访问外网的，虚拟机要连接网络必须保证宿主机能够正常访问网络，虚拟机是否能访问外网可以通过"ping www.baidu.com"命令来验证，如果能 ping 通百度，则说明外网访问成功。

```
[root@Master001 ~]# ping www.baidu.com
    PING www.baidu.com (180.97.33.107) 56(84) bytes of data.
    64 bytes from 180.97.33.107: icmp_seq=1 ttl=128 time=36.2 ms
    ...
    64 bytes from 180.97.33.107: icmp_seq=6 ttl=128 time=41.2 ms
    ^C
    --- www.baidu.com ping statistics ---
    6 packets transmitted, 6 received, 0% packet loss, time 5729ms
    rtt min/avg/max/mdev = 36.278/38.147/41.229/1.858 ms
```

7）创建普通用户

计算机的操作难免会有失误，如果关于内核的操作不当，则会对系统造成重大破坏，如一些工具不能使用、系统无法启动等。为了减少误操作对系统造成的伤害，出于安全性需要建立普通用户。

(1) 创建用户名叫 hadoop 的用户。

```
[root@Master001 ~]# adduser hadoop
```

(2) 给 hadoop 用户指定密码（密码：123456）。

```
[root@Master001 ~]# passwd hadoop
    更改用户 hadoop 的密码。
    新的密码：
    无效的密码：过于简单化/系统化
    无效的密码：过于简单
    重新输入新的密码：
    passwd：所有的身份验证令牌已经成功更新。
```

(3) 验证用户是否创建成功，如果能成功切换表示用户创建成功。

```
[root@Master001 ~]# su -l hadoop
[hadoop@Master001 ~]#
```

8）安装 Xshell

Xshell 是系统的用户界面，提供了用户与内核进行交互操作的一种接口，它接收用户输入的命令并把它送入内核去执行。可以把 Xshell 理解为一个客户端，可以通过这个客户端来远程操作 Linux 系统，就像用 Navicat 去连接 MySQL 服务器一样，可以远程操作 MySQL 数据库。

2. 安装 Xshell

（1）在安装文件目录中找到 Xme4.exe 文件，并双击安装 Xme4.exe。

（2）选中 I accept the terms of the license agreement 单选按钮，然后单击 Next 按钮，如图 5.3 所示。

（3）输入名字、公司和密钥，单击 Next 按钮，如图 5.4 所示。

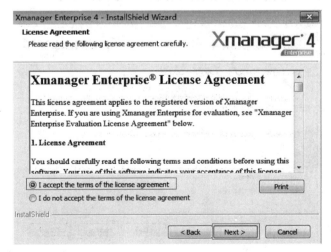

图 5.3

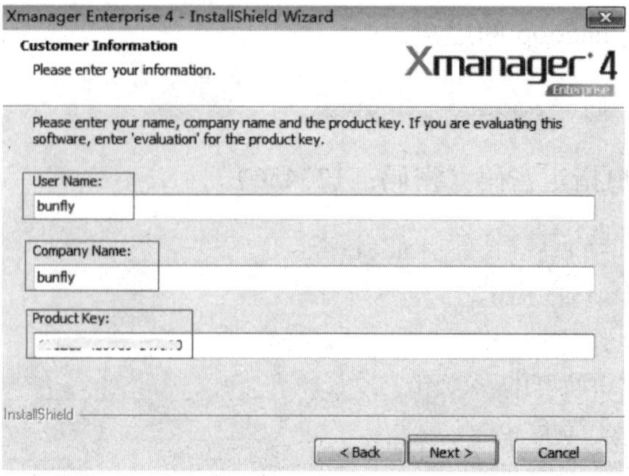

图 5.4

(4) 修改安装地址,单击 Next 按钮,如图 5.5 所示。

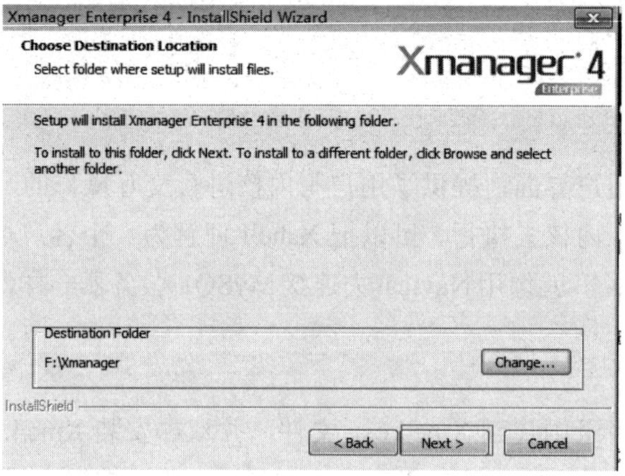

图 5.5

(5) 选择 Typical 安装模式,单击 Next 按钮,如图 5.6 所示。

图 5.6

（6）以后操作均为默认选项，当出现 InstallShield Wizard Complete 提示后单击 Finish 按钮即可。

3. 连接 Xshell

（1）双击 Xshell 图标，单击"新建连接"按钮，打开 Xshell 终端，如图 5.7 所示。

图 5.7

（2）配置需要连接的虚拟机 IP 地址、用户名和密码。

这里使用 hadoop 用户登录，连接成功后将进入 hadoop 用户家目录，如果是 root 用户登录连接成功将进入 root 用户家目录，如图 5.8 和图 5.9 所示。

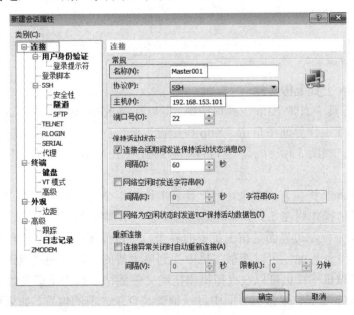

图 5.8

图 5.9

（3）选择连接 Xshell，如果信息栏出现"[hadoop@Master001 ~]$"，表示连接成功。

（4）切换到家目录，在家目录下创建一个名字叫 software 的文件夹，用于管理安装文件。

```
[hadoop@Master001 ~]$ cd ~
[hadoop@Master001 ~]$ mkdir software
```

（5）进入到 softeware 目录。

```
[hadoop@Master001 ~]$ cd software
[hadoop@Master001 software]$
```

4. 利用 Xftp 工具上传文件

（1）Xshell 工具中自带 Xftp 工具快捷键，可以利用 Xftp 快捷键进入 Xftp 工具中，Xftp 工具可以从 Xshell 工作界面中单击"Xftp 快捷键"按钮登录，登录的用户与 Xshell 登录的用户为同一用户，如图 5.10 所示。也可以单独通过双击 Xftp 工具输入 IP 地址、用户名和密码单独登录。

如果使用 Xshell 快捷方式登录，用户登录上传的那个文件权限将属于该用户也是经常失误的地方。具体操作将在 5.3 节"常见问题汇总"中详细讲解。

（2）图 5.11 中，左边界面是宿主机中的界面，右边界面是虚拟机中的界面，下面界面是传输数据的进度条界面。可以在宿主机中找到要上传的文件，通过双击或者拖动的方式将文件上传到虚拟机中；也可以在虚拟机中拖动文件到宿主机中下载文件，通过 Xshell 快捷方式登录到 Xftp 工具。虚拟机界面中目录位置是登录之前的位置，如果这个位置不是想要的位置，可在 Xftp 中通过选择栏进行选择。

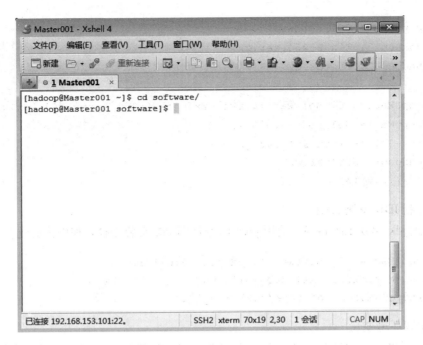

图 5.10

将 hadoop-2.6.5.tar.gz 和 jdk-8u131-linux-x64.tar.gz 安装包文件上传到虚拟机 software 文件夹中，如图 5.11 所示。

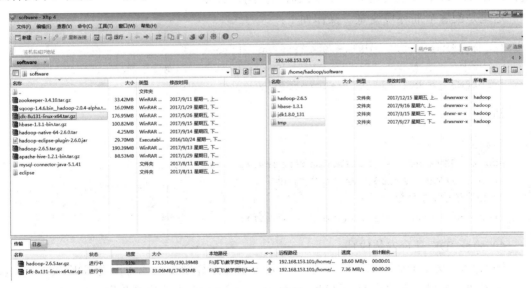

图 5.11

5. 安装 JDK

1）查看 Xftp 上传的文件内容并解压安装文件

在 Xshell 中输入 ls 命令可以查看 Xftp 上传的文件内容，通过 tar 命令解压 jdk-8u131-linux-x64.tar.gz 压缩文件，操作如下：

```
[hadoop@Master001 software]$ ls
    jdk-8u131-linux-x64.tar.gz
    hadoop-2.6.5.tar.gz

[hadoop@Master001 software]$ tar -zxf jdk-8u131-linux-x64.tar.gz
[hadoop@Master001 software]$ ls
    jdk-8u131-linux-x64.tar.gz
    hadoop-2.6.5.tar.gz
    jdk1.8.0_131
```

2）复制 JDK 安装目录

进入到 jdk1.8.0_131 目录，使用 pwd 命令打印 jdk 安装路径，利用鼠标选择复制路径。

```
[hadoop@Master001 software]$ cd jdk1.8.0_131/
[hadoop@Master001 jdk1.8.0_131]$ pwd
    /home/hadoop/software/jdk1.8.0_131
```

3）配置环境变量

Linux 系统中环境变量分为两种：全局变量和局部变量。profile 文件是全局变量配置文件，只有管理员用户对 profile 文件才有写入权限，所以要编写 profile 文件需要切换到 root 用户，因为在全局变量中配置的环境变量对所有用户都有效。.bashrc 文件是局部变量配置文件，在 .bashrc 文件配置的环境变量只对当前用户有效。

这里是配置全局环境变量。操作如下：

```
[hadoop@Master001 jdk1.8.0_131]$ su -l root
    密码：
[root@Master001 ~]# vi /etc/profile
```

插入：

```
#java
export JAVA_HOME=/home/hadoop/software/jdk1.8.0_131
export PATH=$PATH:$JAVA_HOME/bin
```
4）使用环境变量生效
```
[root@Master001 ~]# source /etc/profile
```

4）验证 JDK 是否安装成功

输入 java 或者 java -version，如果出现 java 命令的详细说明或者出现 JDK 版本号，表示安装成功，如果出现 "-bash: dddd: command not found" 表示安装失败。

6. 安装 Hadoop

1）查看上传的文件内容并解压安装文件

切换到 hadoop 用户，并进入到 software 目录，使用 ls 命令可以查看 Xftp 上传的文件内容，通过 tar 命令解压 hadoop-2.6.5.tar.gz 压缩文件。操作如下：

```
[root@Master001 ~]# su -l hadoop
```

```
[hadoop@Master001 ~]$ cd software/
[hadoop@Master001 software]$ ls
    jdk-8u131-linux-x64.tar.gz       jdk1.8.0_131
    hadoop-2.6.5.tar.gz
[hadoop@Master001 software]$ tar -zxf hadoop-2.6.5.tar.gz
[hadoop@Master001 software]$ ls
    jdk-8u131-linux-x64.tar.gz       jdk1.8.0_131
    hadoop-2.6.5.tar.gz              hadoop-2.6.5
```

2）复制 Hadoop 安装目录

进入到 hadoop-2.6.5 目录，使用 pwd 命令打印 Hadoop 安装路径，利用鼠标选择复制路径。

```
[hadoop@Master001 software]$ cd hadoop-2.6.5
[hadoop@Master001 hadoop-2.6.5]$ pwd
    /home/hadoop/software/hadoop-2.6.5
```

3）配置 Hadoop 环境变量

切换到 root 用户，编辑 profile 文件，并插入 Hadoop 配置文件。操作如下：

```
[hadoop@Master001 hadoop-2.6.5]$ su -l root
```

密码：

```
[root@Master001 ~]# vi /etc/profile
```

插入：

```
#hadoop
export HADOOP_HOME=/home/hadoop/software/hadoop-2.6.5
export PATH=$PATH:$HADOOP_HOME/bin
export PATH=$PATH:$HADOOP_HOME/sbin
export HADOOP_MAPRED_HOME=$HADOOP_HOME
export HADOOP_COMMON_HOME=$HADOOP_HOME
export HADOOP_HDFS_HOME=$HADOOP_HOME
export YARN_HOME=$HADOOP_HOME
export HADOOP_COMMON_LIB_NATIVE_DIR=$HADOOP_HOME/lib/native
export HADOOP_OPTS="-Djava.library.path=$HADOOP_HOME/lib"
export JAVA_LIBRARY_PATH=$HADOOP_HOME/lib/native:$JAVA_LIBRARY_PATH
```

4）使环境变量生效

```
[root@Master001 ~]# source /etc/profile
```

5）验证 Hadoop 安装是否成功

输入 Hadoop 命令，如果出现 Hadoop 命令相关的详细信息，表示安装成功；如果出现"-bash: dddd: command not found"，表示安装失败。

6）配置 core-site.xml 文件

切换到 hadoop 用户，进入到 hadoop-2.6.5/etc/hadoop/ 目录，编辑 core-site.xml 文件。

```
[root@Master001 ~]# su -l hadoop
[hadoop@Master001 ~]$ cd software/hadoop-2.6.5/etc/hadoop/
[hadoop@Master001 hadoop]$ ls
    core-site.xml           mapred-site.xml         salves
    hadoop-env.cmd          hdfs-site.xml           yarn-site.xml
    ...
[hadoop@Master001 hadoop]$ vi core-site.xml
```

插入：

```
<configuration>
    <!--指定Zookeeper地址-->
    <property>
        <name>ha.zookeeper.quorum</name>
        <value>Slave001:2181,Slave002:2181,Slave003:2181</value>
    </property>
    <!--指定Hadoop临时目录-->
    <property>
        <name>hadoop.tmp.dir</name>
        <value>/home/hadoop/software/hadoop-2.6.5/tmp</value>
    </property>
    <!--指定Eclipse访问端口-->
    <property>
        <name>fs.defaultFS</name>
        <value>hdfs://mycluster</value>
    </property>
</configuration>
```

7）配置 hadoop-env.sh 文件

编辑 hadoop-env.sh 文件，修改 java_home 地址，java_home 地址是解压的 JDK 地址，配置 java_home 是为了使用 Java 的环境。

修改：

```
# The java implementation to use.
export JAVA_HOME=/home/hadoop/software/jdk1.8.0_131
```

8）配置 hdfs-site.xml 文件

hdfs-site.xml 文件是 Hadoop 2.0 以后版本的必备配置文件之一，可以在 hdfs-site.xml 中配置集群名字空间、访问端口、URL 地址、故障转移等。

```
[hadoop@Master001 hadoop]# vi hdfs-site.xml
```

插入：

```xml
<configuration>
    <!--设置设备备份数量-->
    <property>
        <name>dfs.replication</name>
        <value>3</value>
    </property>
    <!--指定高可用集群的名字空间-->
    <property>
        <name>dfs.nameservices</name>
        <value>mycluster</value>
    </property>
    <!--指定NameNode节点的名字空间-->
    <property>
        <name>dfs.ha.namenodes.mycluster</name>
        <value>nn1,nn2</value>
    </property>
    <!--rpc：远程调用,设置第一个远程调用的地址和端口-->
    <property>
        <name>dfs.namenode.rpc-address.mycluster.nn1</name>
        <value>Master001:9000</value>
    </property>
    <property>
        <name>dfs.namenode.rpc-address.mycluster.nn2</name>
        <value>Master002:9000</value>
    </property>
    <!--设置网页访问的地址和端口-->
    <property>
        <name>dfs.namenode.http-address.mycluster.nn1</name>
        <value>Master001:50070</value>
    </property>
    <property>
        <name>dfs.namenode.http-address.mycluster.nn2</name>
        <value>Master002:50070</value>
    </property>
    <!--设置共享edits的存放地址，将共享edits文件存放在QJournal集群中的
    QJCluster目录下-->
    <property>
        <name>dfs.namenode.shared.edits.dir</name>
        <value>qjournal://Slave001:8485;Slave002:8485;Slave003:8485/QJCluster
        </value>
    </property>
    <!--设置JournalNode节点的edits文件本地存放路径，是QJournal真实数据存放路径-->
    <property>
        <name>dfs.journalnode.edits.dir</name>
        <value>/home/hadoop/software/hadoop-2.6.5/QJEditsData</value>
    </property>
    <!--开启自动故障转移-->
    <property>
        <name>dfs.ha.automatic-failover.enabled</name>
```

```xml
        <value>true</value>
    </property>
    <!--设置HDFS客户端用来与活动的namenode节点联系的java类-->
    <property>
        <name>dfs.client.failover.proxy.provider.mycluster</name>
        <value>org.apache.hadoop.hdfs.server.namenode.ha.ConfiguredFailoverProxyProvider</value>
    </property>
    <!-- 用于停止活动NameNode节点的故障转移期间的脚本。保证任何时候只有一个NameNode处于活动状态；
    SShfence参数：采用ssh方式连接活动NameNode节点，并杀掉进程；
    shell脚本：表示如果sshfence执行失败，在执行自定义的shell脚本，确定只能有一个NameNode处于活动状态 -->
    <property>
        <name>dfs.ha.fencing.methods</name>
        <value>
            sshfence
            shell(/home/hadoop/software/hadoop-2.6.5/ensure.sh)
        </value>
    </property>
    <!--为了实现SSH登录杀掉进程，还需要配置免密码登录的SSH密匙信息-->
    <property>
        <name>dfs.ha.fencing.ssh.private-key-files</name>
        <value>/home/hadoop/.ssh/id_rsa</value>
    </property>
    <!--设置ssh连接超时时间-->
    <property>
        <name>dfs.ha.fencing.ssh.connect-timeout</name>
        <value>30000</value>
    </property>
</configuration>
```

9）配置 mapred-site.xml 文件

在 Hadoop 包里是没有 mapred-site.xml 文件的，需要通过 mapred-site.xml.template 模板文件复制出 mapred-site.xml 文件。操作如下：

```
[hadoop@Master001 hadoop]# cp mapred-site.xml.template mapred-site.xml
[hadoop@Master001 hadoop]# vi mapred-site.xml
```

插入：

```xml
<configuration>
    <!--设置jar程序启动Runner类的main方法运行在yarn集群中-->
    <property>
        <name>mapreduce.framework.name</name>
        <value>yarn</value>
    </property>
</configuration>
```

10）配置 slaves 文件

Hadoop 集群中所有的 DataNode 节点都需要写入 slaves 文件中，因为它是用来指定存储数据的节点文件，Master 会读取 slaves 文件来获取存储信息，根据 slaves 文件来做资源平衡。

📢注意：

（1）slaves 文件名全部是小写，有很多初学者使用 vi Slaves 来编辑 slaves 文件，它将会在 hadoop 目录中重新创建一个首字母为大写的 slaves 文件，这样是错误的。

（2）slaves 文件打开后里面有一个"localhost"，这个 localhost 需要删除，如果没有删除，集群会把 Master 也当做 DataNode 节点，这样会造成 Master 节点负载过重。

```
[hadoop@Master001 hadoop]# vi slaves
```

删除：

```
localhost
```

插入：

```
Slave001
Slave002
Slave003
```

11）配置 yarn-site.xml

yarn-site.xml 文件是 ResourceManager 进程相关的配置参数。

```
[hadoop@Master001 hadoop]# vi yarn-site.xml
```

插入：

```
<configuration>
    <!--设置ResourceManager在哪台节点-->
    <property>
        <name>yarn.resourcemanager.hostname</name>
        <value>Master001</value>
    </property>
    <!-- Reduce取数据的方式是mapreduce_shuffle -->
    <property>
        <name>yarn.nodemanager.aux-services</name>
        <value>mapreduce_shuffle</value>
    </property>
</configuration>
```

7. 安装 SSH

SSH 是一种远程传输通信协议，用于两台或多台节点之间的数据传输。通过 yum 方式在线安装 SSH，yum 是在线安装工具，因此使用 yum 安装时必须连接网络。yum 是一个 Shell 前端软件包管理器，它能够从 yum 服务器自动下载 rpm 包然后安装，一次安装完所有需要的软件包,不必一次次地下载,非常简单方便。

1)在线安装 yum 工具

yum 工具属于 root 用户工具,所以需要切换到 root 用户进行在线安装。

```
[hadoop@Master001 hadoop]$ su -l root
```

密码:

```
[root@Master001 ~]#
```

2)查找 yum 库有哪些 SSH 软件的 rpm 包

在安装 SSH 之前需要先查找 yum 库有哪些 SSH 软件的 rpm 包:

```
[root@Master001 ~]# yum list | grep ssh
    openssh.x86_64              5.3p1-84.1.el6         updates
    openssh-server.x86_64       5.3p1-123.el6_9        updates
    openssh-clients.x86_64      5.3p1-123.el6_9        updates
    ...
```

3)使用 yum 工具在线安装 server 和 clients 软件

```
[root@Master001 ~]# yum install -y openssh-clients.x86_64
[root@Master001 ~]# yum install -y openssh-server.x86_64
```

安装过程如图 5.12 所示。

```
[root@Master001 ~]# yum install -y openssh-clients.x86_64
Loaded plugins: fastestmirror, security
Loading mirror speeds from cached hostfile
 * base: mirrors.sohu.com
 * extras: mirrors.sohu.com
 * updates: mirrors.sohu.com
Setting up Install Process
Resolving Dependencies
--> Running transaction check
---> Package openssh-clients.x86_64 0:5.3p1-123.el6_9 will be installed
--> Finished Dependency Resolution

Dependencies Resolved

================================================================================
 Package              Arch         Version              Repository       Size
================================================================================
Installing:
 openssh-clients      x86_64       5.3p1-123.el6_9      updates          444 k

Transaction Summary
================================================================================
Install       1 Package(s)
```

图 5.12

4)验证 SSH 是否安装成功

验证方法一:输入 ssh 命令,如果出现 ssh 命令的详细信息表示安装成功,如果出现

"-bash: dddd: command not found"则表示安装失败。

```
[root@Master001 ~]# ssh
    usage: ssh [-1246AaCfgKkMNnqsTtVvXxYy] [-b bind_address] [-c cipher_spec]
        [-D [bind_address:]port] [-e escape_char] [-F configfile]
        [-I pkcs11] [-i identity_file]
        [-L [bind_address:]port:host:hostport]
        [-l login_name] [-m mac_spec] [-O ctl_cmd] [-o option] [-p port]
        [-R [bind_address:]port:host:hostport] [-S ctl_path]
        [-W host:port] [-w local_tun[:remote_tun]]
        [user@]hostname [command]
```

验证方法二：使用 rpm 工具验证。输入 rpm -qa | grep ssh 命令查找已经安装的 SSH 相关程序，如果出现 server 和 clients 表示安装成功。

```
[root@Master001 ~]# rpm -qa | grep ssh
    openssh-server-5.3p1-123.el6_9.x86_64
    openssh-clients-5.3p1-123.el6_9.x86_64
    libssh2-1.4.2-1.el6.x86_64
    openssh-5.3p1-123.el6_9.x86_64
```

5.2.3 步骤3——复制虚拟机

现在已经安装好一台节点虚拟机的配置，其他 4 台节点虚拟机可以通过复制的方式来安装。在 2.1 节"系统的安装"中提到选择安装目录，这个安装目录就是整个虚拟机文件目录。复制这个目录就可以创建出另一个虚拟机，但在复制目录之前需要先把虚拟机关机。

1. 关闭虚拟机（halt 命令需要 root 权限）

```
[root@Master001 ~]# halt
```

2. 复制虚拟机

复制出另外 4 台虚拟机，并把复制的文件夹重新命名为 Master001、Master002、Slave001、Slave002、Slave003 方便管理，如图 5.13 所示。

图 5.13

3. 打开虚拟机

1）通过菜单打开虚拟机

通过"文件"→"打开"选择复制的虚拟机来打开虚拟机。为了方便管理需将虚拟名字修改为文件夹名称，如图 5.14 所示。

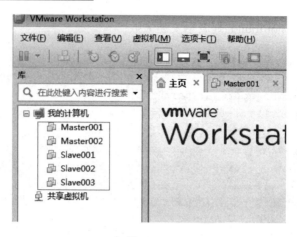

图 5.14

2）启动虚拟机

单击左边虚拟机名称，待右边出现对应的界面后，单击"开启此虚拟机"按钮打开虚拟机，如图 5.15 所示。

图 5.15

3）选择"我已复制该虚拟机"

每一台计算机都有一个唯一的 MAC 地址，虚拟机也一样。虽然它是虚拟状态的，但它同样有内存、处理器、硬盘和 MAC 地址等。虚拟机是通过复制出另一台一模一样的虚拟机，包括 MAC 地址，所以需要在启动副本虚拟机时选择"我已复制该虚拟机"按钮来告诉 VMware 平台"我这台虚拟机需要重新生成一个新的 MAC 地址"。如果单击"我已移动该虚拟机"按钮，VMware 平台将不会为新虚拟机生成新的 MAC 地址，如图 5.16 所示。

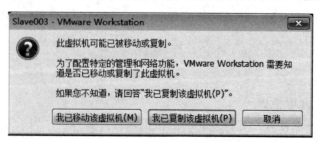

图 5.16

4. 修改虚拟配置

1) 查看 MAC 地址

修改 MAC 地址之前需要到 70-persistent-net.reles 文件去查看最新的 MAC 地址，最后的一条为最新的 MAC 地址，并记住 ATTR 和 NAME 的值，如图 5.17 所示。

```
[root@Master001 ~]# cat /etc/udev/rules.d/70-persistent-net.rules
```

图 5.17

2) 修改 MAC 地址和 IP 地址

需要到 profile 文件中修改最新的 MAC 地址和网络名称，按之前约定的配置规则来修改 IP 地址，如图 5.18 所示。

```
[root@Master001 ~]# vi /etc/sysconfig/network-scripts/ifcfg-eth0
```

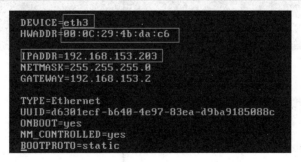

图 5.18

3) 修改主机名

按之前约定的配置规则来修改主机名。

```
[root@Master001 ~]# vi /etc/sysconfig/network
```

改写：

HOSTNAME=Slave003

4）使修改生效

如果只是修改 profile 文件，可以重启网络服务即可使修改生效。如果修改主机名，需要重启虚拟机才能生效。

```
[root@Master001 ~]# reboot
```

5）验证修改是否成功

如果登录主机名变成修改的主机名表示主机名修改成功，如图 5.19 所示。

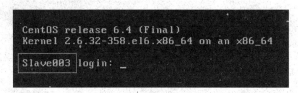

图 5.19

输入用户名和密码登录后，输入"ifconfig"命令，如果出现修改后的网络名称和 IP 地址表示静态 IP 修改成功，如图 5.20 所示。

图 5.20

5. 修改其他虚拟机

依次执行 5.2.3 节中的复制虚拟机操作，修改其他几台虚拟机，当所有虚拟机都修改完成后，可以互相 ping IP 地址或主机名来验证内网是否联通。

```
[hadoop@Slave003 ~]$ ping 192.168.153.101
    PING 192.168.153.101 (192.168.153.101) 56(84) bytes of data.
    64 bytes from 192.168.153.101: icmp_seq=1 ttl=64 time=0.797 ms
    64 bytes from 192.168.153.101: icmp_seq=2 ttl=64 time=0.774 ms
    ^C
    --- 192.168.153.101 ping statistics ---
    2 packets transmitted, 2 received, 0% packet loss, time 1876ms
    rtt min/avg/max/mdev = 0.774/0.785/0.797/0.030 ms
```

```
[hadoop@Slave003 ~]$ ping Master001
    PING Master001 (192.168.153.101) 56(84) bytes of data.
    64 bytes from Master001 (192.168.153.101): icmp_seq=1 ttl=64 time=1.69 ms
    64 bytes from Master001 (192.168.153.101): icmp_seq=2 ttl=64 time=0.703 ms
    ^C
    --- Master001 ping statistics ---
    2 packets transmitted, 3 received, 0% packet loss, time 2391ms
    rtt min/avg/max/mdev = 0.703/1.061/1.697/0.452 ms
```

5.2.4 步骤4——设置免密

安装 Hadoop 之前，由于集群中大量主机进行分布式计算需要相互进行数据通信，服务器之间的连接需要通过 SSH 来进行，所以要安装 SSH 服务。默认情况下通过 SSH 登录服务器需要输入用户名和密码进行连接，如果不配置免密码登录，每次启动 Hadoop 都要输入密码来访问每台机器的 DataNode，因为 Hadoop 集群有上百或者上千台机器，靠人力输入密码工程浩大，所以一般都会配置 SSH 的免密码登录。在 Hadoop 集群中 Master 节点需要对所有节点进行访问，了解每个节点的健康状态，所以只需要对 Master 做免密设置，该集群是高可用集群，有两个 Master。这两个 Master 都需要生成自己的私密，然后对所有节点（包括自己）传输密钥，以 Master001 为例，Master002 只需要执行 Master001 相同操作即可。具体操作如下。

1. 生成密钥

密钥就像是进入一扇门的钥匙，生成密钥就是生成这把钥匙。由于要对 hadoop 用户进行免密设置，所以需要切换到 hadoop 用户，并回到该用户的家目录。

执行 ssh-keygen -t rsa -P 命令后，将在/home/hadoop/.ssh/目录下以 rsa 方式生成 id_rsa 的密钥。

```
[hadoop@Master001 ~]$ cd ~
[hadoop@Master001 ~]$ ssh-keygen -t rsa -P ''
    Generating public/private rsa key pair.
    Enter file in which to save the key (/home/hadoop/.ssh/id_rsa):
    Your identification has been saved in /home/hadoop/.ssh/id_rsa.
    Your public key has been saved in /home/hadoop/.ssh/id_rsa.pub.
    The key fingerprint is:
    2c:a9:91:36:4b:18:e6:09:60:f9:1a:22:23:3b:d6:af hadoop@Master001
    The key's randomart image is:
    +-------[ RSA 2048]-------+
    |...                      |
    |o.                       |
    |. +                      |
    |== = . o                 |
    |+oB * o S                |
    |oo + = .                 |
    |.. +                     |
    |    .                    |
    |    E.                   |
    +-------------------------+
```

2. 对所有节点进行免密

将密钥分发给集群中所有节点（包括自己），就免去输入密码去访问其他虚拟机。执行 ssh-copy-id 命令后，会将 id_rsa 中的密钥传输到目标虚拟机的/home/hadoop/.ssh/authorized_keys 文件中。

```
[hadoop@Master001 ~]$ ssh-copy-id Master001
    hadoop@master001's password:
    Now try logging into the machine, with "ssh 'Master001'", and check in:
    .ssh/authorized_keys
    to make sure we haven't added extra keys that you weren't expecting.
[hadoop@Master001 ~]$ ssh-copy-id Master002
    hadoop@master002's password:
    Now try logging into the machine, with "ssh 'Master002'", and check in:
    .ssh/authorized_keys
    to make sure we haven't added extra keys that you weren't expecting.
[hadoop@Master001 ~]$ ssh-copy-id Slave001
    hadoop@slave001's password:
    Now try logging into the machine, with "ssh 'Slave001'", and check in:
    .ssh/authorized_keys
    to make sure we haven't added extra keys that you weren't expecting.
[hadoop@Master001 ~]$ ssh-copy-id Slave002
    hadoop@slave002's password:
    Now try logging into the machine, with "ssh 'Slave002'", and check in:
    .ssh/authorized_keys
    to make sure we haven't added extra keys that you weren't expecting.
[hadoop@Master001 ~]$ ssh-copy-id Slave003
    hadoop@slave003's password:
    Now try logging into the machine, with "ssh 'Slave003'", and check in:
    .ssh/authorized_keys
    to make sure we haven't added extra keys that you weren't expecting.
```

3. 验证免密设置是否成功

验证免密是免密设置最关键的一步，如果不输入密码就能访问到目标虚拟机，表示免密设置成功。

```
[hadoop@Master001 ~]$ ssh Master001
    Last login: Tue Dec 19 14:44:02 2017 from 192.168.153.1
[hadoop@Master001 ~]$ exit
    logout
    Connection to Master001 closed.
[hadoop@Master001 ~]$ ssh Master002
    Last login: Fri Dec 15 08:38:51 2017 from 192.168.153.1
[hadoop@Master002 ~]$ exit
    logout
    Connection to Master002 closed.
[hadoop@Master001 ~]$ ssh Slave001
    Last login: Fri Dec 15 08:38:54 2017 from 192.168.153.1
[hadoop@Slave001 ~]$ exit
```

```
        logout
        Connection to Slave001 closed.
[hadoop@Master001 ~]$ ssh Slave002
        Last login: Fri Dec 15 08:38:56 2017 from 192.168.153.1
[hadoop@Slave002 ~]$ exit
        logout
        Connection to Slave002 closed.
[hadoop@Master001 ~]$ ssh Slave003
        Last login: Tue Dec 19 14:44:05 2017 from 192.168.153.1
[hadoop@Slave003 ~]$ exit
        logout
        Connection to Slave003 closed.
[hadoop@Master001 ~]$
```

5.2.5 步骤 5——安装 Zookeeper

关于 Zookeeper 在大数据集群中的作用和安装的讲解视频可扫描二维码观看。

ZooKeeper 是一个分布式的，开放源码的分布式应用程序协调服务，它是一个为分布式应用提供一致性服务的软件，提供的功能包括：配置维护、域名服务、分布式同步、组服务等。Zookeeper 在 Hadoop 大数据集群中提供 Master 管理和元数据存储。Zookeeper 是以集群的形式出现，一般为 3 台或者 5 台，它可以通过选举机制来选取 Master，保证两个 Master 中一台为 active 状态，另外一台为 standby 状态。它还存储 NameNode 的 eidlogs（快照），用来减轻 Master 负担，当 active 状态 Master 死机后，Zookeeper 通过选举组成员选出一个新的 Master，新的 Master 通过实时更新的 fsimage 来恢复所有数据。集群采用 3 台 Zookeeper，分别安装在 Slave001、Slave002、Slave003 中。

1. 上传 Zookeeper 安装文件

利用 Xftp 工具将 zookeeper-3.4.10.tar.gz 安装文件分别上传到 Slave003 的 hadoop 用户的 software 目录中，如图 5.21 所示。

图 5.21

2. 配置 Zookeeper

（1）进入到 hadoop 用户 software 目录下，解压 zookeeper-3.4.10.tar.gz 文件。

```
[hadoop@Slave003 software]$ tar -zxf zookeeper-3.4.10.tar.gz
```

（2）进入 Zookeeper 的 conf 目录，将 zoo_sample.cfg 模板复制出 zoo.cfg 文件，并编辑 zoo.cfg 文件。

```
[hadoop@Slave003 ~]$ cd software/zookeeper-3.4.10/conf/
[hadoop@Slave003 conf]$ ls
   configuration.xsl    log4j.properties    zoo_sample.cfg
[hadoop@Slave003 conf]$ cp zoo_sample.cfg zoo.cfg
[hadoop@Slave003 conf]$ ls
   configuration.xsl    log4j.properties    zoo.cfg zoo_sample.cfg
[hadoop@Slave003 conf]$ vi zoo.cfg
```

改写：

```
#快照存放目录
dataDir=/home/hadoop/software/zookeeper-3.4.10/tmp/zookeeper
```

插入：

```
#服务器名称与地址：集群信息（服务器编号，服务器地址，LF通信端口，选举端口），这个配置项的书写格式比较特殊，规则如下：server.N=YYY:A:B
server.1=Slave001:2888:3888
server.2=Slave002:2888:3888
server.3=Slave003:2888:3888
```

（3）创建/home/hadoop/software/zookeeper-3.4.10/tmp/zookeeper 文件夹。

```
[hadoop@Slave003 conf]$ mkdir -p /home/hadoop/software/zookeeper-3.4.10/tmp/zookeeper
[hadoop@Slave003 conf]$ cd /home/hadoop/software/zookeeper-3.4.10/ tmp/zookeeper
[hadoop@Slave003 zookeeper]$
```

（4）创建 myid 文件。

```
[hadoop@Slave003 zookeeper]$ vi myid
```

插入：

```
3
```

3. 传输到 Zookeeper 配置

（1）将 Slave003 中的 zookeeper-3.4.10 目录传送到 Slave001 和 Slave002。

```
[hadoop@Slave003 conf]$ cd ~/software/
```

```
[hadoop@Slave003 software]$ ls
    hadoop-2.6.5    jdk1.8.0_131    zookeeper-3.4.10
[hadoop@Slave003 software]$ scp -r zookeeper-3.4.10/ Slave003:~/software/
```

(2)修改 myid。

将 Slave001 和 Slave002 的/software/zookeeper-3.4.10/tmp/zookeeper 目录下的 myid 分别修改为 1 和 2。

```
[hadoop@Slave001 ~]$ cd ~/software/zookeeper-3.4.10/tmp/zookeeper
[hadoop@Slave001 zookeeper]$ vi myid
```

改写：

```
1
 [hadoop@Slave002 ~]$ cd ~/software/zookeeper-3.4.10/tmp/zookeeper
[hadoop@Slave002 zookeeper]$ vi myid
```

改写：

```
2
```

4. 启动 Zookeeper

由于 Zookeeper 没有配置环境变量，要想启动 Zookeeper 需要进入到 Zookeeper 的 bin 目录，找到 zkServer.sh 文件。zkServer.sh 文件是 Zookeeper 服务文件，可以通过 zkServer.sh 文件查看 Zookeeper 服务状态、启动 Zookeeper 集群和停止 Zookeeper 集群。

```
[hadoop@Slave001 ~]$ cd ~/software/zookeeper-3.4.10/bin/
[hadoop@Slave001 bin]$ ls
    zkCli.cmd      …   zkServer.cmd         zkServer.sh       zookeeper.out
[hadoop@Slave001 bin]$ ./zkServer.sh start
    ZooKeeper JMX enabled by default
    Using config: /home/hadoop/software/zookeeper-3.4.10/bin/../conf/zoo.cfg
    Starting zookeeper … STARTED
[hadoop@Slave001 bin]$ jps
    1782    QuorumPeerMain
    1302    Jps

[hadoop@Slave002 ~]$ cd ~/software/zookeeper-3.4.10/bin/
[hadoop@Slave002 bin]$ ls
    zkCli.cmd      …   zkServer.cmd         zkServer.sh       zookeeper.out
[hadoop@Slave002 bin]$ ./zkServer.sh start
    ZooKeeper JMX enabled by default
    Using config: /home/hadoop/software/zookeeper-3.4.10/bin/../conf/zoo.cfg
    Starting zookeeper … STARTED
[hadoop@Slave002 bin]$ jps
    3788    QuorumPeerMain
    1206    Jps
```

```
[hadoop@Slave003 ~]$ cd ~/software/zookeeper-3.4.10/bin/
[hadoop@Slave003 bin]$ ls
    zkCli.cmd         …    zkServer.cmd    zkServer.sh      zookeeper.out
[hadoop@Slave003 bin]$ ./zkServer.sh start
    ZooKeeper JMX enabled by default
    Using config: /home/hadoop/software/zookeeper-3.4.10/bin/../conf/zoo.cfg
    Starting zookeeper … STARTED
[hadoop@Slave003 bin]$ jps
    1788    QuorumPeerMain
    1806    Jps
```

5.2.6 步骤6——启动 Hadoop 集群

关于启动 Hadoop 集群的讲解视频可扫描二维码观看。

1. 启动 journalnode 进程

高可用集群格式化 NameNode 时需要到 Zookeeper 每个节点上分别启动 journalnode 进程。

```
[hadoop@Slave001 bin]$ hadoop-daemon.sh start journalnode
    starting journalnode, logging to /home/hadoop/software/hadoop-2.6.5/
    logs/hadoop-hadoop-journalnode-Slave001.out
[hadoop@Slave001 bin]$ jps
    1895    Jps
    1849    JournalNode
    1788    QuorumPeerMain

[hadoop@Slave002 bin]$ hadoop-daemon.sh start journalnode
    starting journalnode, logging to /home/hadoop/software/hadoop-2.6.5/
    logs/hadoop-hadoop-journalnode-Slave002.out
[hadoop@Slave002 bin]$ jps
    1895    Jps
    1849    JournalNode
    1788    QuorumPeerMain

[hadoop@Slave003 bin]$ hadoop-daemon.sh start journalnode
    starting journalnode, logging to /home/hadoop/software/hadoop-2.6.5/
    logs/hadoop-hadoop-journalnode-Slave003.out
[hadoop@Slave003 bin]$ jps
    1895    Jps
    1849    JournalNode
    1788    QuorumPeerMain
```

2. 在 Master001 中格式化 NameNode

在 Master001 中格式化 NameNode 会生成~/software/hadoop-2.6.5/tmp 目录，该目录中

存放版本号和元数据等相关信息。

```
[hadoop@Master001 ~]$ hdfs namenode -format
```

3. 传送 tmp 文件到其他节点

```
[hadoop@Master001 ~]$ cd ~/software/hadoop-2.6.5/
[hadoop@Master001 hadoop-2.6.5]$ ls
    bin    etc  ...    sbin   share   tmp
[hadoop@Master001 hadoop-2.6.5]$ scp -r tmp/ Master002:~/software/ hadoop-2.6.5/
    VERSION                              100%   206     0.2KB/s   00:00
    fsimage_0000000000000000000          100%   323     0.3KB/s   00:00
    fsimage_0000000000000000000.md5      100%   62      0.1KB/s   00:00
    seen_txid                            100%   2       0.0KB/s   00:00
[hadoop@Master001 hadoop-2.6.5]$ scp -r tmp/ Slave001:~/software/ hadoop-2.6.5/
    VERSION                              100%   206     0.2KB/s   00:00
    ...
[hadoop@Master001 hadoop-2.6.5]$ scp -r tmp/ Slave002:~/software/ hadoop-2.6.5/
    VERSION                              100%   206     0.2KB/s   00:00
    ...
[hadoop@Master001 hadoop-2.6.5]$ scp -r tmp/ Slave003:~/software/ hadoop-2.6.5/
    VERSION                              100%   206     0.2KB/s   00:00
    ...
```

4. 启动 hdfs

启动 hdfs 只需要在 Master001 中执行 start-dfs.sh 即可。它分别会在 Master001 和 Master002 中启动 namenode 进程，在 Slave001、Slave002 和 Slave003 中启动 datanode 进程。

```
[hadoop@Master001 ~]$ start-dfs.sh
    Starting namenodes on [Master001 Master002]
    Master001: starting namenode, logging to /home/hadoop/software/hadoop-
    2.6.5/logs/hadoop-hadoop-namenode-Master001.out
    Master002: starting namenode, logging to /home/hadoop/software/hadoop-
    2.6.5/logs/hadoop-hadoop-namenode-Master002.out
    Slave002: starting datanode, logging to /home/hadoop/software/hadoop-
    2.6.5/logs/hadoop-hadoop-datanode-Slave002.out
    Slave003: starting datanode, logging to /home/hadoop/software/hadoop-
    2.6.5/logs/hadoop-hadoop-datanode-Slave003.out
    Slave001: starting datanode, logging to /home/hadoop/software/hadoop-
    2.6.5/logs/hadoop-hadoop-datanode-Slave001.out
    Starting journal nodes [Slave001 Slave002 Slave003]
    Slave003: starting journalnode, logging to /home/hadoop/software/
    hadoop-2.6.5/logs/hadoop-hadoop-journalnode-Slave003.out
    Slave001: starting journalnode, logging to /home/hadoop/software/
    hadoop-2.6.5/logs/hadoop-hadoop-journalnode-Slave001.out
    Slave002: starting journalnode, logging to /home/hadoop/software/
```

```
hadoop-2.6.5/logs/hadoop-hadoop-journalnode-Slave002.out
Starting ZK Failover Controllers on NN hosts [Master001 Master002]
Master002: starting zkfc, logging to /home/hadoop/software/hadoop-
2.6.5/logs/hadoop-hadoop-zkfc-Master002.out
Master001: starting zkfc, logging to /home/hadoop/software/hadoop-
2.6.5/logs/hadoop-hadoop-zkfc-Master001.out
```

5. 启动 MapReduce

启动 MapReduce 只需要在 Master001 中执行 start-yarn.sh 即可。它分别会在 Master001 中启动 resourcemanager 进程，在 Slave001、Slave002 和 Slave003 中启动 nodemanager 进程。

```
[hadoop@Master001 ~]$ start-yarn.sh
    starting yarn daemons
    starting resourcemanager, logging to /home/hadoop/software/hadoop-
2.6.5/logs/yarn-hadoop-resourcemanager-Master001.out
    Slave003: starting nodemanager, logging to /home/hadoop/software/hadoop-
2.6.5/logs/yarn-hadoop-nodemanager-Slave003.out
    Slave001: starting nodemanager, logging to /home/hadoop/software/hadoop-
2.6.5/logs/yarn-hadoop-nodemanager-Slave001.out
    Slave002: starting nodemanager, logging to /home/hadoop/software/hadoop-
2.6.5/logs/yarn-hadoop-nodemanager-Slave002.out
```

6. 验证集群是否成功启动

1）方法一：查看进程

当集群启动成功后每个节点中都有一些必须存在的进程。具体进程如下：

```
[hadoop@Master001 ~]$ jps
    3766    Jps
    3510    ResourceManager
    3399    DFSZKFailoverController
    3118    NameNode

[hadoop@Master002 ~]$ jps
    2129    NameNode
    2289    Jps
    2230    DFSZKFailoverController

[hadoop@Slave001 ~]$ jps
    2610    NodeManager
    2500    JournalNode
    2358    QuorumPeerMain
    2745    Jps

[hadoop@Slave002 ~]$ jps
    2675    Jps
    2549    NodeManager
```

```
    2439        JournalNode
    2297        QuorumPeerMain

[hadoop@Slave003 ~]$ jps
    2737        Jps
    2611        NodeManager
    2501        JournalNode
    1788        QuorumPeerMain
```

2）方法二：查看 HDFS 文件

在任意节点输入"hadoop fs -ls /"命令查看 hdfs 文件。如果没有报错（警告可忽略），说明集群启动成功。

```
[hadoop@Slave002 ~]$ hadoop fs -ls /
    17/12/19 17:11:42 WARN util.NativeCodeLoader: Unable to load native-
    hadoop library for your platform... using builtin-java classes where
    applicable
[hadoop@Slave002 ~]$
```

5.2.7 正常启动顺序

格式化 NameNode 只是在第一次启动和非法关机导致版本号不一致时使用。下面将讲解正常时大数据集群的启动顺序。

1. 启动 Zookeeper

分别进入到 Zookeeper 的 bin 目录启动 Zookeeper 服务。

```
[hadoop@Slave001 ~]$ cd software/zookeeper-3.4.10/bin/
[hadoop@Slave001 bin]$ ./zkServer.sh start
    ZooKeeper JMX enabled by default
    Using config: /home/hadoop/software/zookeeper-3.4.10/bin/../conf/zoo.cfg
    Starting zookeeper ... STARTED
[hadoop@Slave002 ~]$ cd software/zookeeper-3.4.10/bin/
[hadoop@Slave002 bin]$ ./zkServer.sh start
    ZooKeeper JMX enabled by default
    Using config: /home/hadoop/software/zookeeper-3.4.10/bin/../conf/zoo.cfg
    Starting zookeeper ... STARTED
[hadoop@Slave003 ~]$ cd software/zookeeper-3.4.10/bin/
[hadoop@Slave003 bin]$ ./zkServer.sh start
    ZooKeeper JMX enabled by default
    Using config: /home/hadoop/software/zookeeper-3.4.10/bin/../conf/zoo.cfg
    Starting zookeeper ... STARTED
```

2. 启动 HDFS

在 Master001 节点下启动 start-dfs.sh。

```
[hadoop@Master001 ~]$ start-dfs.sh
```

```
Starting namenodes on [Master001 Master002]
Master001: starting namenode, logging to /home/hadoop/software/hadoop-
2.6.5/logs/hadoop-hadoop-namenode-Master001.out
Master002: starting namenode, logging to /home/hadoop/software/hadoop-
2.6.5/logs/hadoop-hadoop-namenode-Master002.out
Slave002: starting datanode, logging to /home/hadoop/software/hadoop-
2.6.5/logs/hadoop-hadoop-datanode-Slave002.out
Slave003: starting datanode, logging to /home/hadoop/software/hadoop-
2.6.5/logs/hadoop-hadoop-datanode-Slave003.out
Slave001: starting datanode, logging to /home/hadoop/software/hadoop-
2.6.5/logs/hadoop-hadoop-datanode-Slave001.out
Starting journal nodes [Slave001 Slave002 Slave003]
Slave003: starting journalnode, logging to /home/hadoop/software/hadoop-
2.6.5/logs/hadoop-hadoop-journalnode-Slave003.out
Slave001: starting journalnode, logging to /home/hadoop/software/hadoop-
2.6.5/logs/hadoop-hadoop-journalnode-Slave001.out
Slave002: starting journalnode, logging to /home/hadoop/software/hadoop-
2.6.5/logs/hadoop-hadoop-journalnode-Slave002.out
Starting ZK Failover Controllers on NN hosts [Master001 Master002]
Master002: starting zkfc, logging to /home/hadoop/software/hadoop-
2.6.5/logs/hadoop-hadoop-zkfc-Master002.out
Master001: starting zkfc, logging to /home/hadoop/software/hadoop-
2.6.5/logs/hadoop-hadoop-zkfc-Master001.out
```

3. 启动 MapReduce

在 Master001 节点下启动 start-yarn.sh。

```
[hadoop@Master001 ~]$ start-yarn.sh
    starting yarn daemons
    starting resourcemanager, logging to /home/hadoop/software/hadoop-
    2.6.5/logs/yarn-hadoop-resourcemanager-Master001.out
    Slave003: starting nodemanager, logging to /home/hadoop/software/hadoop-
    2.6.5/logs/yarn-hadoop-nodemanager-Slave003.out
    Slave001: starting nodemanager, logging to /home/hadoop/software/hadoop-
    2.6.5/logs/yarn-hadoop-nodemanager-Slave001.out
    Slave002: starting nodemanager, logging to /home/hadoop/software/hadoop-
    2.6.5/logs/yarn-hadoop-nodemanager-Slave002.out
```

4. 启动成功后出现的进程

启动成功后，在各个节点分别会出现以下进程。

```
[hadoop@Master001 ~]$ jps
    3766    Jps
    3510    ResourceManager
    3399    DFSZKFailoverController
    3118    NameNode
```

```
[hadoop@Master002 ~]$ jps
    2129    NameNode
    2289    Jps
    2230    DFSZKFailoverController

[hadoop@Slave001 ~]$ jps
    2610    NodeManager
    2500    JournalNode
    2358    QuorumPeerMain
    2745    Jps

[hadoop@Slave002 ~]$ jps
    2675    Jps
    2549    NodeManager
    2439    JournalNode
    2297    QuorumPeerMain

[hadoop@Slave003 ~]$ jps
    2737    Jps
    2611    NodeManager
    2501    JournalNode
    1788    QuorumPeerMain
```

5.3 常见问题汇总

（1）Hadoop 安全模式异常。

报错：mkdir: Cannot create directory /one. Name node is in safe mode.

原因：Hadoop 处于安全模式下。

解决：bin/hadoop dfsadmin -safemode leave（离开安全模式）。

（2）Hadoop 大数据集群两个 Master 均为 standby 状态。

报错：Operation category READ is not supported in state standby.

原因：Hadoop 集群异常，两个 NameNode 全部为 standBy 状态。DFSZKFailoverController 进程没有启动，因为没有做 hdfs zkfc-formatZK 格式化，在 Zookeeper 集群中没有创建 hadoop-ha 节点。正常情况 Master 有两种状态：active（活跃状态）和 standby（备用状态），其中正常的集群是一台 active，另一台必然是 standby。可以到 Zookeeper 节点上通过 "./zkCli.sh -server 192.168.xx.xx" 命令登录 Zookeeper 服务器，通过 ls /命令查看 Zookeeper 集群中是否有 hadoop-ha 节点。当检查到有 hadoop-ha 节点后，再到任意一台 Master 节点启动服务。

解决：在任意一台 Master 节点上执行 "hdfs zkfc –formatZK"（注：执行 "hdfs zkfc –formatZK" 命令时，Zookeeper 集群必须在启动状态）。

（3）Zookeeper 异常。

异常：Starting zookeeper…already running as process 1490.

原因：在临时文件下有个 zookeeper_server.pid 文件，这个文件是用来记录进程 id 的。由于机器意外断电异常关闭，导致 pid 文件残留。

解决：删除 zookeeper_server.pid 文件即可。

（4）集群中 namenode 和 datanode 启动后又挂掉。

异常：集群中 NameNode 或 DataNode 启动后，过一会儿又找不到进程。

原因：由于非法关机或其他误操作导致版本不一致。

解决：删除每个节点中存放版本号的 tmp 文件，在 Master001 节点执行"hdfs namenode –format"命令格式化 NameNode，然后将格式化后新生成的 tmp 拷贝用 scp 方式传输到各个节点。

```
[hadoop@Master001 hadoop-2.6.5]$ scp -r tmp/ Master002:~/software/hadoop-2.6.5/
[hadoop@Master001 hadoop-2.6.5]$ scp -r tmp/ Slave001:~/software/hadoop-2.6.5/
[hadoop@Master001 hadoop-2.6.5]$ scp -r tmp/ Slave002:~/software/hadoop-2.6.5/
[hadoop@Master001 hadoop-2.6.5]$ scp -r tmp/ Slave003:~/software/hadoop-2.6.5/
```

5.4　习题与思考

1．为什么要用 Hadoop 技术？它解决了哪些问题？Hadoop 技术有哪些优缺点？

2．尝试重新搭建 Hadoop 高可用集群，尝试解决非正常关机导致集群启动失败的问题。

第 6 章 HDFS 技术

HDFS（Hadoop Distributed File System）是 Hadoop 项目的核心子项目，是分布式计算中数据存储管理的基础，为了满足基于硬盘迭代模式访问和处理超大离线文件的需求而开发，可以运行于廉价的商用服务器或 PC 上。

HDFS 有以下优点。

（1）高容错性。上传的数据自动保存多个副本（默认 3 个副本）。它是通过增加副本的数量，来增加它的容错性的。如果某一个副本丢失，HDFS 机制会复制其他机器上的副本，而不必关注它的实现。

（2）适合大数据的处理。能够处理 GB、TB、甚至 PB 级别的数据和能够处理百万规模的数据。

（3）基于硬盘迭代的 I/O 写入。一次写入，多次读取。文件一旦写入，就不能修改，只能增加，这样可以保证数据的一致性。

（4）可以装在廉价的机器上。

HDFS 有以下缺点。

（1）低延时数据访问。它在低延时的情况下是不行的，它适合高吞吐率的场景，即在某一时间内写入大量的数据。对低延时要求高的情况一般使用 Spark 来完成，具体将在第 3 篇中详细讲解。

（2）小文件的存储。如果存放大量的小文件，它会大量占用 NameNode 的内存存储文件、目录、块信息，这样会对 NameNode 节点造成负担，小文件存储的寻道时间会超过文件的读取时间。这违背了 HDFS 的设计目标。

（3）不能并发写入，文件不能随机修改。一个文件只能一个线程写，不能多个线程同时写。仅支持文件的追加，不支持文件的随机修改。

6.1 HDFS 架构

HDFS 的高可用是 HDFS 持续对各类客户端提供读写服务的能力，因为客户端对 HDFS 的读写操作之前都要访问 NameNode 服务器，客户端只有从 NameNode 获取元数据之后才能继续进行读写。所以 HDFS 的高可用关键在于 NameNode 上元数据的持续可用。Hadoop 官方提供了一种 QuorumJournalManager 来实现高可用。在高可用配置下，editlog 存放在一个共享存储的地方，这个共享存储由若干个 JournalNode 组成，一般是 3 个节点（JN 集群），每个 JournalNode 专门用于存放来自 NameNode 的编辑日志（editlog），编辑日志由活跃状态的名称节点写入。

要有两个 NameNode 节点，二者之中只能有一个处于活跃状态（active），另一个是待命状态（standby）。只有 active 状态的节点才能对外提供读写 HDFS 服务，也只有 active 状态的节点才能向 JournalNode 写入编辑日志；standby 的名称节点只负责从 JN 集群中的 JournalNode 节点拷贝数据到本地存放。另外，各个 DataNode 会定期向两个 NameNode 节点报告自己的状态（心跳信息、块信息）。

一主一从的两个 NameNode 节点同时和 3 个 JournalNode 构成的组保持通信，活跃的 NameNode 节点负责往 JournalNode 集群写入编辑日志，待命的 NameNode 节点负责观察 JournalNode 组中的编辑日志，并且把日志拉取到待命节点，再加入到两节点各自的 fsimage 镜像文件，这样一来就能确保两个 NameNode 的元数据保持同步。一旦 active 不可用，提前配置的 Zookeeper 会把 standby 节点自动变为 active，继续对外提供服务。

对于 HA 群集的正确操作至关重要，因此一次只能有一个 NameNodes 处于活动状态。否则，命名空间状态将在两者之间迅速分歧，出现数据丢失或其他不正确的结果。为了确保这个属性并防止所谓的"分裂大脑情景"，JournalNodes 将只允许一个 NameNode 作为"领导者"。在故障切换期间，变为活动状态的 NameNode 将简单地接管写入 JournalNodes 的角色，这将有效地防止其他 NameNode 继续处于活动状态，允许新的 Active 安全地进行故障切换。HDFS 高可用架构如图 6.1 所示。

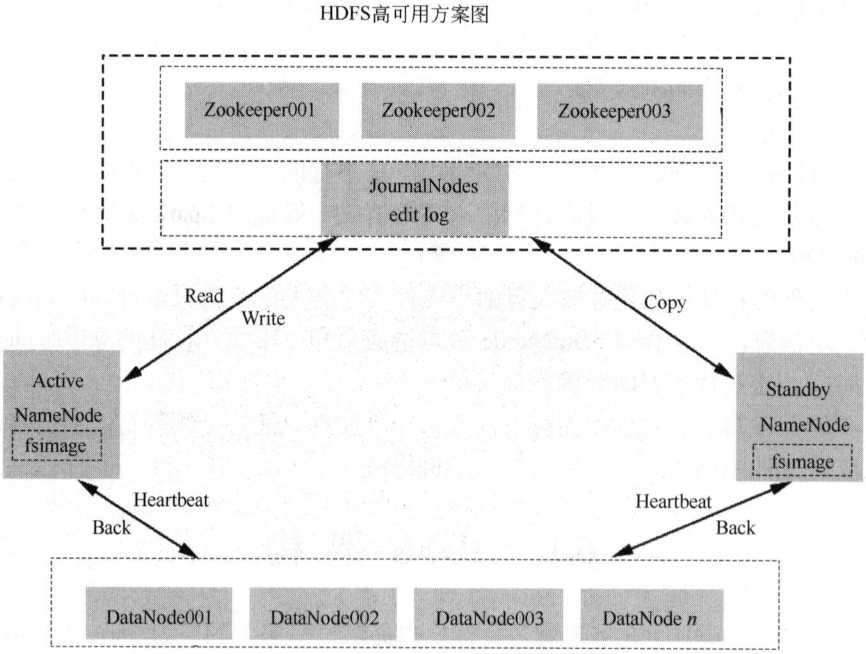

图 6.1

6.2 HDFS 命令

在 HDFS 中所有的 Hadoop 命令均由 bin/hadoop 脚本引发，不指定参数地运行 Hadoop

脚本会打印所有命令的描述。本节将介绍常用的 HDFS 命令的操作。

6.2.1 version 命令

用法：

```
hadoop version
```

version 命令可以打印 Hadoop 版本详细信息，示例如下：

```
[hadoop@Slave001 ~]$ hadoop version
    Hadoop 2.6.5
    Subversion https://github.com/apache/hadoop.git -r e8c9fe0b4c252caf
2ebf1464220599650f119997
    Compiled by sjlee on 2016-10-02T23:43Z
    Compiled with protoc 2.5.0
    From source with checksum f05c9fa095a395faa9db9f7ba5d754
    This command was run using /home/hadoop/software/hadoop-2.6.5/share/
hadoop/common/hadoop-common-2.6.5.jar
```

6.2.2 dfsadmin 命令

dfsadmin 命令可以查看集群存储空间使用情况和各个节点存储空间使用情况，示例如下：

```
[hadoop@Slave001 ~]# hadoop dfsadmin -report
    DEPRECATED: Use of this script to execute hdfs command is deprecated.
    Instead use the hdfs command for it.
    Configured Capacity: 37139136512 (34.59 GB)
    Present Capacity: 30914732032 (28.79 GB)
    DFS Remaining: 30734471168 (28.62 GB)
    DFS Used: 180260864 (171.91 MB)
    DFS Used%: 0.58%
    Under replicated blocks: 99
    Blocks with corrupt replicas: 0
    Missing blocks: 0

    -------------------------------------------------
    Live datanodes (2):

    Name: 192.168.153.201:50010 (Slave001)
    Hostname: Slave001
    Decommission Status : Normal
    Configured Capacity: 18569568256 (17.29 GB)
```

```
            DFS Used: 90128384 (85.95 MB)
            Non DFS Used: 3115909120 (2.90 GB)
            DFS Remaining: 15363530752 (14.31 GB)
            DFS Used%: 0.49%
            DFS Remaining%: 82.73%
            Configured Cache Capacity: 0 (0 B)
            Cache Used: 0 (0 B)
            Cache Remaining: 0 (0 B)
            Cache Used%: 100.00%
            Cache Remaining%: 0.00%
            Xceivers: 1
            Last contact: Tue Dec 26 22:33:14 CST 2017

            Name: 192.168.153.202:50010 (Slave002)
            Hostname: Slave002
            Decommission Status : Normal
            Configured Capacity: 18569568256 (17.29 GB)
            DFS Used: 90132480 (85.96 MB)
            Non DFS Used: 3108495360 (2.90 GB)
            DFS Remaining: 15370940416 (14.32 GB)
            DFS Used%: 0.49%
            DFS Remaining%: 82.77%
            Configured Cache Capacity: 0 (0 B)
            Cache Used: 0 (0 B)
            Cache Remaining: 0 (0 B)
            Cache Used%: 100.00%
            Cache Remaining%: 0.00%
            Xceivers: 1
            Last contact: Tue Dec 26 22:33:12 CST 2017
```

6.2.3 jar 命令

jar 命令是运行 jar 包文件命令。用户可以把它们的 MapReduce 代码捆绑到 jar 文件中，使用 jar 命令使程序运行起来。

格式：hadoop jar <jar> [mainClass]。

<jar>：jar 包。

[mainClass]：可选选项，指定运行主类。

使用 "adoop jar" 命令在 Hadoop 集群中运行 WordCount.jar 程序，示例如下：

```
[hadoop@Slave001 ~]# hadoop jar WordCount.jar
```

6.2.4 fs 命令

fs 参数是运行通用文件系统参数，在 Hadoop 后面跟上 fs 参数，表示是对 HDFS 中的文件进行操作。

格式：

```
hadoop fs [GENERIC_OPTIONS] [COMMAND_OPTIONS]
```

[GENERIC_OPTIONS]：通用选项。

[COMMAND_OPTIONS]：命令选项。

fs 常用的基本选项如下。

1. cat

cat 命令可以在 HDFS 中查看指定文件或指定文件夹下所有文件内容。

格式：

```
hadoop fs -cat <hdfs:pathFile>
```

例如，查看 HDFS 中 input 目录下所有文件内容。

```
[hadoop@Slave001 ~]# hadoop fs -cat /input/*
```

查看 HDFS 中 input 目录下 part-r-00000 文件中内容。

```
[hadoop@Slave001 ~]# hadoop fs -cat /input/part-r-00000
```

注：

（1）/input/*中的"*"代表所有，/input/*代表 input 目录下的所有文件。

（2）/input/part*代表 input 目录中文件名以 part 开始的所有文件。

2. copyFromLocal

copyFromLocal 命令类似于 put 命令，与 put 命令不同的之处，copyFromLocal 命令拷贝的源地址必须是本地文件地址。

格式：

```
hadoop fs -copyFromLocal <local:pathFile> <hdfs:pathDirectory>
```

3. copyToLocal

copyFromLocal 命令的作用与 put 命令很像，都是上传文件到 HDFS 中，它们的区别在于 copyToLocal 的源路径只能是一个本地文件，而 put 的源路径可能是多个文件，也可能是标准输入。

格式：

```
hadoop fs -copyToLocal <hdfs:pathFile> <local:pathDirectory>
```

除了限定目标路径是一个本地文件外，和 get 命令类似。

4. cp

格式：

```
hadoop fs -cp <hdfs:pathFile> <hdfs:pathDirectory>
```

cp 命令可以将 HDFS 中的指定文件复制到 HDFS 中目标路径。这个命令允许有多个源路径，此时目标路径必须是一个目录。

示例如下：

```
[hadoop@Slave001 ~]# hadoop fs -cp /user/hadoop/file1 /user/hadoop/file2
[hadoop@Slave001 ~]# hadoop fs -cp /user/hadoop/file1 /user/hadoop/file2
/user/hadoop/dir
```

5. du

格式：

```
hadoop fs -du <hdfs:pathDirectory>
```

du 命令是显示文件或文件夹属性的命令，可以显示指定文件的大小、多个指定文件的大小、指定目录中所有文件大小和指定多个目录中所有文件的大小。

示例如下：

```
[hadoop@Slave001 ~]# hadoop fs -du /input /output
    73962           /output/relation
    8829            /output/word
    14215           /output/word_count
    53              /input/2017-12-22.1514227013843
    41              /input/2017-12-26.1514227103154
    3297            /input/choose_column
```

6. dus

格式：

```
hadoop fs -dus <hdfs:pathDirectory>
```

dus 命令可以显示指定文件目录的大小或者指定多个文件的目录大小。

示例如下：

```
[hadoop@Slave001 ~]$ hadoop fs -dus /input
    3169072         /input
[hadoop@Slave001 ~]$ hadoop fs -dus /input /ouput
    3169072         /input
    8829            /ouput
```

7. expunge

expunge 命令的字面意思是"清除"，它在 Hadoop 中的作用是清空回收站。

格式：

```
hadoop fs -expunge
```

示例如下：

```
[hadoop@Slave001 ~]$ hadoop fs -expunge
```

```
17/12/26 22:31:55 INFO fs.TrashPolicyDefault: Namenode trash configuration:
Deletion interval = 0 minutes, Emptier interval = 0 minutes.
```

8. get

格式：

```
hadoop fs -get <hdfs:pathFile> <local:pathDirectory>
```

get 命令从 HDFS 中复制指定文件、指定目录下所有文件和指定多个文件到本地文件目录，执行 get 之前，本地文件目录必须事先存在。get 是一种常用的下载命令。

示例如下：

（1）在 HDFS 中复制 input 目录下 word_count 文件到本地 file 目录中。

```
[hadoop@Slave001 file]$ hadoop fs -get /input/word_count ~/file
```

（2）复制 HDFS 中 input 和 output 目录所有文件到本地 file 目录中。

```
[hadoop@Slave001 file]$ hadoop fs -get /input /output ~/file
[hadoop@Slave001 file]$ ls
    input    output
[hadoop@Slave001 file]$ cd input/
[hadoop@Slave001 input]$ ls
    2017-12-22.1514227013843    2017-12-26.1514227103154    choose_column
    city_data    major    monitor_data    score    word
```

9. getmerge

格式：

```
hadoop fs -getmerge <hdfs:pathDirectory> <hdfs:localFile>
```

将 HDFS 中指定目录下的所有文件加载到本地文件中。如果文件名不存在将在本地新创建文件，如果文件名存在，则覆盖文件内所有内容。

示例如下：

```
[hadoop@Slave001 file]$ hadoop fs -getmerge /output/word_count ~/file/output
[hadoop@Slave001 file]$ ls
    file    output
[hadoop@Slave001 file]$ cat output
    Just    出现：1次
    ...
    Not 出现：1次
    Noticing    出现：1次
```

10. ls

格式：

```
hadoop fs -ls <hdfs:pathDirectory>
```

在 HDFS 中显示指定文件的详细内容。如果是目录，则返回它直接子文件的列表。

详细内容包括：权限、用户、文件所在组、文件大小、创建日期和路径等信息。

示例如下：

（1）显示 HDFS 中 word 文件的详细属性。

```
[hadoop@Slave001 file]$ hadoop fs -ls /input/word
    -rw-r--r--   3 hadoop      supergroup 8829 2017-12-25 22:15 /input/word
```

（2）显示 HDFS 中 input 目录下所有文件的详细属性。

```
[hadoop@Slave001 file]$ hadoop fs -ls /input
    Found 4 items
    -rw-r--r--   3 hadoop supergroup      482 2017-12-26 00:19 /input/major
    -rw-r--r--   3 hadoop supergroup  2954128 2017-12-25 04:33 /input/monitor
    -rw-r--r--   3 hadoop supergroup    56812 2017-12-25 23:40 /input/score
    -rw-r--r--   3 hadoop supergroup     8829 2017-12-25 22:15 /input/word
```

11. lsr

格式：

```
hadoop fs -lsr <hdfs:pathDirectory>
```

lsr 命令是 ls -R 简写，是用来递归显示 HDFS 中指定目录下的所有子文件。

示例如下：

```
[hadoop@Slave001 file]$ hadoop fs -lsr /input
    lsr: DEPRECATED: Please use 'ls -R' instead.
    -rw-r--r--   3 hadoop supergroup      53 2017-12-26 02:36 /input/1514227013843
    -rw-r--r--   3 hadoop supergroup      41 2017-12-26 02:38 /input/1514227103154
    -rw-r--r--   3 hadoop supergroup    3297 2017-12-25 23:05 /input/choose
    -rw-r--r--   3 hadoop supergroup  145430 2017-12-25 04:52 /input/city_data
    -rw-r--r--   3 hadoop supergroup     482 2017-12-26 00:19 /input/major
    -rw-r--r--   3 hadoop supergroup 2954128 2017-12-25 04:33 /input/monitor
    -rw-r--r--   3 hadoop supergroup   56812 2017-12-25 23:40 /input/score
    -rw-r--r--   3 hadoop supergroup    8829 2017-12-25 22:15 /input/word
```

12. mkdir

格式：

```
hadoop fs -mkdir <paths>
```

mkdir 命令可以在 HDFS 中创建新目录，但它只能创建一级目录。创建多级目录上一级目录必须先存在，或者使用-p 参数。

示例如下：

（1）使用 mkdir 命令在 HDFS 的 input 目录下创建一个 file 目录。

```
[hadoop@Slave001 ~]$ hadoop fs -mkdir /input/file
[hadoop@Slave001 ~]$ hadoop fs -ls /input
    Found 4 items
    drwxr-xr-x   - hadoop supergroup       0 2017-12-26 23:30 /input/file
    -rw-r--r--   3 hadoop supergroup     482 2017-12-26 00:19 /input/major
```

```
-rw-r--r--   3 hadoop supergroup      56812 2017-12-25 23:40 /input/score
-rw-r--r--   3 hadoop supergroup       8829 2017-12-25 22:15 /input/word
```

（2）使用 mkdir 命令在 HDFS 中分别在 input 目录下创建一个 file2 目录，在 output 目录下也创建一个 file2 目录。

```
[hadoop@Slave001 ~]$ hadoop fs -mkdir /input/file2 /output/file2
[hadoop@Slave001 ~]$ hadoop fs -ls /input /output
Found 6 items
    drwxr-xr-x   - hadoop supergroup          0 2017-12-26 23:30 /input/file
    drwxr-xr-x   - hadoop supergroup          0 2017-12-26 23:32 /input/file2
    ...
    -rw-r--r--   3 hadoop supergroup       8829 2017-12-25 22:15 /input/word
Found 10 items
    drwxr-xr-x   - hadoop supergroup          0 2017-12-25 23:41 /output/averager
    drwxr-xr-x   - hadoop supergroup          0 2017-12-25 23:06 /output/choose
    ...
    drwxr-xr-x   - hadoop supergroup          0 2017-12-25 23:32 /output/file2
```

（3）使用 mkdir 命令在 HDFS 中创建一个多级目录/file/file1/file2/file3。

```
[hadoop@Slave001 ~]$ hadoop fs -mkdir -p /file/file1/file2/file3
[hadoop@Slave001 ~]$ hadoop fs -ls -R /file
    drwxr-xr-x   - hadoop supergroup          0 2017-12-26 23:35 /file/file1
    drwxr-xr-x   - hadoop supergroup          0 2017-12-26 23:35 /file/file1/file2
    drwxr-xr-x   - hadoop supergroup          0 2017-12-26 23:35 /file/file1/file2/file3
```

13. mv

格式：

```
hadoop fs -mv <hdfs:sourcepath> [hdfs:sourcepath ...] <hdfs:targetPath>
```

mv 命令可以在 HDFS 中将文件从源路径移动到目标路径，这个命令允许有多个源路径，此时目标路径必须是一个目录。

示例如下：

（1）将 HDFS 中的/input/major 文件移动到/file/file1/file2 中。

```
[hadoop@Slave001 ~]$ hadoop fs -mv /input/major /file/file1/file2
[hadoop@Slave001 ~]$ hadoop fs -ls /file/file1/file2
    Found 2 items
    drwxr-xr-x - hadoop supergroup     0 2017-12-26 23:35 /file/file1/file2/file3
    -rw-r--r-- 3 hadoop supergroup   482 2017-12-26 00:19 /file/file1/file2/major
```

（2）使用 mv 命令将 HDFS 中的/input/major 文件和/input/word 文件移动到/file/file1/file2/file3 目录中。

```
[hadoop@Slave001 ~]$ hadoop fs -mv /input/score /input/word /file/file1/file2/file3
[hadoop@Slave001 ~]$ hadoop fs -ls /file/file1/file2/file3
    Found 2 items
    -rw-r--r-- 3 hadoop supergroup 56812 2017-12-25 23:40 /file/file1/file2/file3/score
```

```
            -rw-r--r-- 3 hadoop supergroup 8829 2017-12-25 22:15 /file/file1/file2/file3/word
[hadoop@Slave001 ~]$ hadoop fs -ls /input
    Found 7 items
    -rw-r--r-- 3 hadoop supergroup 145430 2017-12-25 04:52 /input/city_data
    ...
    drwxr-xr-x - hadoop supergroup 0 2017-12-26 23:32 /input/file2
```

14. put

格式:

```
hadoop fs -put <local:pathFile> [local:pathFile] <hdfs:pathDirctory>
```

put 命令可以从本地文件系统中复制单个或多个源路径到目标文件系统。HDFS 中接收文件的目录必须事先存在。

示例: 从本地上传 city_data 文件和 monitor_data 文件到 HDFS 的 test 目录中。

```
[hadoop@Slave001 ~]$ ls
    aaaip9.jar  city_data  file    monitor_data   rrwquy.jar
[hadoop@Slave001 ~]$ hadoop fs -mkdir /test
[hadoop@Slave001 ~]$ hadoop fs -put city_data monitor_data /test
[hadoop@Slave001 ~]$ hadoop fs -ls /test
    Found 2 items
    -rw-r--r-- 3 hadoop supergroup 145430 2017-12-26 23:48 /test/city_data
    -rw-r--r-- 3 hadoop supergroup 2954128 2017-12-26 23:48 /test/monitor_data
```

15. rm

格式:

```
hadoop fs -rm <hdfs:pathFile> [hdfs:pathFile]
```

rm 命令是用于删除一个指定的文件或多个指定文件的命令,并且加上"-r"参数可以删除指定目录。

示例如下:

```
[hadoop@Slave001 ~]$ hadoop fs -ls /test
Found 2 items
-rw-r--r-- 3 hadoop supergroup 145430 2017-12-26 23:48 /test/city_data
-rw-r--r-- 3 hadoop supergroup  2954128 2017-12-26 23:48 /test/monitor_data
[hadoop@Slave001 ~]$ hadoop fs -rm /test/city_data /test/monitor_data
17/12/26 23:52:19 INFO fs.TrashPolicyDefault: Namenode trash configuration:
Deletion interval = 0 minutes, Emptier interval = 0 minutes.
Deleted /test/city_data
17/12/26 23:52:20 INFO fs.TrashPolicyDefault: Namenode trash configuration:
Deletion interval = 0 minutes, Emptier interval = 0 minutes.
Deleted /test/monitor_data
```

16. rmr

格式:

```
hadoop fs -rmr <hdfs:pathDirectory>[hdfs:pathDirectory]
```

rmr 命令可以删除目录或递归删除子文件，如果使用-rmr 命令删除一个目录时，不管目录下是否有其他文件，均将一并删除。

示例如下：

```
[hadoop@Slave001 ~]$ hadoop fs -ls -R /test /file
    drwxr-xr-x   - hadoop supergroup  0 2017-12-26 23:35 /file/file1
    drwxr-xr-x   - hadoop supergroup  0 2017-12-26 23:40 /file/file1/file2
    ...
    -rw-r--r--   3 hadoop supergroup 482 2017-12-26 00:19 /file/file1/file2/major
    -rw-r--r--   3 hadoop supergroup  36365 2017-12-20 21:51 /file/input
[hadoop@Slave001 ~]$ hadoop fs -rmr /test /file
    rmr: DEPRECATED: Please use 'rm -r' instead.
    17/12/26 23:55:49 INFO fs.TrashPolicyDefault: Namenode trash configuration:
    Deletion interval = 0 minutes, Emptier interval = 0 minutes.
    Deleted /test
    17/12/26 23:55:49 INFO fs.TrashPolicyDefault: Namenode trash configuration:
    Deletion interval = 0 minutes, Emptier interval = 0 minutes.
    Deleted /file
[hadoop@Slave001 ~]$ hadoop fs -ls -R /test /file
    ls: `/test': No such file or directory
    ls: `/file': No such file or directory
```

17. tail

格式：

```
hadoop fs -tail [-f] <hdfs:pathFile>
```

tail 命令可以将文件尾部 1KB 的内容输出到标准输出。并且 tail 命令支持 "-f" 选项，加上 "-f" 选项表示实时显示文件内容。

示例如下：

```
[hadoop@Slave001 ~]$ hadoop fs -tail /input/city_data
    90100751295        PMS_淳溪长一村4#   淳溪供电所      南京
    90100751318        PMS_新杨4#（加工厂边）淳溪供电所   南京
    ...
    90100796714        PMS_固城湖佳苑#1   南京市高淳区供电公司   南京
    90100796715        PMS_固城湖佳苑#2   南京市高淳区供电公司   南京
```

18. text

格式：

```
hadoop fs -text <hdfs:pathFile>
```

text 命令可以将 HDFS 中的源文件以文本格式输出。

19. touchz

格式：

```
hadoop fs -touchz <hdfs:newFile>
```

touchz 命令可以在 HDFS 中创建一个 0 字节的空文件。

示例如下：

```
[hadoop@Slave001 ~]$ hadoop fs -touchz /newfile
[hadoop@Slave001 ~]$ hadoop fs -ls /newfile
   -rw-r--r--   3 hadoop supergroup       0 2017-12-27 00:01 /newfile
```

6.3 API 的使用

关于 HDFS API 开发配置的讲解视频可以扫描以下两个二维码观看。

HDFS 是一个分布式文件系统。既然是文件系统，就可以对其文件进行操作，例如新建文件、删除文件、读取文件内容等。下文将讲解如何使用 Java API 对 HDFS 中的文件进行操作。

在 HDFS 中的文件操作涉及以下几个分类。

Configuration 类：该类的对象封装了客户端或者服务器的配置。

FileSystem 类：这类对象是一个文件系统对象，可以用该对象的一些方法来对文件进行操作。例如"FileSystem fs = FileSystem.get(conf);"通过 FileSystem 的静态方法 get 获得该对象。

FSDataInputStream 和 FSDataOutputStream：这两个类是 HDFS 中的输入输出流。分别通过 FileSystem 的 open 方法和 create 方法获得。

对 HDFS 文件进行操作的示例如下。

例 6.1 利用 HDFS API 在 HDFS 中创建新文件（讲解视频可扫二维码观看）。

```java
//入口
public static void main(String[] args)
    throws IOException {

String s = "hello world & 世界你好";
HdfsFile.createFile("/input", s.getBytes());

}
//在HDFS中创建新文件的方法
public static void createFile(String dst, byte[] contents)
    throws IOException {
    Configuration conf = new Configuration();
    FileSystem fs = FileSystem.get(conf);
    Path dstPath = new Path(dst); // 目标路径
    // 打开一个输出流
    FSDataOutputStream outputStream = fs.create(dstPath);
    outputStream.write(contents);
```

```
        outputStream.close();
        fs.close();
        System.out.println("文件创建成功！");
    }
```

例 6.2 利用 HDFS API 将本地中的文件上传到 HDFS 中（讲解视频可扫描二维码观看）。

```
//入口
public static void main(String[] args)
        throws IOException {

//从本地上传文件到HDFS中
//file: 是目录文件自动生成; input: 是文件名, 数据将放到这个文件名下
    HdfsFile.uploadFile("/home/hadoop/usernumber.txt","/file/input");
}
// 从本地上传文件到HDFS中
public static void uploadFile(String src, String dst)
        throws IOException {
    Configuration conf = new Configuration();
    FileSystem fs = FileSystem.get(conf);
    Path srcPath = new Path(src);  // 原路径
    Path dstPath = new Path(dst);  // 目标路径
// 调用文件系统的文件复制函数, 前面参数指是否删除原文件, true为删除, 默认为false
    fs.copyFromLocalFile(false, srcPath, dstPath);

    FileStatus[] fileStatus = fs.listStatus(dstPath);

    for (FileStatus file : fileStatus) {
        //查看目录文件路径
        System.out.println("HDFS中目标路径是: "+file.getPath());
        System.out.println("上传文件成功！");
    }
    fs.close();
}
```

例 6.3 利用 HDFS API 将 HDFS 中的指定文件删除。

```
//入口
public static void main(String[] args)
        throws IOException {
    //删除文件
    HdfsFile.delete("/input");
}
// 删除文件
public static void delete(String filePath)
        throws IOException {
    Configuration conf = new Configuration();
    FileSystem fs = FileSystem.get(conf);
    Path path = new Path(filePath);
    boolean isok = fs.deleteOnExit(path);
```

```java
    if (isok) {
        System.out.println("文件已经被删除");
    } else {
        System.out.println("操作失败");
    }
    fs.close();
}
```

例 6.4 利用 HDFS API 读取 HDFS 中指定的数据内容（讲解视频可扫描二维码观看）。

```java
//入口
public static void main(String[] args)
        throws IOException {
    //读取文件内容
    HdfsFile.readFile("/file/input");
}
```

```java
// 读取文件的内容
public static void readFile(String filePath)
        throws IOException {
    Configuration conf = new Configuration();
    FileSystem fs = FileSystem.get(conf);
    Path srcPath = new Path(filePath);
    InputStream in = null;
    try {
        in = fs.open(srcPath);
        // 复制到标准输出流，4096是缓冲大小
        IOUtils.copyBytes(in, System.out, 4096, false);
    } finally {
        IOUtils.closeStream(in);
    }
}
```

6.4 习题与思考

1. 尝试在 HDFS 中创建文件夹、上传文件、查询文件、拷贝文件和下载文件。

2. 尝试使用 HDFS API 在 HDFS 中创建文件夹、上传文件、查询文件、复制文件和下载文件。

第 7 章　MapReduce 技术

7.1　MapReduce 工作原理

7.1.1　MapReduce 作业运行流程

MapReduce 作业运行流程如图 7.1 所示。

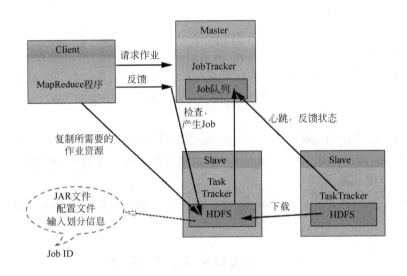

图　7.1

MapReduce 作业流程分析如下，相关讲解视频可扫描二维码观看。

（1）在客户端启动一个作业。

（2）向 JobTracker 请求一个 Job ID。

（3）将运行作业所需要的资源文件复制到 HDFS 上，包括 MapReduce 程序打包的 JAR 文件、配置文件和客户端计算所得的输入划分信息。这些文件都存放在 JobTracker 专门为该作业创建的文件夹中，文件夹名为该作业的 Job ID。JAR 文件默认会有 10 个副本（mapred.submit.replication 属性控制），输入划分信息告诉了 JobTracker 应该为这个作业启动多少个 map 任务等信息。

（4）JobTracker 接收到作业后，将其放在一个作业队列里，等待作业调度器对其进行调度，当作业调度器根据自己的调度算法调度到该作业时，会根据输入划分信息为每个划分创建一个 map 任务，并将 map 任务分配给 TaskTracker 执行。对于 map 和 reduce 任务，

TaskTracker 根据主机核的数量和内存的大小有固定数量的 map 槽和 reduce 槽。这里需要强调的是：map 任务不是随随便便地分配给某个 TaskTracker 的，这里有个概念：数据本地化（Data-Local）。意思是将 map 任务分配给含有该 map 处理的数据块的 TaskTracker，同时将程序 JAR 包复制到该 TaskTracker 来运行，这是"运算移动，数据不移动"，分配 reduce 任务时并不考虑数据本地化。

（5）TaskTracker 每隔一段时间会给 JobTracker 发送一个心跳，告诉 JobTracker 它依然在运行，同时心跳中还携带着很多信息，比如当前 map 任务完成的进度等信息。当 JobTracker 收到作业的最后一个任务完成信息时，便把该作业设置成"成功"。当 JobClient 查询状态时，它将得知任务已完成，便显示一条消息给用户。

7.1.2 早期 MapReduce 架构存在的问题

早期的 MapReduce 架构非常简单明了，在最初几年出现了众多的成功案例，获得业界广泛的支持和肯定，但随着分布式系统集群的规模和其工作负荷的增长，原框架的问题逐渐浮出水面，主要集中在如下几方面。

（1）JobTrack 单点故障问题。由于 JobTrack 是 MapReduce 的集中处理点，如果出单点故障，集群将不能使用，集群的高可用性也就得不到保障。

（2）JobTrack 任务过重。JobTracker 节点完成了太多的任务，会造成过多的资源消耗，当 Job 任务非常多的时候，会造成很大的内存开销，潜在地增加了 JobTracker 节点死机的风险。

（3）容易造成 TaskTracker 端内存溢出。在 TaskTracker 端，是以 map 或 reduce 的任务数量作为资源的，没有考虑到内存的占用情况，如果两个大内存消耗的任务被调度到一块，这样很容易出现内存溢出。

（4）容易造成资源浪费。在 TaskTracker 端，把资源强制划分为 map 任务和 reduce 任务，如果当系统中只有 map 任务或者只有 reduce 任务时，会造成资源的浪费。

7.2 YARN 运行概述

7.2.1 YARN 模块介绍

从业界使用分布式系统的变化趋势和 Hadoop 框架的长远发展来看，MapReduce 的 JobTracker 和 TaskTracker 机制需要大规模的调整来修复它在可扩展性、内存消耗、可靠性和性能上的缺陷。在过去的几年中，Hadoop 开发团队做了一些 bug 的修复，但是最近这些修复的成本越来越高，这表明对原框架做出改变的难度越来越大。

为了从根本上解决旧 MapReduce 框架的性能瓶颈，促进 Hadoop 框架的更长远发展，从 Hadoop 0.23.0 版本开始，Hadoop 的 MapReduce 框架完全重构，发生了根本的变化，新的 Hadoop MapReduce 框架命名为 YARN。

YARN 是一个资源管理、任务调度的框架，主要包含三大模块：ResourceManager（RM）、NodeManager（NM）、ApplicationMaster（AM）。其中，ResourceManager 负责所有资源的监控、分配和管理；ApplicationMaster 负责每一个具体应用程序的调度和协调；NodeManager

负责每一个节点的维护；对于所有的 applications，ResourceManager 拥有绝对的控制权和资源的分配权，而每个 ApplicationMaster 则会和 ResourceManager 协商资源，同时和 NodeManager 通信来执行和监控 task。

1. ResourceManager

ResourceManager 负责整个集群的资源管理和分配，是一个全局的资源管理系统。NodeManager 以心跳的方式向 ResourceManager 汇报资源使用情况（目前主要是 CPU 和内存的使用情况）。ResourceManager 只接收 NodeManager 的资源回报信息，对于具体的资源处理则交给 NodeManager 处理。YARN Scheduler 根据 application 的请求为其分配资源，不负责 Application Job 的监控、追踪、运行状态反馈、启动等工作。

2. NodeManager

NodeManager 是每个节点上的资源和任务管理器，是管理这台机器的代理，负责该节点程序的运行，以及该节点资源的管理和监控。YARN 集群每个节点都运行一个 NodeManager。NodeManager 定时向 ResourceManager 汇报本节点资源（CPU、内存）的使用情况和 Container 的运行状态。当 ResourceManager 死机时 NodeManager 自动连接 ResourceManager 备用节点。NodeManager 接收并处理来自 ApplicationMaster 的 Container 启动、停止等各种请求。

3. ApplicationMaster

用户提交的每个应用程序均包含一个 ApplicationMaster，它可以运行在 ResourceManager 以外的机器上，有以下作用：

（1）负责与 ResourceManager 调度器协商以获取资源（用 Container 表示）。

（2）将得到的任务进一步分配给内部的任务（资源的二次分配）。

（3）与 ResourceManager 通信以启动/停止任务。

（4）监控所有任务运行状态，并在任务运行失败时重新为任务申请资源以重启任务。

（5）当前 YARN 自带了两个 ApplicationMaster 实现，一个是用于演示 ApplicationMaster 编写方法的实例程序 DistributedShell，它可以申请一定数目的 Container 以并行运行一个 Shell 命令或者 Shell 脚本；另一个是运行 MapReduce 应用程序的 AM—MRAppMaster。

7.2.2 YARN 工作流程

运行在 YARN 上的应用程序主要分为两类：短应用程序和长应用程序。其中，短应用程序是指一定时间内可运行完成并正常退出的应用程序，比如 MapReduce 作业、Tez DAG 作业等；长应用程序是指不出意外，永不终止运行的应用程序，通常是一些服务，比如 Storm Service（主要包括 Nimbus 和 Supervisor 两类服务），HBase Service（包括 Hmaster 和 RegionServer 两类服务）等，它们本身作为一个框架提供编程接口供用户使用。尽管这两类应用程序作用不同，一类直接运行数据处理程序，一类用于部署服务（服务之上再运行数据处理程序），但运行在 YARN 上的流程是相同的。

当用户向 YARN 中提交一个应用程序后，YARN 将分两个阶段运行该应用程序：第一个阶段是启动 ApplicationMaster；第二个阶段是由 ApplicationMaster 创建应用程序，为它申请资源，并监控它的整个运行过程，直到运行完成，YARN 的工作流程如图 7.2 所示。

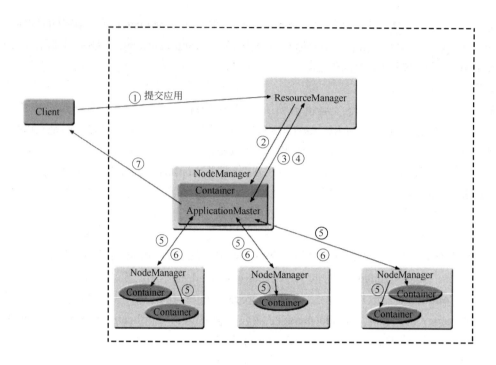

图 7.2

（1）Client 向 ResourceManager 提交应用程序，包括启动该应用的 ApplicationMaster 的必需信息，例如 ApplicationMaster 程序、启动 ApplicationMaster 的命令、用户程序等。

（2）ResourceManager 在 NodeManager 中启动一个 Container 用于运行 ApplicationMaster。

（3）启动中的 ApplicationMaster 向 ResourceManager 注册自己，启动成功后与 Resource Manager 保持心跳。

（4）ApplicationMaster 向 ResourceManager 发送请求，申请相应数目的 Container。

（5）ResourceManager 返回 ApplicationMaster 申请的 Container 信息。申请成功的 Container 由 ApplicationMaster 进行初始化。Container 的启动信息初始化后，ApplicationMaster 与对应的 NodeManager 通信，要求 NodeManager 启动 Container。

（6）ApplicationMaster 与 NodeManager 保持相同，从而对 NodeManager 上运行的任务进行监控和管理。Container 运行期间 ApplicationMaster 对 Container 进行监控。Container 通过 RPC 协议向对应的 ApplicationMaster 汇报自己的进度和状态等信息。

（7）应用运行期间 Client 直接与 ApplicationMaster 通信获取应用的状态、进度更新等信息。

（8）应用程序运行完成后，ApplicationMaster 向 ResourceManager 注销并关闭自己，并允许属于它的 Container 被收回。

7.3 MapReduce 编程模型

从 MapReduce 自身的命名特点可以看出，MapReduce 由两个阶段组成：map 和 reduce。

用户只需编写 map()和 reduce()两个函数，即可完成简单的分布式程序的设计。

map()函数以 key/value 对作为输入，产生另外一系列 key/value 对作为中间输出写入本地磁盘。MapReduce 框架会自动将这些中间数据按照 key 值进行聚集，且 key 值相同（用户可设定聚集策略，默认情况下是对 key 值进行哈希取模）的数据统一交给 reduce()函数处理。

reduce()函数以 key 及对应的 value 列表作为输入。经合并 key 相同的 value 值后，产生另外一系列 key/value 对作为最终输出写入 HDFS。

下面以 MapReduce 中的"hello world"程序——WordCount 为例介绍程序设计方法。

"hello world"程序是学习任何一门编程语言时编写的第一个程序。它简单且易于理解，能够帮助读者快速入门。同样，分布式处理框架也有自己的"hello world"程序：WordCount。它完成的功能是统计输入文件中的每个单词出现的次数。

其中 Map 部分代码如图 7.3 所示。

```
// key:字符串偏移量
// value:一行字符串内容
map(String key, String value) :
  // 将字符串分割成单词
  words = SplitIntoTokens(value);
  for each word w in words:
    EmitIntermediate(w, "1");

Reduce部分如下：

// key:一个单词
// values:该单词出现的次数列表
reduce(String key, Iterator values):
  int result = 0;
  for each v in values:
    result += StringToInt(v);
  Emit(key, IntToString(result));
```

图 7.3

用户编写完 MapReduce 程序后，按照一定规则指定程序的输入和输出目录，并提交到 Hadoop 集群中。Hadoop 将输入数据切分成若干个输入分片（input split，后面简称 split），并将每个 split 交给一个 Map Task 处理，Map Task 不断地从对应的 split 中解析出一个个 key/value，并调用 map()函数处理，处理完之后根据 Reduce Task 个数将结果分成若干个分片（partition）写到本地磁盘。同时，每个 Reduce Task 从每个 Map Task 上读取属于自己的那个 partition，然后使用基于排序的方法将 key 相同的数据聚集在一起，调用 reduce()函数处理，并将结果输出到文件中。

上面的程序还缺少三个基本的组件，功能分别如下：

（1）指定输入文件格式。将输入数据切分成若干个 split，且将每个 split 中的数据解析成一个个 map()函数要求的 key/value 对。

（2）确定 map()函数产生的每个 key/value 对发给哪个 Reduce Task 函数处理。

（3）指定输出文件格式，即每个 key/value 对以何种形式保存到输出文件中。

在 Hadoop MapReduce 中，这三个组件分别是 InputFormat、Partitioner 和 OutputFormat，它们均需要用户根据自己的应用需求配置。而对于上面的 WordCount 例子，默认情况下 Hadoop 采用的默认实现正好可以满足要求，因而不必再提供。

综上所述，Hadoop MapReduce 对外提供了 5 个可编程组件，分别是 InputFormat、Mapper、Partitioner、Reducer 和 OutputFormat。

7.4 MapReduce 数据流

Hadoop 的核心组件工作流水线如图 7.4 所示。

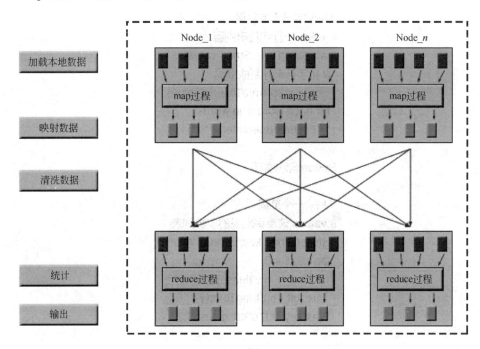

图 7.4

MapReduce 的输入一般来自 HDFS 中的文件，这些文件分别存储在集群内的节点上。运行一个 MapReduce 程序会在集群的许多节点甚至所有节点上运行 mapping 任务。每一个 mapping 任务都是平等的，Mapper 没有特定"标识物"与其关联。因此，任意的 mapper 都可以处理任意的输入文件。每一个 mapper 会加载一些存储在运行节点本地的文件集来进行处理（这是移动计算，把计算移动到数据所在节点可以避免额外的数据传输开销）。

当 Mapping 阶段完成后，这阶段所生成的中间键值对数据必须在节点间进行交换，把具有相同键的数值发送到同一个 reducer。reduce 任务在集群内的分布节点同 mapper 的一样。这是 MapReduce 中唯一的任务节点间的通信过程。map 任务间不会进行任何的信息交换，也不会去关心其他 map 任务的存在。同理，不同的 reduce 任务之间也不会有通信。用户不能显式地从一台机器发送信息到另外一台机器；所有数据传送都是由 Hadoop MapReduce 平台自身去做的，这些是通过关联到数值上的不同键来隐式引导的，这是 Hadoop MapReduce 的可靠性的基础元素。如果集群中的节点失效了，任务会被重新启动。如果任务已经执行了有副作用的操作，比如跟外面进行通信，共享状态必须存在可以重启的任务上。消除了通信和副作用，重启就会更优雅。

在图 7.5（a）中，描述了 Hadoop MapReduce 的高层视图。可以看到 mapper 和 reducer 组件是如何用到词频统计程序中的，是如何完成它们的目标的。如图 7.5（b）所示是近距离看这个系统的细节。

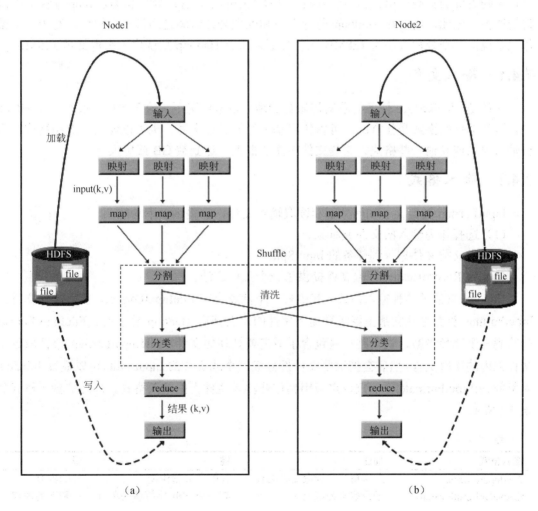

图 7.5

从图 7.5 中可以看出当执行一个 job 时，是从当地的 HDFS 存储加载文件。Hadoop 会将输入数据划分成等长的数据块，成为数据分片。Hadoop 会为每个分片构建一个 map 任务，然后进入到 Shuffle 阶段。Shuffle 是 map 和 reduce 之间的过程，包含了 partition（分割）和 combine（分类）两部分，它只是理论上存在，属于 MapReduce 框架，主要作用是把 map 任务的每个分片拆分，然后按相同内容进行分组，是以<key,value>形式将结果存在一个临时空间，reduce 从临时空间里读取<key,value>结果，通过 key 进行分组，相同 key 的 value 值作统计计算，把统计出来的结果利用 contest.write(key,value)语句写入到 HDFS 中，以一个新的文件形式存在。

可以把 MapRedue 过程看成一条 SQL 语句"select key, sum(value) from talbe group by key"。map 部分就像是没有加 group by 的 SQL 语句，只是对字段的映射，reduce 部分像是加了 group by 的 SQL 语句，先要对 key 进行分组，然后对相同 key 的 value 进行统计。

注：并行的处理分片时间肯定会少于处理整个大数据块的时间，但由于各个节点性能及作业运行情况的不同，每个分片的处理时间可能不一样。因此，把数据分片切分得更细可以得到更好的负载均衡。但另一方面，分片太小的话，管理分片和构建 map 任务的时间将会增多。因此，需要在 Hadoop 分片大小和处理分片时间之间做一个权衡。对于大多数作业来说，一个分片大小为 128MB 比较合适，所有 Hadoop 默认块大小也就是 128MB。

7.4.1 输入文件

文件是 MapReduce 任务的数据初始存储地。正常情况下，输入文件一般是存在 HDFS 中。这些文件的格式是任意的，可以使用基于行的日志文件，也可以使用二进制格式，多行输入记录或其他一些格式。这些文件往往会很大，达到数 GB 或更大。

7.4.2 输入格式

InputFormat 类定义了如何分割和读取输入文件，它提供以下功能：
（1）选择作为输入的文件或对象。
（2）定义把文件划分到任务的 InputSplits。
（3）为 RecordReader 读取文件提供了一个工厂方法。

Hadoop 自带了多种输入格式。其中有一个抽象类叫 FileInputFormat，所有操作文件的 InputFormat 类都是从它那里继承功能和属性的。当开启 Hadoop 作业时，FileInputFormat 会得到一个路径参数，这个路径内包含了所需要处理的文件。FileInputFormat 会读取这个文件夹内的所有文件，然后会把这些文件拆分成一个或多个的 InputSplit。可以通过 JobConf 对象的 setInputFormat()方法来设定应用到作业输入文件上的输入格式。标准的输入格式如表 7.1 所示。

表 7.1

输入格式	描述	键	值
TextInputFormat	默认格式，读取文件的行	行的字节偏移量	行的内容
KeyValueInputFormat	把行解析为键值对	第一个 Tab 字符前的所有字符	行剩下的内容
SequenceFileInputFormat	Hadoop 定义的高性能二进制格式	用户自定义	用户自定义

默认的输入格式是 TextInputFormat。它把输入文件每一行作为单独的一个记录，但不做解析处理。这对那些没有被格式化的数据或是基于行的记录来说是很有用的，比如日志

文件。KeyValueInputFormat 这个格式也是把输入文件每一行作为单独的一个记录。然而不同的是，TextInputFormat 把整个文件行当做值数据，KeyValueInputFormat 则是通过搜寻 Tab 字符来把行拆分为键值对，这在把一个 MapReduce 的作业输出作为下一个作业的输入时显得特别有用，因为默认输出格式正是按 KeyValueInputFormat 格式输出数据。

最后来讲解 SequenceFileInputFormat。它会读取特殊的特定于 Hadoop 的二进制文件，这些文件包含了很多能让 Hadoop 的 mapper 快速读取数据的特性，Sequence 文件是块压缩的并提供了对几种数据类型直接的序列化与反序列化操作。Squence 文件可以作为 MapReduce 任务的输出数据，并且用它做一个 MapReduce 作业到另一个作业的中间数据是很高效的。

7.4.3 数据片段

一个输入块描述了构成 MapReduce 程序中单个 map 任务的一个单元。把一个 MapReduce 程序应用到一个数据集上，即是指一个作业，会由多个任务组成，Map 任务可能会读取整个文件，但一般是读取文件的一部分。默认情况下，FileInputFormat 及其子类会以 128MB 为基数来拆分文件。可以在 mapred-default.xml 文件内设定 mapred.min.split.size 参数来控制具体划分大小，或者在具体 MapReduce 作业的 JobConf 对象中重写这个参数。通过以块形式处理文件，可以让多个 map 任务并行地操作一个文件。如果文件非常大的话，这个特性可以通过并行处理大幅地提升性能。更重要的是，因为多个块（Block）组成的文件可能会分散在集群内的好几个节点上，这样就可以把任务调度在不同的节点上，因此所有的单个块都是本地处理的，而不是把数据从一个节点传输到另外一个节点。

输入格式定义了组成 mapping 阶段的 map 任务列表，每一个任务对应一个输入块。接着根据输入文件块所在的物理地址，这些任务会被分派到对应的系统节点上，可能会有多个 map 任务被分派到同一个节点上。任务分派好后，节点开始运行任务，尝试去最大并行化执行。节点上的最大任务并行数由 mapred.tasktracker.map.tasks.maximum 参数控制。

7.4.4 记录读取器

RecordReader 类是用来加载数据并把数据转换为适合 mapper 读取的键值对。RecordReader 实例是由输入格式定义的，默认的输入格式 TextInputFormat，提供了一个 LineRecordReader。这个类会把输入文件的每一行作为一个新的值，关联到每一行的键则是该行在文件中的字节偏移量。RecordReader 会在输入块上被重复地调用直到整个输入块被处理完毕，每一次调用 RecordReader 都会调用 Mapper 的 map()方法。

而 SequenceFileInputFormat 对应的 RecordReader 是 SequenceFileRecordReader。LineRecordReader 是每行的偏移量作为读入 map 的 key，每行的内容作为读入 map 的 value。很多时候 hadoop 内置的 RecordReader 并不能满足需求，比如在读取记录时，希望 map 读入的 key 值不是偏移量而是行号或者是文件名，这时候可以自定义 RecordReader。

7.4.5 Mapper

Mapper 执行了 MapReduce 程序第一阶段中有趣的用户定义的工作。给定一个键值对，map()方法会生成一个或多个键值对，这些键值对会被送到 Reducer 那里。对于整个作业输入部分的每一个 map 任务（输入块），每一个新的 Mapper 实例都会在单独的 Java 进程中被初始化，Mapper 之间不能进行通信。这就使得每一个 map 任务的可靠性不受其他 map 任务的影响，只由本地机器的可靠性来决定。

map()方法除了键值，对外还会接收额外的两个参数。

（1）OutputCollector 对象有一个叫 collect()的方法，它可以利用该方法把键值对送到作业的 reduce 阶段。

（2）Reporter 对象提供当前任务的信息，它的 getInputSplit()方法会返回一个描述当前输入块的对象，并且还允许 map 任务提供关于系统执行进度的额外信息。setStatus()方法允许生成一个反馈给用户的状态消息，incrCounter()方法允许递增共享的高性能计数器，除了默认的计数器外，还可以定义更多的想要的计数器。每一个 Mapper 都可以递增计数器，JobTracker 会收集由不同处理得到的递增数据并把它们聚集在一起以供作业结束后的读取。

7.4.6 Shuffle

Shuffle 的本义是洗牌、混洗，把一组有一定规则的数据尽量转换成一组无规则的数据，越随机越好。MapReduce 中的 Shuffle 更像是洗牌的逆过程，把一组无规则的数据尽量转换成一组具有一定规则的数据。

为什么 MapReduce 计算模型需要 Shuffle 过程？MapReduce 计算模型一般包括两个重要的阶段：map 是映射，负责数据的过滤分发；reduce 是归约，负责数据的计算归并。reduce 的数据来源于 map，map 的输出即是 reduce 的输入，reduce 需要通过 Shuffle 来获取数据。

从 map 输出到 reduce 输入的整个过程可以广义地称为 Shuffle。Shuffle 横跨 map 端和 reduce 端，在 map 端包括 split 过程，在 reduce 端包括 copy 和 sort 过程，如图 7.6 所示。

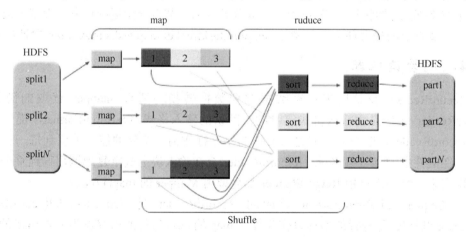

图 7.6

在 map 端的 Shuffle 过程是对 map 的结果进行分区（partition）、排序（sort）和分割（split），然后将属于同一个划分的输出合并在一起（merge）并写在硬盘上，同时按照不同的划分将结果发送给对应的 reduce。reduce 端又会将各个 map 送来的属于同一个划分的输出进行合并（merge），然后对 merge 的结果进行排序，最后交给 reduce 处理。通俗地讲，就是对

map 输出结果先进行分区（partition），如 "aaa" 经过 Partitioner 后返回 0，也就是这对值应当交由第一个 reducer 来处理。接下来，需要将数据写入内存缓冲区中。缓冲区的作用是批量收集 map 结果，减少磁盘 IO 的影响。key/value 对以及 Partition 的结果都会被写入缓冲区。当然写入之前，key 与 value 值都会被序列化成字节数组。这个内存缓冲区是有大小限制的，默认为 100MB（80%）。当 map task 的输出结果很多时，需要在一定条件下将缓冲区中的数据临时写入磁盘，然后重新利用这块缓冲区。这个从内存往磁盘写数据的过程被称为 spill。spill 可以认为是一个包括 sort 和 combiner（Combiner 是可选的，用户如果定义就有）的过程。先进行 sort 可以把缓冲区中一段范围 key 的数据排在一起，（如果数据多的时候，多次刷新往内存缓冲区中写入的数据可能会有属于相同范围的 key，也就是说，多个 spill 文件中可能会有统一范围的 key，这就是需要下面 map 端 merge 的原因）。具体的介绍可以看下面的详细过程。执行过 sort 之后，如果用户定义了 combiner 就会执行 combine，然后执行 merge 操作，接着就是 reduce 端。

7.4.7 排序

每一个 reduce 任务负责归约（reduceing）关联到相同键上的所有数值，每一个节点收到的中间键集合在被送到具体的 reducer 那里前已经被 Hadoop 自动排序。

7.4.8 归约

每个 reduce 任务都会创建一个 reducer 实例。这是一个用户自定义代码的实例，负责执行特定作业的第二个重要的阶段。对于每一个已赋予到 reducer 的 partition 内的键来说，reducer 的 reduce() 方法只会调用一次。它会接收一个键和关联到键的所有值的一个迭代器，迭代器会以一个未定义的顺序返回关联到同一个键的值。reducer 同时要接收一个 OutputCollector 和 Report 对象，它们像在 Map() 方法中那样被使用。

7.4.9 输出格式

提供给 OutputCollector 的键值对会被写到输出文件中，写入的方式由输出格式控制。OutputFormat 的功能与前面描述的 InputFormat 类相似。Hadoop 提供的 OutputFormat 的实例会把文件写在本地磁盘或 HDFS 上，它们都是继承自公共的 FileInputFormat 类。每一个 reducer 会把结果输出写在公共文件夹中一个单独的文件内。这些文件的命名一般是 part-xxxxx，xxxxx 是关联到某个 reduce 任务的 partition 的 id，输出文件夹通过 FileOutputFormat.setOutputPath() 来设置。可以通过具体 MapReduce 作业的 JobConf 对象的 setOutputFormat() 方法来设置具体用到的输出格式，输出格式如表 7.2 所示。

表 7.2

输出格式	描述
TextOutputFormat	默认的输出格式，以 "key \t value" 的方式输出行
SequenceFileOutputFormat	输出二进制文件，适合读取为 MapReduce 作业的输入
NullOutputFormat	忽略收到的数据，即不做输出

Hadoop 提供了一些 OutputFormat 实例用于写入文件，基本的（默认的）实例是

TextOutputFormat，它会以一行一个键值对的方式把数据写入一个文本文件。这样接下来的 MapReduce 任务就可以通过 KeyValueInputFormat 类简单地重新读取所需的输入数据，而且也适合人们阅读。更适合于在 MapReduce 作业间使用的中间格式是 SequenceFileOutputFormat，它可以通过快速地序列化任意的数据类型到文件中，而对应 SequenceFileInputFormat 则会把文件反序列化为相同的类型并提交为下一个 Mapper 的输入数据，方式和前一个 Reducer 的生成方式一样。NullOutputFormat 不会生成输出文件并丢弃任何通过 OutputCollector 传递给它的键值对。如果想要 reduce()方法中显示输出文件并且不想被 Hadoop 框架输出额外的空输出文件，这个类可以满足要求。

recordWriter：与 InputFormat 中通过 RecordReader 读取单个记录的实现相似，OutputFormat 类是 RecordWriter 对象的工厂方法，用来把单个的记录写到文件中，如同 OuputFormat 直接写入。

Reducer 输出的文件会留在 HDFS 上供其他应用使用。比如另外一个进行 MapReduce 作业，或一个给人工检查的单独程序。

7.5 MapReduce API 编程

7.5.1 词频统计

词频分析是对文章中重要词汇出现的次数进行统计与分析，是文本挖掘的重要手段。它是文献计量学中传统的和具有代表性的一种内容分析方法，基本原理是通过词出现频次多少的变化，来确定热点及其变化趋势。

词频统计是 Hadoop 编程中最为经典的例子，其思路简单：首先将 HDFS 中的文件读取出来，然后在 Mapper 类中使用 Stringokenizer 对象将字符串分隔，用 st.nextToken()方法循环读取被分隔的字符以 key 形式写入到 HDFS 中；并给每个字符串作上标记，表示这个字符串出现过一次；最后在 Reduce 类中对相同的 key 分组，对 values 值求和。由于 values 值是以字符形式传到 Reduce 类，在做求和计算前需要先将字符类型转为 int 类型。

配置类：

```java
package com.bunfly.word_count;
import org.apache.hadoop.conf.Configuration;
import org.apache.hadoop.fs.Path;
import org.apache.hadoop.io.Text;
import org.apache.hadoop.mapreduce.Job;
import org.apache.hadoop.mapreduce.lib.input.FileInputFormat;
import org.apache.hadoop.mapreduce.lib.output.FileOutputFormat;

public class Word_Count {
    public static void main(String[] args) throws Exception {
        //获取配置文件中的参数
        Configuration conf = new Configuration();
        //通过参数配置获取一个job对象
```

```java
        Job job = Job.getInstance(conf,"Word_Count");
        //启动job需要知道的入口
        job.setJarByClass(Word_Count.class);
        //启动Map入口
        job.setMapperClass(Word_Count_Mapper.class);
        //启动Reduce入口
        job.setReducerClass(Word_Count_Reduce.class);
        //设置Reduce输出的数量
        job.setNumReduceTasks(1);
        //key值输出类型
        job.setOutputKeyClass(Text.class);
        //value值输出类型
        job.setOutputValueClass(Text.class);
        //设置输入源
        String inl = "/input/word";
        //设置输出路径
        String out = "/output/word_count";
        Path input = new Path(inl);
        Path output = new Path(out);
        //将路径加入到文件格式中
        FileInputFormat.addInputPath(job, input);
        FileOutputFormat.setOutputPath(job, output);
        System.out.println(job.waitForCompletion(true) ? 0 : 1);
    }
}
```

Mapper 类：

```java
package com.bunfly.word_count;
import java.io.IOException;
import java.util.StringTokenizer;
import org.apache.hadoop.io.Text;
import org.apache.hadoop.mapreduce.Mapper;

public class Word_Count_Mapper extends Mapper<Object, Text, Text, Text> {
    public void map (Object key, Text value, Context contex) throws
    IOException, InterruptedException {

        String s = value.toString();
        StringTokenizer st = new StringTokenizer(s);
        while (st.hasMoreElements()) {
            String wd = st.nextToken();
            //过滤符号
            String word = wd.replace(",", "").replace(".", "").replace(":",
            "").replace("“", "").replace("”", "").replace(";", "").replace
```

```
                ("?", "").replace("(", "").replace(")", "");
                contex.write(new Text(word), new Text("1"));
            }
        }
    }
```

Reduce 类：

```
package com.bunfly.word_count;
import java.io.IOException;
import org.apache.hadoop.io.Text;
import org.apache.hadoop.mapreduce.Reducer;

public class Word_Count_Reduce extends Reducer<Text, Text, Text, Text> {
    public void reduce (Text key, Iterable<Text> values, Context context)
    throws IOException, InterruptedException {
        int sum=0;
        for (Text val : values) {
            String[] q = val.toString().split("\t");
            int num = Integer.parseInt(q[0]);
            sum = sum + num;
        }
        context.write(key, new Text("出现: "+sum+"次"));
    }
}
```

7.5.2 指定字段

指定字段是 Hadoop 编程中非常适用的编写方式。它只在 Mapper 类中编写，将文件字段映射到 HDFS 中。如果一个文件字段特别多，而且只需要其中几个字段，可以只在 Mapper 类中编写，列出需要的字段并写入到 HDFS 中即可。

例如，房产登记数据只需要购买人姓名、电话和楼盘名称，就可以在 Mapper 类中实现，数据结构如图 7.7 所示。

0	1	2	3	4	5	6	7	8	9	10	11
houses_name	deal_time	house_type_id	area	unit_price	price	payment_method	district_id	name	phone		
楼盘名称	交易时间	房屋类型代码	建筑面积	单价	交易总价	付款方式	区域编码	购买人	联系电话	证件类型	证件号
天祥尚府	2017/1/1	201	85.49	12113	1035540	全款	510105	袁善	13163260059	201	130925197805141579
环球汇蔚然	2017/4/3	201	57.68	12466	719038.9	商贷	510104	姬蓉	13102709059	201	130823196108255263
绿地中心星朗	2017/3/6	201	154.5	13380	2067210	公积金	510104	宇栽	13102893689	201	510104197009255491
朗御	2017/6/27	201	131.84	18708	2466463	全款	510104	甄品	13103006776	201	511304196805236437
华瑞中心	2017/1/1	201	146.26	12149	1776913	公积金	510107	卓艳	13719858228	201	510132199808232927
九龙仓擎天半岛	2017/4/1	201	122.57	17698	2169244	商贷	510104	吴惭	13103711420	201	513435198402186869
绿地锦天府	2017/3/27	201	123.6	15317	1893181	公积金	510104	于座	13104479059	201	513435197803174887
吉宝凌云峰阁	2017/7/12	201	126.69	17410	2205673	公积金	510104	宇乔谢	13106624915	201	510182197403247818
伊泰天骄	2017/7/20	201	123.6	15331	1894912	全款	510104	戴牲	13107756757	201	131103199004131645
星汇广场	2017/9/25	201	133.9	13267	1776451	公积金	510104	卫伸	13107832450	201	511001197302243973
阿玛尼艺术公寓	2017/2/13	201	132.87	13606	1807829	公积金	510104	周怨	13108384303	201	513225196209165549
昆仑华庭二期	2017/8/7	201	95.79	20870	1999137	商贷	510104	车城	13108813560	201	139002198602143748
俊发星雅俊园Park公馆	2017/4/21	201	150.38	17926	2695712	全款	510104	傅祟	13109054920	201	130801195708275794
望江名门	2017/9/2	201	60.77	17196	1045001	商贷	510104	廉厕	13109105358	201	130684197507121599
仁恒滨河湾	2017/2/5	201	125.66	10555	1326341	商贷	510104	孙或	13109636729	201	511001196002123491
环球汇天誉	2017/1/8	201	151.41	24804	3755574	公积金	510104	庄盏	13110390622	201	130681198207289676
中港CCPARK	2017/7/23	201	107.12	16797	1799295	全款	510104	宋断舱	13110440271	201	130984197408154516

图 7.7

配置类:

```java
package com.bunfly.choose_column;
import org.apache.hadoop.conf.Configuration;
import org.apache.hadoop.fs.Path;
import org.apache.hadoop.io.Text;
import org.apache.hadoop.mapreduce.Job;
import org.apache.hadoop.mapreduce.lib.input.FileInputFormat;
import org.apache.hadoop.mapreduce.lib.output.FileOutputFormat;

public class Choose_Column {
    public static void main(String[] arg) throws Exception {
        Choose_Column.Conf();
    }
    public static void Conf() throws Exception {
        // 获取配置文件中的参数
        Configuration conf = new Configuration();
        // 通过参数配置获取一个job对象
        Job job = Job.getInstance(conf, "Choose_Column");
        // 启动job需要知道的入口
        job.setJarByClass(Choose_Column.class);
        // 启动Map入口
        job.setMapperClass(Choose_Column_Mapper.class);
        // 关闭Reduce输出
        job.setNumReduceTasks(0);
        // key值输出类型
        job.setOutputKeyClass(Text.class);
        // value值输出类型
        job.setOutputValueClass(Text.class);
        // 设置输入源
        String inl = "/input/choose_column";
        // 设置输出路径
        String out = "/output/choose_column";
        Path input = new Path(inl);
        Path output = new Path(out);
        // 将路径加入到文件格式中
        FileInputFormat.addInputPath(job, input);
        FileOutputFormat.setOutputPath(job, output);
        System.out.println(job.waitForCompletion(true) ? 0 : 1);
    }
}
```

Mapper 类:

```java
package com.bunfly.choose_column;
import java.io.IOException;
```

```java
import org.apache.hadoop.io.Text;
import org.apache.hadoop.mapreduce.Mapper;

public class Choose_Column_Mapper extends Mapper<Object, Text, Text, Text> {
    public void map (Object key, Text value, Context context) throws
        IOException, InterruptedException {
            //每一条字符串按空格分隔,并存储到数组中
            String[] q = value.toString().split("\t");
            context.write(new Text(q[8] +"\t"+ q[9] +"\t"+ q[0]), new Text(""));
    }
}
```

7.5.3 求平均数

在 SQL 基本函数中,聚合函数对一组值执行计算,并返回单个值。聚合函数经常与 Select 语句的 Group by 子句一起使用。在 Hadoop 编程中求和、求平均等也是经常用到的。接下来求全班学生语文、数学、英语的平均成绩。

score 表的结构如图 7.8 所示。

0	1	2	3	4	5	6	7
学号 studentid	姓名 name	性别 sex	出生日期 birth	语文 chinese	数学 mathematics	英语 english	班级编号 classid
20093333001	刘岳	男	1995/7/15	66	51	64	dz1955001001
20093333002	周啊娜	女	1996/8/7	78	42	42	dz1955001001
20093333003	刘玉	女	1996/4/16	52	84	43	dz1955001001
20093333004	李凤扬	女	1996/2/5	87	98	43	dz1955001001
20093333005	田文	女	1996/3/10	null	65	76	dz1955001001
20093333006	倪婧	女	1995/9/17	81	75	97	dz1955001001
20093333007	解晶	男	1995/8/9	94	66	72	dz1955001001
20093333008	谢岩	男	1995/12/17	76	null	76	dz1955001001
20093333009	董凤睿	男	1996/1/22	97	61	58	dz1955001001
20093333010	任菊	女	1996/11/15	82	80	56	dz1955001001
20093333011	顾庆敏	女	1995/5/15	61	99	69	dz1955001001
20093333012	柳媚	女	1996/2/20	99	65	93	dz1955001001
20093333013	廖文	女	1995/10/16	82	57	85	dz1955001001
20093333014	丁莹	女	1995/1/7	73	75	89	dz1955001001
20093333015	蒋庆盈	男	1996/11/23	77	61	85	dz1955001001

图 7.8

配置类:

```java
package com.bunfly.average;
import org.apache.hadoop.conf.Configuration;
import org.apache.hadoop.fs.Path;
import org.apache.hadoop.io.Text;
import org.apache.hadoop.mapreduce.Job;
import org.apache.hadoop.mapreduce.lib.input.FileInputFormat;
import org.apache.hadoop.mapreduce.lib.output.FileOutputFormat;
```

```java
public class Average {
    public static void main(String[] arg) throws Exception {
        Average.Conf();
    }
    public static void Conf() throws Exception {
        //获取配置文件中的参数
        Configuration conf = new Configuration();
        //通过参数配置获取一个job对象
        Job job = Job.getInstance(conf,"Average");
        //启动job需要知道的入口
        job.setJarByClass(Average.class);
        //启动Map入口
        job.setMapperClass(Average_Mapper.class);
        //启动Reduce入口
        job.setReducerClass(Average_Reduce.class);
        //设置Reduce输出的数量
        job.setNumReduceTasks(1);
        //key值输出类型
        job.setOutputKeyClass(Text.class);
        //value值输出类型
        job.setOutputValueClass(Text.class);
        //设置输入源
        String inl = "/input/score";
        //设置输出路径
        String out = "/output/averager_score";
        Path input = new Path(inl);
        Path output = new Path(out);
        //将路径加入到文件格式中
        FileInputFormat.addInputPath(job, input);
        FileOutputFormat.setOutputPath(job, output);
        System.out.println(job.waitForCompletion(true) ? 0 : 1);
    }
}
```

Mapper 类：

```java
package com.bunfly.average;
import java.io.IOException;
import org.apache.hadoop.io.Text;
import org.apache.hadoop.mapreduce.Mapper;

public class Average_Mapper extends Mapper<Object, Text, Text, Text> {
    public void map (Object key, Text value, Context context) throws
    IOException, InterruptedException {
        String s = value.toString();
```

```
            String[] q = s.split("\t");
            //q[7]：班级编号 q[4]：语文   q[5]：数学   q[6]：英语
            context.write(new Text(q[7]), new Text(q[4] +"\t"+ q[5] +"\t"+
            q[6]));
        }
    }
}
```

Reduce 类：

```
package com.bunfly.average;
import java.io.IOException;
import org.apache.hadoop.io.Text;
import org.apache.hadoop.mapreduce.Reducer;

public class Average_Reduce extends Reducer<Text, Text, Text, Text> {
    public void reduce (Text key, Iterable<Text> values, Context context)
    throws IOException, InterruptedException {
        //用来接收转换后的字符串，c:语文   m:数学   e:英语
        double c=0, m=0, e=0;
        double c_sum=0, m_sum=0, e_sum=0;
        //计数器，记录全班多少人
        int count = 0;
        for (Text val : values) {
            String[] q = val.toString().split("\t");
            c = Double.valueOf(q[0]);
            m = Double.valueOf(q[1]);
            e = Double.valueOf(q[2]);
            c_sum = c_sum + c;
            m_sum = m_sum + m;
            e_sum = e_sum + e;
            count++;
        }
        //以班级编码分组，分别统计语文、数学、英语平均成绩
        context.write(key, new Text(c_sum/count +"\t"+ m_sum/count +"\t"+
        e_sum/count));
    }
}
```

7.5.4 关联

关联业务是通过两个文件的共有外键关联拼出完整信息的方式。在 7.5.3 节中求平均数业务，是通过班级 id 来求平均数的。由于无法得知班级的名称，就需要两个文件关联编程。一个文件同样利用 score 表，另一个文件对应 major 表，表结构如图 7.9 所示。

	0	1	2
	班级编号	班级名称	系名称
	classid	classname	deptname
	dz1955001001	数据挖掘	数学系
	dz1955001002	数学计算	数学系
	dz1955002001	生物工程	生物系
	dz1955002002	环境监测	生物系
	dz1955003001	刑侦	公安系
	dz1955003002	犯罪侦察	公安系
	dz1955004001	电子工程	电子系
	dz1955004002	电子应用	电子系
	dz1955004003	电子技术	电子系
	dz1955005001	Web前端	计算机系
	dz1955005002	java应用	计算机系
	dz1955005003	大数据	计算机系
	dz1955005004	网络工程	计算机系

图 7.9

配置类：

```java
package com.bunfly.relation;
import org.apache.hadoop.conf.Configuration;
import org.apache.hadoop.fs.Path;
import org.apache.hadoop.io.Text;
import org.apache.hadoop.mapreduce.Job;
import org.apache.hadoop.mapreduce.lib.input.FileInputFormat;
import org.apache.hadoop.mapreduce.lib.output.FileOutputFormat;

public class Relation {
    public static void main(String[] arg) throws Exception {
        Relation.Conf();
    }
    public static void Conf() throws Exception {
        //获取配置文件中的参数
        Configuration conf = new Configuration();
        //通过参数配置获取一个job对象
        Job job = Job.getInstance(conf,"Relation");
        //启动job需要知道的入口
        job.setJarByClass(Relation.class);
        //启动Map入口
        job.setMapperClass(Relation_Mapper.class);
        //启动Reduce入口
        job.setReducerClass(Relation_Reduce.class);
        //设置Reduce输出的数量
        job.setNumReduceTasks(1);
        //key值输出类型
        job.setOutputKeyClass(Text.class);
```

```java
        //value值输出类型
        job.setOutputValueClass(Text.class);
        //设置输入源
        String in1 = "/input/score";
        String in2 = "/input/major";
        //设置输出路径
        String out = "/output/relation";
        Path input = new Path(in1);
        Path input2 = new Path(in2);
        Path output = new Path(out);
        //将路径加入到文件格式中
        FileInputFormat.addInputPath(job, input);
        FileInputFormat.addInputPath(job, input2);
        FileOutputFormat.setOutputPath(job, output);
        System.out.println(job.waitForCompletion(true) ? 0 : 1);
    }
}
```

Mapper类：

```java
package com.bunfly.relation;
import java.io.IOException;
import org.apache.hadoop.io.Text;
import org.apache.hadoop.mapreduce.InputSplit;
import org.apache.hadoop.mapreduce.Mapper;
import org.apache.hadoop.mapreduce.lib.input.FileSplit;

public class Relation_Mapper extends Mapper<Object, Text, Text, Text> {
    public void map (Object key, Text value, Context context) throws
    IOException, InterruptedException {
        InputSplit t = context.getInputSplit();
        String path = ((FileSplit) t).getPath().getName();
        /*
         * 通过路径判断出哪个是score文件，哪个是major文件，以班级id分组，并给value
         值加上标记便于区分
         */
        if (path.contains("score")) {
            String[] q = value.toString().split("\t");
            context.write(new Text(q[7]),
                new Text("AAAAA" +"\t"+ q[0] +"\t"+q[1] +"\t"+q[2] +
                "\t"+q[3] +"\t"+q[4] +"\t"+q[5] +"\t"+ q[6]));
        } else if (path.contains("major")) {
            String[] q = value.toString().split("\t");
            context.write(new Text(q[0]),
```

```java
                    new Text("BBBBB" +"\t"+ q[1] +"\t"+q[2]));
        }
    }
}
```

Reduce 类:

```java
package com.bunfly.relation;
import java.io.IOException;
import java.util.ArrayList;
import org.apache.hadoop.io.Text;
import org.apache.hadoop.mapreduce.Reducer;

public class Relation_Reduce extends Reducer<Text, Text, Text, Text> {
    public void reduce (Text key, Iterable<Text> values, Context context)
    throws IOException, InterruptedException {
        ArrayList<String> list1 = new ArrayList<String>();
        ArrayList<String> list2 = new ArrayList<String>();
        for (Text val :values) {
            String[] q = val.toString().split("\t");
            //通过对q[0]判断区分文件
            if (q[0].equals("AAAAA")) {
                //将属于AAAAA部分的文件加入list1集合
                list1.add(q[1] +"\t"+q[2] +"\t"+q[3] +"\t"+q[4] +"\t"+q[5]
                +"\t"+ q[6] +"\t"+ q[7]);
            } else if (q[0].equals("BBBBB")) {
                list2.add(q[1] +"\t"+q[2]);
            }
        }
        //将集合通过迭代器循环定稿到HDFS中
        for (String i : list1) {
            for (String j : list2) {
                context.write(key, new Text(i +"\t"+ j));
            }
        }
    }
}
```

7.6 习题与思考

1. 根据表 7.3 和表 7.4，通过 MapReduce 框架编写程序，统计出每个学生的考试总成绩。

表 7.3

姓名	性别	出生年份	院系	家庭住址	学号
刘一	男	1994/11/26	数学系	上海	20157259
陈二	男	1993/6/11	数学系	北京	20153174
张三	女	1994/9/21	数学系	北京	20157824
李四	男	1993/1/26	信息工程系	云南	20155367

表 7.4

学号	课程名	分数
20157259	语文	90
20157259	数学	58
20157259	英语	39
20153174	语文	91
20153174	数学	95
20153174	英语	75
20157824	语文	60
20157824	数学	58
20157824	英语	53
20155367	语文	62
20155367	数学	43
20155367	英语	74

2. 根据表 7.3 和表 7.4，通过 MapReduce 框架编写程序，统计出每个系的英语平均成绩在所有系中的占比。

第 8 章　Hive 数据仓库

8.1　Hive 模型

Hive 是基于 Hadoop 构建的一套数据仓库分析工具，它提供了丰富的 SQL 查询方式来分析存储在 Hadoop 分布式文件系统中的数据。可以将结构化的数据文件映射为一张数据库表，并提供完整的 SQL 查询功能；也可以将 SQL 语句转换为 MapReduce 任务运行，通过 SQL 去查询分析需要的内容。这套类 SQL 简称 HQL，使对 MapReduce 不熟悉的用户利用 HQL 语言查询、汇总、分析数据，简化 MapReduce 代码，从而使用 Hadoop 集群。而 MapReduce 开发人员可以把已写的 Mapper 和 Reducer 作为插件来支持 Hive 做更复杂的数据分析。

8.1.1　Hive 架构与基本组成

Hive 的架构图如图 8.1 所示。

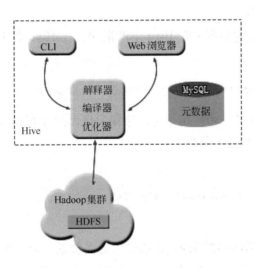

图　8.1

Hive 的体系结构可以分为以下几部分。

（1）用户接口主要有三个：CLI、Java Client 和 Web WUI。其中最常用的是 CLI，CLI 启动时会同时启动一个 Hive 副本。Client 是 Hive 的客户端，用户连接至 Hive Server，在启动 Client 模式的时候，需要指出 Hive Server 所在节点，并且在该节点启动 Hive Server，该访问方式基本不用。Web WUI（Web 浏览器）通过浏览器访问 Hive。

（2）Hive 将元数据存储在数据库中，如 MySQL、derby。Hive 中的元数据包括表的名字、表的列和分区及其属性、表的属性（是否为外部表等）以及表的数据所在目录等。

（3）解释器、编译器、优化器完成 HQL 查询语句从词法分析、语法分析、编译、优化以及查询计划的生成。生成的查询计划存储在 HDFS 中，并在随后由 MapReduce 调用执行。

（4）Hive 的数据存储在 HDFS 中，大部分的查询、计算由 MapReduce 完成。

Hive 将元数据存储在 RDBMS 中，有如下三种模式可以连接到数据库。

（1）单用户模式。此模式连接到一个 In-memory 的数据库 Derby，一般用于 Unit Test。

（2）多用户模式。通过网络连接到一个数据库中，是最经常使用的模式。

（3）远程服务器模式。用于非 Java 客户端访问元数据库，在服务器端启动 MetaStoreServer，客户端利用 Thrift 协议通过 MetaStoreServer 访问元数据库。

关于数据存储，Hive 没有专属的数据存储格式，也没有为数据建立索引。用户可以自由组织 Hive 中的表，只需要在创建表的时候通过 Hive 数据中的列分隔符和行分隔符，Hive 就可以解析数据。Hive 中所有的数据都存储在 HDFS 中，存储结构主要包括数据库、文件、表和视图。Hive 中包含以下数据模型：Table 内部表、External Table 外部表、Partition 分区、Bucket 桶。Hive 默认可以直接加载文本文件，同时支持 Sequence File、RCFile。

8.1.2　Hive 的数据模型

1. 创建 Hive 数据库

创建数据库是用来创建数据库在 Hive 中的语句。在 Hive 数据库中是一个命名空间或表的集合。语法声明如下：

```
create database|schema [if not exists] <database name>
```

在这里，if not exists 是一个可选子句，通知用户已经存在相同名称的数据库。可以使用 schema 在 database 中的这个命令。

简单示例如下：

```
命令行 hive > create database test_database;
```

2. 内部表

Hive 的内部表与数据库中的 Table 在概念上是类似的。每一个 Table 在 Hive 中都有一个相应的目录存储数据。例如一个表 pvs，它在 HDFS 中的路径为/wh/pvs，其中 wh 是在 hive-site.xml 中由${hive.metastore.warehouse.dir} 指定的数据仓库的目录，所有的 Table 数据（不包括 External Table）都保存在这个目录中。删除表时，元数据与数据都会被删除。

简单示例如下：

创建数据文件：test_inner_table.txt。

创建表：create table test_inner_table (key string)。

加载数据：load data local inpath 'filepath' into table test_inner_table。

查看数据：select * from test_inner_table; select count(*) from test_inner_table。

删除表：drop table test_inner_table。

3. 外部表

外部表指向已经在 HDFS 中存在的数据，可以创建 Partition。它和内部表在元数据的组织上是相同的，而实际数据的存储则有较大的差异。内部表的创建过程和数据加载过程可以分别独立完成，也可以在同一个语句中完成。在加载数据的过程中，实际数据会被移动到数据仓库目录中，之后对数据的访问将会直接在数据仓库目录中完成。删除表时，表中的数据和元数据将会被同时删除。而外部表只有一个过程，加载数据和创建表同时完成（create external table…location）。实际数据是存储在 location 后面指定的 HDFS 路径中，并不会移动到数据仓库目录中。当删除一个 external table 时，仅删除该链接。

简单示例如下：

创建数据文件：test_external_table.txt。

创建表：create external table test_external_table (key string)。

加载数据：load data inpath 'filepath' into table test_inner_table。

查看数据：select * from test_external_table; •select count(*) from test_external_table。

删除表：drop table test_external_table。

8.2 Hive 的安装

8.2.1 Hive 的基本安装

关于 Hive 的安装讲解视频可扫描二维码观看。

（1）在 hadoop 用户状态下，将 Hive 的安装文件拷贝到安装目录下并解压。

（2）配置 Hive 的环境变量（需要 root 用户配置，因为 porfile 文件属于 root 用户）：

```
vi /etc/porfile
添加：
    export HIVE_HOME=/xx/xx/hive.xx.xx
    export PATH=$PATH:$HIVE_HOME/bin
```

（3）使环境变量生效：source /etc/porfile。

（4）验证 Hive： hive。

在这一步 Hive 已经安装成功,但 Hive 用来存储元数据的数据库是自带数据库(derby)。这个数据库不能实现远程操作，这就会产生不必要的麻烦。所以需要安装另一个其他关系型数据库来代替它自身的数据库，以消除这个不足。

8.2.2 MySQL 的安装

1. 安装 MySQL

（1）查看 Linux 系统中是否存在自带数据库。

```
rpm -qa | grep mysql
```

（2）卸载 Linux 系统集成的 MySQL 数据库，卸载分为普通模式和强力模式。强力模式是针对提示有依赖的其他文件时使用。

普通删除模式：rpm -e mysql.xx.xx。

强力删除模式：rpm -e --nodeps mysql.xx.xx。

（3）通过 yum 来安装 MySQL，安装前，可以通过 yum list | grep mysql 命令查看 yum 上提供哪些 MySQL 数据库可下载的版本。找到 mysql-server、mysql、mysql-devel 服务并进行安装。

查看 yum 上的程序：yum list | grep mysql。

多个程序安装：yum install -y mysql-server mysql mysql-deve。

（4）验证 MySQL 是否安装成功：

```
rpm -qa mysql-server
```

2. MySQL 的初始化

（1）启动 MySQL 服务：service mysqld start。

或者启动服务：service mysqld restart。

（2）在使用 MySQL 数据库时，需首先启动 MySQL 服务。可以通过 chkconfig --list | grep mysql 命令来查看 MySQL 服务是不是开机自动启动,如果出现全部关闭状态表示没有开启自动启动。

```
chkconfig --list | grep mysqld
```

结果：mysql 0:关闭 1:关闭 2:关闭 3:关闭 4:关闭 5:关闭 6:关闭。

（3）开启自动启动：chkconfig mysqld on，再执行 chkconfig --list|grep mysql。

如果出现 2、3、4、5 为启用状态，表示自动启动已经开启。

结果：mysql 0:关闭 1:关闭 2:启用 3:启用 4:启用 5:启用 6:关闭。

（4）MySQL 数据库安装完以后只会有一个 root 管理员账号，此时的 root 账号并没有设置密码。在第一次启动 MySQL 服务时，会进行数据库的一些初始化工作，在输出的一大串信息中，会看到 "/usr/bin/mysqladmin -u root password 'new-password'" 信息，这条信息告诉我们需要用 "root password 'new-password'" 命令为 root 账号设置密码。

所以，可以通过该命令给 root 账号设置密码（这个 root 账号是 MySQL 的 root 账号，非 Linux 的 root 账号）。如下：

```
mysqladmin -u root password '密码'
```

（5）登录 MySQL 数据库：

```
mysql -u root -p
```

8.2.3　Hive 配置

1. 配置 MySQL 为 Hive 元数据存储数据库

1）使用 root 用户进入 MySQL

```
mysql -u root -p
```

2）创建 hadoop 账户

```
create user 'hadoop'@'localhost' identified by 'hadoop'
```

3）为 hadoop 账户添加权限

```
grant all privileges on *.* to 'hadoop'@'localhost' with grant option
```

4）检查账户是否创建成功

```
select User,Host from mysql.user
```

5）退出 MySQL，并重启 MySQL 服务

```
exit;
service mysqld restart;
```

6）使用 hadoop 账户登录

```
mysql -u hadoop -p
```

7）创建 Hive 元数据库

```
create database hive_metadata
```

8）配置 Hive

进入 Hive 的 conf 目录，把 hive-default.xml.template 复制一个副本，并重命名为 hive-site.xml。如下：

```
cp hive-default.xml.template hive-site.xml
```

9）打开 hive-site.xml 进行相关参数配置

```
vi hive-site.xml
```

更改：

```
//指定Hive连接的数据库的数据库连接字符串
javax.jdo.option.ConnectionURL
jdbc:mysql://localhost:3306/hive_metadata?createDatabaseIfNotExist=true
javax.jdo.option.ConnectionDriverName
com.mysql.jdbc.Driver
javax.jdo.option.ConnectionUserName
hadoop

javax.jdo.option.ConnectionPassword
```

```
hadoop
//设置Hive作业的本地临时空间，iotmp地址需要自己创建
hive.exec.local.scratchdir
/xx/xx/hive/iotmp
hive.downloaded.resources.dir
/xx/xx/hive/iotmp/${hive.session.id}_resources

//指定Hive的数据存储目录，指定的是HDFS上的位置
hive.metastore.warehouse.dir
/user/hive/warehouse
```

10）配置 MySQL 插件

下载 mysql-connector-java-5.1.10-bin.jar, 把 mysql-connector-java-5.1.10-bin.jar 放到 Hive 安装目录的 lib 下。

2. 配置 hosts（只在 Hive 客户端所在服务器配置）

1）进入 root 用户

```
su -l root
```

2）编辑/etc/hosts

```
vi /etc/hosts
```

插入：

```
192.168.153.201 Hive
```

8.3 HQL 详解

关于 HQL 使用的讲解视频可分别扫描以下三个二维码观看。

8.3.1 Hive 数据管理方式

Hive 是建立在 Hadoop 上的数据仓库基础构架。它提供了一系列的工具，用来进行数据提取、转化、加载。这是一种可以存储、查询和分析存储在 Hadoop 中的大规模数据的机制。Hive 定义了简单的类 SQL 查询语言，称为 HQL，它允许熟悉 SQL 的用户查询数据。作为一个数据仓库，Hive 的数据管理按照使用层次可以从元数据存储、数据存储和数据交换三个方面来介绍。Hive 客户端所安装的软件如图 8.2 所示。

1. 元数据存储

Hive 将元数据存储在 RDBMS 中，有以下三种模式可以连接到数据库。

（1）单用户模式。此模式连接到一个 In-memory 的数据库 Derby。一般用于 Unit Test。

（2）多用户模式。通过网络连接到一个数据库中，是最经常使用的模式。

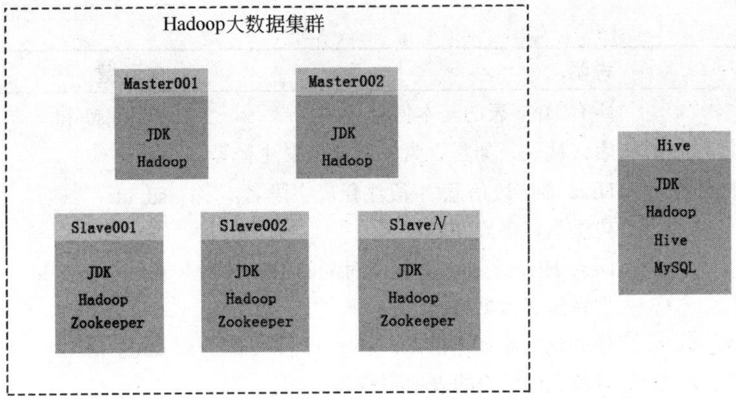

图 8.2

（3）远程服务器模式。用于非 Java 客户端访问元数据库，在服务器端启动 MetaStoreServer，客户端利用 Thrift 协议通过 MetaStoreServer 访问元数据库。

2. 数据存储

首先，Hive 没有专属的数据存储格式，也没有为数据建立索引。用户可以自由组织 Hive 中的表，只需要在创建表的时候通告 Hive 数据中的列分隔符和行分隔符，就可以解析数据。

其次，Hive 中所有的数据都存储在 HDFS 中，Hive 中包含 4 种数据模型：Database、Table、Partition 和 Bucket。

1）Database（数据库）

这种模型相当于关系数据库中的命名空间（namespace）。其作用是将用户和数据库的应用隔离到不同的数据库或模式中。该模型在 Hive 0.6.0 之后的版本支持，Hive 提供了 create database dbname、use dbname 以及 drop database dbname 这样的语句。

2）Table（表）

Hive 的表逻辑上由存储的数据和描述表格中的数据形式的相关元数据组成。元数据存储在关系数据库中。表存储的数据存放在 Hive 的数据仓库中，这个数据仓库是 HDFS 上的一个目录，该目录是在 hive-site.xml 中由${Hive.metastore.warehouse.dir}指定的，这里假定为/user/hive/warehouse/。创建一张 Hive 的表，即在 HDFS 的仓库目录下创建一个文件夹。表分为内部表和外部表两种。

Hive 元数据对应的表约有 20 个，其中和表结构信息有关的有 9 张，其余的十几张或为空，或只有简单的几条记录。表的简要说明如表 8.1 所示。

从表 8.1 的内容来看，Hive 整个创建表的过程较为清晰。

（1）解析用户提交的 Hive 语句，对其进行解析，分解为表、字段、分区等 Hive 对象。

（2）根据解析到的信息构建对应的表、字段、分区等对象，从 sequence_table 中获取构建对象的最新 ID，与构建对象信息（名称，类型等）一同通过 DAO 方法写入到元数据表中，成功后将 sequence_table 中对应最新 ID+5。

表 8.1

表名	说明	关联键
tbls	所有 Hive 表的基本信息	tbl_id,sd_id
table_param	表级属性，如是否为外部表、表注释等	tbl_id
columns	Hive 表字段信息(字段注释，字段名，字段类型，字段序号)	sd_id
sds	所有 Hive 表、表分区所对应的 HDFS 数据目录和数据格式	sd_id,serde_id
serde_param	序列化反序列化信息，如行分隔符、列分隔符、NULL 的表示字符等	serde_id
partitions	Hive 表分区信息	part_id,sd_id,tbl_id
partition_keys	Hive 分区表分区键	tbl_id
partition_key_vals	Hive 表分区名(键值)	part_id

3）Partition（分区表）

Hive 中分区的概念是根据"分区列"的值对表的数据进行粗略划分的机制。在 Hive 存储上就体现在表的主目录下的一个子目录，这个文件夹的名字就是定义的"分区列+值"。

分区以字段的形式存在表结构中，通过 describe table 命令可以查看到字段。但并不是对应着数据文件中某个列的字段，它不存放实际的数据内容，仅仅是分区的表示（伪列）。

用户存储的每个数据文档要放到哪个分区，由用户决定，这是单纯的数据文档的移动。即用户在加载数据的时候必须显示的指定该部分数据放到哪个分区。

进行分区的优点是提高了查询效率。在 Hive Select 查询中一般会扫描整个表内容，这样就会消耗很多时间成本。有时候我们只需扫描表中关心的一部分数据，因此建表时引入了 partition 概念。如当前互联网应用每天都要存储大量的日志文件，数 GB、数十 GB 甚至更大都有可能。存储日志中必然有个属性是日志产生的日期。在产生分区时，就可以按照日志产生的日期列进行划分，把每一天的日志当作一个分区。

示例如下。

（1）创建一个分区表，以 time 为分区列。

```
create table partition_table (id int, name string)
partitioned by (time string)
row format delimited
fields terminated by '\t';
```

（2）将数据添加到时间为 2017-01-16 的这个分区中。

```
load data local inpath '/home/hadoop/software/data.txt' overwrite into
table invites partition (time='2017-01-16');
```

（3）从一个分区中查询数据。

```
select * from partition_table where time ='2017-01-16';
```

（4）往一个分区表的某一个分区中添加数据。

```
insert overwrite table partition_table
partition (time='2017-01-16')
select id,max(name) from test group by id;
```

（5）使用以下命令可以查看分区的具体情况。

```
hadoop fs -ls /home/hadoop.hive/warehouse/partition_table;
```

或查看分区内容：

```
hadoop fs -cat /home/hadoop.hive/warehouse/partition_table/p*;
```

4）Bucket（桶）

对于每一个表（Table）或者分区，Hive 可以进一步组织成桶，即桶是更为细粒度的数据范围划分。它是对数据源数据文件本身进行拆分数据，使用桶的表会将源数据文件按一定规律拆分成多个文件。物理上，每个桶就是表(或分区)目录中的一个文件，Hive 是针对某一列进行桶的组织。这里的列字段是对应于数据文件中具体某个列的 Hive 采用对列值哈希，然后除以桶的个数求余的方式决定该条记录存放在哪个桶当中。

把表（或者分区）组织成桶的好处如下。

（1）获得更高的查询处理效率。桶为表加上了额外的结构，Hive 在处理有些查询时能利用这个结构。具体而言，连接两个在(包含连接列的)相同列上划分桶的表，可以使用 Map 端连接（Map-Side Join）高效地实现。比如对于 JOIN 操作两个表有一个相同的列，如果对这两个表都进行桶操作，那么只需保存相同列值的桶进行 JOIN 操作，大大减少 JOIN 的数据量。

（2）使取样（Sampling）更高效。在处理大规模数据集的开发和修改查询阶段，如果能在数据集的一小部分数据上试运行查询，将带来很大便利。

示例如下：

（1）创建桶表。

```
create table bucketed_user(id int,name string)
clustered by (id) sorted by(name) into 4 buckets
row format delimited fields terminated by '\t' stored as textfile;
```

（2）往桶表中插入数据。

```
insert overwrite table bucketed_user
select * from users;
```

（3）对桶中的数据进行采样（查询一半返回的桶数）。

```
select * from bucketed_user tablesample (bucket 1 out of 2 on id);
```

8.3.2 HQL 操作

1. 前期准备

前期需要准备两个数据文件:Socre 和 Unit_name。Socre 文件中记录的是学生的语、数、英考试成绩和学生信息。Unit_name 文件中记录的是学院的组织结构,包括班级 id、班级名称和系统名称信息。两个文件可以通过共有的 classid 字段关联组合出自己所需要的内容。接下来将通过对 Socre 和 Unit_name 两个文件在 Hive 中创建表、加载数据、操作数据、删除数据、删除表等多方面讲解 HQL 的用法。

Socre 文件表结构如图 8.3 所示。

studentid	name	sex	birth	chinese	mathematics	english	classid
20093333001	刘岳	男	1995/7/15	66	51	64	dz1955001001
20093333002	周啊娜	女	1996/8/7	78	42	42	dz1955001001
20093333003	刘玉	女	1996/4/16	52	84	43	dz1955001001
20093333004	李凤扬	女	1996/2/5	87	98	43	dz1955001001
20093333005	田文	女	1996/3/10	null	65	76	dz1955001001
20093333006	倪婧	女	1995/9/17	81	75	97	dz1955001001
20093333007	解晶	男	1995/8/9	94	66	72	dz1955001001
20093333008	谢岩	男	1995/12/17	76	null	76	dz1955001001
20093333009	董凤睿	男	1996/1/22	97	61	58	dz1955001001
20093333010	任菊	女	1996/11/15	82	80	56	dz1955001001
20093333011	顾庆敏	女	1995/5/15	61	99	69	dz1955001001
20093333012	柳媚	女	1996/2/20	99	65	93	dz1955001001
20093333013	廖文	女	1995/10/16	82	57	85	dz1955001001
20093333014	丁莹	女	1995/1/7	73	75	89	dz1955001001
20093333015	蒋庆盈	男	1996/11/23	77	61	85	dz1955001001
20093333016	沈沉	男	1995/11/24	93	66	57	dz1955001001

图 8.3

Unit_name 文件表结构如图 8.4 所示。

classid	classname	deptname
dz1955001001	数据挖掘	数学系
dz1955001002	数学计算	数学系
dz1955002001	生物工程	生物系
dz1955002002	环境监测	生物系
dz1955003001	刑侦	公安系
dz1955003002	犯罪侦察	公安系
dz1955004001	电子工程	电子系
dz1955004002	电子应用	电子系
dz1955004003	电子技术	电子系
dz1955005001	Web前端	计算机系
dz1955005002	Java应用	计算机系
dz1955005003	大数据	计算机系
dz1955005004	网络工程	计算机系

图 8.4

2. 创建数据库

Hive 数据库是一个命名空间或表的集合。语法声明如下:

```
create database [if not exists] <database_name>
```

[if not exists]是一个可选子句,用以通知用户已经存在相同名称的数据库。
示例如下:

```
create database if not exists bunfly;
```

可以使用 show 查看数据库是否创建成功。以下命令将显示出所有的数据库:

```
show database;
```

利用 use 命令可以切换到想要的数据库:

```
use bunfly;
```

3. 删除数据库

删除一个空的数据库:

```
drop database [if exists] databasename;
```

删除一个有内容的数据库:

```
drop database [if exists] databasename cascade;
```

删除数据库可以用 drop database dataname 语句来删除,但这个方式只能删除空的数据库。如果数据库中有表要删除,数据库可以加"cacade"关键字,或者先删除数据库中的所有表,然后再使用 drop database dataname 语句进行删除。if exists 关键字判断 Hive 中是否存在此数据库名,如果没有将通知用户不存在此数据库名。

4. 创建内部表(管理表)

Hive 由 SQL 语法演变而来,其数据类型与 SQL 基本相似。Hive 常用的基本数据类型如表 8.2 所示。

表 8.2

数据类型	说明
int	整型,4B 整数
bigint	整型,8B 整数
boolean	布尔类型,true 或 false
float	单精度浮点数
double	双精度浮点数
string	字符串类型

创建内部表的示例如下:

```
create table bunfly.score (
studentId    string,
name         string,
sex          string,
birth        string,
```

```
chinese      double,
mathematics double,
english      double,
classid string
)
row format delimited fields terminated by '\t';
```

上面语句在 bunfly 数据库中创建了一张名为 score 的内部表，并以\t 制表符（空格）分隔数据。

```
create table bunfly.unit_name(
classid          string,
classname        string,
deptname         string
)
row format delimited fields terminated by '\t';
```

上面语句在 bunfly 数据库中创建了一张名为 unit_name 的内部表，并以\t 制表符（空格）分隔数据。

5. 修改表

1）重命名表

语法：

```
alter table <ago_tablename> rename to <new_tablename>
```

ago_tablename 是现在的表文件，new_tablename 是修改后的表名，当执行 alter table <ago_tablename> rename to <new_tablename>语句后，Hive 将修改表的名称和在 HDFS 中的文件目录名。

2）添加列

语法：

```
alter table tablename add columns (
    columns1 type,
    columns2 type,
    ...
)
```

coulumnsX 是需要增加的列名（也称字段名），实现在现有的列后新增列。

3）修改列顺序

语法：

```
alter table tablename change column columns1 columns1 string after columns2;
```

新增列的位置不符合要求，是因为原数据的内容位置和所创建的表字段位置不对应，这时就要使用修改列的顺序的语句。alter table tablename change column columns1 columns1 string after columns2 语句中的 tablename 是要修改列的表（也可以指定数据库），columns1

是要移动的列（注意这里是两个 columns1 字段），columns2 代表要将 columns1 移动到 columns2 之后。

4）删除列

语法：

```
alter table tablename replace columns(columns type);
```

columns type 是需要删除的字段和字段的类型。

6. 加载数据到内部

从本地加载数据到 score 表中：

```
load data local inpath '/home/user/inputfile/score.txt' overwrite into table bunfly.score;
```

从 HDFS 加载数据到 unit_name 表中：

```
load data inpath '/input/unit_name.txt' overwrite into table bunfly.unit_name;
```

如果语句加上"local"关键字，inpath 后的路径是本地路径，加载到 Hive 表中的数据将从本地复制数据到 Hive 表对应的 HDFS 中。如果语句没有加上"local"关键字，inpath 后面的路径是 HDFS 中的路径，那么加载到 Hive 表中的数据将从 HDFS 源数据位置移动到 Hive 表对应的 HDFS 目录下，HDFS 源数据将被删除。

"overwrite"关键字是可选项，加上代表覆盖原文件中的所有内容。如果不加"overwrite"关键字，将会在原文件的基础上追加新的内容。

7. 插入数据

语法：

```
insert overwrite table to_tablename [partiton (partcol1=val1, partcol2=val2,…)]
select column1,column2,…,[val1,val2] from from_tablename
```

partiton 动态插入数据到分区表，其分区字段需要与插入字段最尾部字段对应。例如，val1 与 select 中的 val1 对应，val2 与 select 中的 val2 对应。如果需要创建非常多的分区，用户就需要写入非常多的 SQL，而动态插入可以基于查询参数推断出需要创建的分区名称。动态插入需要与分区表字段一致，分区字段可以有多个，并且对应最后一个字段。例如，如果有两个分区字段，分区表第一个分区字段对应的是查询的倒数第二个字段，分区表第二个分区字段对应的是查询的倒数第一个字段。

在执行动态插入前必须先开启动态插入功能：

```
set hive.exec.dynamic.partition.mode=nonstrict;
```

示例如下：

```
insert overwrite table bunfly.score2
select * from bunfly.score;
```

将 bunfly 数据库中的 score 表内容全部查询出来，插入到 bunfly 数据库中的 score2 表中，并重写 score2 表的内容。执行上面语句相当于数据备份，将数据复制出新的副本。

通过查询将数据保存到本地文件。语法如下：

```
insert overwrite [local] directory path_directory
select * from tablename;
```

示例如下：

```
insert overwrite local directory '/home/user/hive'
select * from bunfly.score;
```

将 bunfly 数据库中查询出来的 score 表的结果写入到本地 user 用户的家目录下的 Hive 文件中。产生的文件会覆盖指定目录中的其他文件，即将目录中已经存在的文件删除。

```
insert overwrite directory '/user/hive'
select * from bunfly.score;
```

将 bunfly 数据库中查询出来的 score 表的结果写入到 HDFS 中 user 目录下的 Hive 文件中。产生的文件会覆盖指定目录中的其他文件，即将目录中已经存在的文件删除。

8. 创建外部表

在创建表的时候可以指定 external 关键字创建外部表，外部表对应的文件存储在 location 指定的目录下。向该目录添加新文件的同时，该表会读取到该文件，但删除外部表不会删除 location 指定目录下的文件。

创建外部表：

```
create external table bunfly.score2 (
studentId string,
name string,
sex string,
birth string,
chinese double,
math double,
english double,
classId string
)
row format delimited fields terminated by '\t';
load data inpath '/input/score.txt' overwrite into table bunfly.score2;
```

上面语句在 bunfly 数据库中创建了一张 score2 的外部表，并以\t 制表符（空格）分隔数据。从 HDFS 中 input 目录下移动 score.txt 文件到 bunfly 数据库中 score2 表在 HDFS 对应的目录下。

内部表与外部表除了创建方式不同，其他使用均相同。可以通过 desc formatted bunfly.score2 查询表是内部表还是外部表。

内部表与外部表有以下区别：

- 创建内部表时，会将数据移动到数据仓库指向的路径。内部表的数据属于自己，而外部表的数据不属于自己。

- 在删除内部表的时候，Hive 将会把属于表的元数据和数据全部删掉。而删除外部表的时候仅仅删除表的元数据，数据不会被删除。
- 外部表相对来说更加安全，数据组织也更加灵活，方便共享源数据。一般情况下，如果所有处理都需要由 Hive 完成，那么应该创建内部表，否则使用外部表。

9. 创建分区表

Hive 的分区表中分区列不是表中的一个实际的字段，而是一个或者多个伪列，即在表的数据文件中实际上并不保存分区列的信息与数据。

创建分区表：

```
create table bunfly.score3 (
studentId string,
name string,
sex string,
birth string,
chinese double,
math double,
english double
)
partitioned by (classId string)
row format delimited fields terminated by '\t';
load data inpath '/input/score.txt' overwrite into table bunfly.score3 artition (classId);
```

上面示例在 bunfly 数据库中创建了一个 score3 分区表，分区字段以 classId 分区，并以\t 制表符分隔。然后从 HDFS 中 input 目录下移动数据到 score3 对应的 HDFS 目录下，其中 score.txt 文件内的最后的一列作为动态分区字段。例如：calssId 是班级 ID，将把相同班级 ID 的所有内容放在一起，并以 calaaId 内容为文件名。

10. HQL 的常用操作

1）语法

```
select column1, column2,…from tablename
```

示例如下：

（1）查询 score 表中所有同学的语、数、英成绩。

```
select name, chinese, mathematics, english from bunfly.score;
```

（2）查询 score 表中所有信息可以通过写入所有字段查询所有信息。如果字段比较多查询输入量会非常大，可以通过"*"通配符来代表所有字段。

```
select * from bunfly.score;
```

（3）使用 limit 命令可以查看若干行的数据，如下命令查看 score 表中前十行数据。

```
select * from bunfly.socre limit 10;
```

(4) 如果要查看多个班，可以通过 distinct 来实现。

```
select distinct classId from bunfly.score;
```

2) Hive 中同样也支持 where、group by、order by 等语句

(1) 查看英语成绩及格的所有同学的信息。

```
select * from bunfly.score where english >= 60;
```

(2) 查看各个班英语成绩总分。

```
select classId, sum(english) from bunfly.score;
```

(3) 查看 dz1955001001 班的学生及英语成绩，并按降序排序。

```
select classId, name, english from bunfly.score where classId= 'dz1955001001' order by desc;
```

(4) 查看各个班英语平均成绩大于 80 的班级。

```
select classId, avg(english) avg_eng from bunfly.score group by classId having avg_eng>80;
```

3) union all 和 union

它们两个都可以把两个或多个表进行合并，每一个 union 子查询都必须具有相同的列。
union：对两个结果集进行并集操作，不包括重复行，同时进行规则的排序。
union all：对两个结果集进行并集操作，包括重复行,同时进行规则的排序。
示例如下：

```
create table score3
as
select name,chinese,mathematics,english from bunfly.score1
union all
select name,chinese,mathematics,english from bunfly.score2;
```

上述操作将 score1 和 score2 表的 name、chinese、mathematics 和 english 字段查询出来，并新创建一张 score3 表把刚刚查询出来的内容插入到 score3 表中。也可以看作将 score1 和 score2 复制到一个文件中并把它重新命名为 score3。

Hive 操作符如图 8.5 所示。

操作符	支持的数据类型	描述
A = B	基本数据类型	如果A等于B则返回true
A <> B, A != B	基本数据类型	如果A不等于B则返回true
A < B	基本数据类型	A小于B返回true
A <= B	基本数据类型	A小于等于B返回true
A > B	基本数据类型	A大于B返回true
A >= B	基本数据类型	A大于等于B返回true
A [not] between B and C	基本数据类型	A在B和C之间返回true
A is null	所有数据类型	A是null返回true
A is not null	所有数据类型	A不是null返回true
A [not] like B	String类型	A与B匹配返回true
A rlike B	String类型	A与B匹配返回true

图 8.5

Hive 算术运算符如图 8.6 所示。
Hive 数据函数如图 8.7 所示。

运算符	类型	描述
A+B	数值	A和B相加
A-B	数值	A减去B
A*B	数值	A和B相乘
A/B	数值	A除以B。返回商数
A%B	数值	A除以B。返回余数

图 8.6

返回值类型	格式	描述
double	Round(double d)	四舍五入，保留整数
double	Round(double d,int n)	四舍五入，保留n位
double	Rand()或rand(int d)	随机数，范围为0~1
double	Ln(double d)	以自然数据为底，d的对数
double	Log(double ba,double d)	以ba为底，d的对数
double	Sqrt(double d)	计算d的平方根
double	abs(double d)	计算d的绝对值
double	Sin(double d)	正弦值
double	Cos(double d)	余弦值
double	Tan(double d)	正切值

图 8.7

Hive 聚合函数如图 8.8 所示。

返回值类型	格式	描述
int	count(*)	计算总行数，包含null
int	count(1)	计算总行数，不含null
double	sum(col)	求和
double	avg(col)	求平均值
double	min(col)	求最小数
double	max(col)	求最大数
double	var_pop(col)	求方差
double	stddev_pop(col)	求标准偏差
double	corr(col1,col2)	两组数值的关系

图 8.8

Hive 内置函数如图 8.9 和图 8.10 所示。

返回值类型	格式	描述
string	concat(string s1,string s2,...)	将s1和s2拼接
string	concat_ws('#','a','b','c')	带分隔符的字符串拼接
int	length(string s)	计算字符串长度
string	lower(string s)	将字符串全部转换成小写字母
string	upper(string s)	将字符串全部转换成大写字母
string	lpad(string s,int len,string p)	从左边开始对字符串进行填充，到len长度为止
string	rpad(string s,int len,string p)	从右边开始对字符串进行填充，到len长度为止
string	ltrim(string s)	去掉左边空格
double	rtrim(string s)	去掉右边空格
double	trim(string s)	去掉前后全部空格

图 8.9

返回值类型	格式	描述
string	regexp_extract(string s,string regex,string replacement)	替换某个字符 将字符串s中符合条件的部分替换成replacement所指定的字符串a
string	Translate(string s,string from,string to)	替换某个字符 绝对匹配替换
string	Repeat(string s,int n)	重复输出n次字符串
string	Reverse(string s)	反转字符串
string	Substr(string s,int I,int j)	对字符串s,从start位置开始截取length长度的字符串，作为子字符串

图 8.10

8.4 习题与思考

1. 根据表 8.3 和表 8.4，通过编写 Hive 脚本创建 Student 表和 Score 表，并加载表 8.3 和表 8.4 的数据。

表 8.3

姓名	性别	出生年份	院系	家庭住址	学号
刘一	男	1994/11/26	数学系	上海	20157259
陈二	男	1993/6/11	数学系	北京	20153174
张三	女	1994/9/21	数学系	北京	20157824
李四	男	1993/1/26	信息工程系	云南	20155367

表 8.4

学号	课程名	分数
20157259	语文	90
20157259	数学	58
20157259	外语	39
20153174	语文	91
20153174	数学	95
20153174	外语	75
20157824	语文	60
20157824	数学	58
20157824	外语	53
20155367	语文	62
20155367	数学	43
20155367	外语	74

2. 尝试统计出每个系所有科目平均成绩在所有系中的占比。
3. 尝试统计出每个系考试不合格学生的占比。

第 9 章　HBase 分布式数据库

HBase 是一个高可靠性、高性能、面向列、可伸缩的分布式存储系统，利用 HBase 技术可在廉价的 PC 服务器上搭建大规模结构化的存储集群。与 MapReduce 的离线批处理计算框架不同，HBase 是一个可以随机访问的存储和检索数据平台，弥补了 HDFS 或 Hive 不能随机访问数据的缺陷，适合实时性要求不是非常高的业务场景。

HBase 从另一个角度处理伸缩性问题，它通过线性方式从下到上增加节点来进行扩展。HBase 不是关系型数据库，也不支持 SQL，但是它有自己的特长，这是 RDBMS 不能处理的，HBase 巧妙地将大而稀疏的表放在商用的服务器集群上。

HBase 是 Google Bigtable 的开源实现，与 Google Bigtable 利用 GFS 作为其文件存储系统类似，HBase 利用 Hadoop HDFS 作为其文件存储系统。Google 运行 MapReduce 来处理 Bigtable 中的海量数据，HBase 同样利用 Hadoop MapReduce 来处理 HBase 中的海量数据，Google Bigtable 利用 Chubby 作为协同服务，HBase 则利用 Zookeeper 作为对应。

HBase 主要有以下几个特点：

（1）大。一个表可以有上亿行，上百万列。

（2）面向列。面向列表（族）的存储和权限控制，列（族）独立检索。

（3）稀疏。对于为空（NULL）的列，并不占用存储空间。因此，表可以设计得非常稀疏。

（4）无模式。每一行都有一个可以排序的主键和任意多的列，列可以根据需要动态增加，同一张表中不同的行可以有截然不同的列。

（5）数据多版本。每个单元中的数据可以有多个版本，默认情况下，版本号自动分配，版本号就是单元格插入时的时间戳。

（6）数据类型单一。HBase 中的数据都是字符串，没有类型。

9.1　HBase 工作原理

9.1.1　HBase 表结构

HBase 表结构如图 9.1 所示。

1. Row Key

与 NoSQL 数据库一样，Row Key 是用来检索记录的主键。访问 HBase Table 中的行只有以下三种方式。

- 通过单个 Row Key 访问；

- 通过 Row Key 的 Range；
- 全表扫描。

Row Key	Column family 001		Column family 002			Column family N
	Column1	Column2	Column1	Column2	Column3	Column1
key1	80					
	100					
key1						
keyN						

图 9.1

Row Key 行键可以是任意字符串（最大长度是 64KB，实际应用中长度一般为 10～100B），在 HBase 内部，Row Key 保存为字节数组。

存储时，数据按照 Row Key 的字典序（Byte Order）排序存储。设计 Key 时，要充分排序存储这个特性，将经常一起读取的行存储到一起（位置相关性）。

注意：字典序对 int 排序的结果是 1,10,100,11,12,13,14,15,16,17,18,19,2,20,21,…,9,91,92,93,94,95,96,97,98,99。要保持整型的自然序，行键必须用 0 作左填充。

2. 列族

HBase 表中的每个列，都归属于某个列族。列族是表的 chema 的一部分（而列不是），必须在使用表之前定义。列名都以列族作为前缀。例如 courses:history，courses:math 都属于 courses 这个列族。

访问控制、磁盘和内存的使用统计都是在列族层面进行的。实际应用中，列族上的控制权限能帮助管理不同类型的应用：允许一些应用可以添加新的基本数据、读取基本数据并创建继承的列族，但一些应用则只允许浏览数据（由于隐私可能无法浏览所有数据）。

3. 时间戳

HBase 中通过 row 和 columns 确定的存储单元称为 cell。每个 cell 都保存着同一份数据的多个版本，版本通过时间戳来索引。时间戳的类型是 64 位整型。时间戳可以由 HBase（在数据写入时自动）赋值，此时时间戳是精确到毫秒的当前系统时间。时间戳也可以由客户显式赋值。如果应用程序要避免数据版本冲突，就必须生成具有唯一性的时间戳。每个 cell 中，不同版本的数据按照时间倒序排序，即最新的数据排在最前面。

为了避免数据存在过多版本造成的管理（包括存储和索引）负担，HBase 提供了两种数据版本回收方式：一是保存数据的最后 n 个版本，二是保存最近一段时间内的版本（比如最近七天）。用户可以针对每个列族进行设置。

4. cell

由 {row key, column(=<family> + <label>), version} 唯一确定的单元。cell 中的数据是没有类型的，全部是字节码形式存储。

9.1.2 体系结构

HBase 体系结构如图 9.2 和图 9.3 所示，HBase 读写流程如图 9.4 所示。

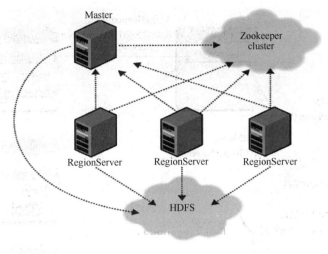

图 9.2

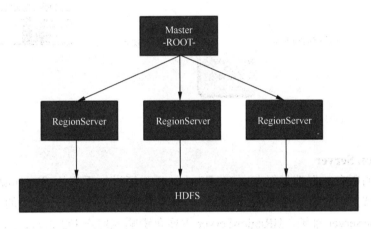

图 9.3

1. **client**

包含访问 HBase 的接口，client 维护着一些 cache 来加快对 HBase 的访问，比如 Region 的位置信息。

2. **Zookeeper**

（1）保证任何时候，集群中只有一个 HMaster。

（2）实时监控 HRegion Server 的上线和下线信息，并实时通知给 HMasert。

（3）存储 HBase 的 Table 表数据和存储所有 Region 的寻址入口。

3. **HMaster**

理论上 HMaster 可以启动多个，但是 Zookeeper 有 MasterElection 机制保证只有一个 HMaster 在运行，来负责 Table 和 Region 的管理工作。

（1）管理 HRegionServer 的负载均衡，调整 Region 分布。

（2）Region Split 后，负责新的 Region 的分布。

（3）在 HRegionServer 停机后，负责失效 HRegionServer 上 Region 的迁移工作。

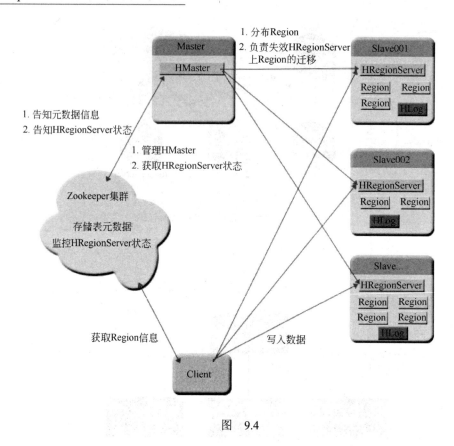

图 9.4

4. HRegion Server

HBase 中的所有数据都是保存在 HDFS 中，用户通过一系列的 HRegionServer 获取这些数据。一个节点上只有一个 HRegionServer 和多个 HRegion,每一个区段的 HRegion 只会被一个 HRegionServer 维护，HRegionServer 主要负责响应用户 I/O 请求，向 HDFS 文件系统读写数据。

5. HRegion

HRegion 是用来存储实际数据的。当表的大小超过预设时，HBase 会自动生成多个 Region 来存储数据。

6. HLog

每个 HRegionServer 中都有一个 HLog 对象，实现用户操作记录。在用户操作将数据写入 MemStore 的时候，同时会写一份数据到 HLog 文件中，HLog 文件会定期滚动刷新，并删除旧的文件。

HLog 的作用：当 HRegionServer 意外终止，HMaster 会通过遗留的 HLog 文件，把不同 HRegion 的 HLog 数据进行拆分，分别放到相应的 HRegion 的目录下，然后再将失效的 HRegion 重新分配到其他的 HRegionServer 中，完成数据恢复。

9.1.3 物理模型

HBase 的物理模型如图 9.5 所示。

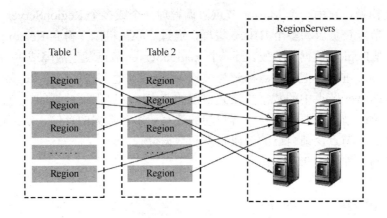

图 9.5

Region 是按大小分割的,每个表开始只有一个 Region。随着数据增多,Region 不断增大,当增大到一个阈值的时候,Region 就会分出一个新的 Region,之后会有越来越多的 Region。

Region 是 HBase 中分布式存储和负载均衡的最小单元,不同 Region 分布到不同 RegionServer 上,如图 9.6 所示。

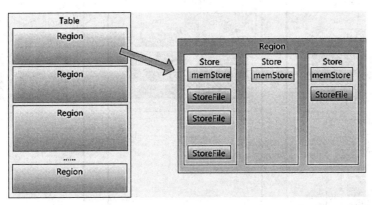

图 9.6

Region 虽然是分布式存储的最小单元,但并不是存储的最小单元。Region 是由一个或多个 Store 组成的,每个 Store 由一个 MemStore 和多个 StoreFile 组成。数据写入时首先进入到 MemStor 中并存储在内存中。MemStore 中的数据是排序的,当 MemStore 累计到一定阈值时,就会创建一个新的 MemStore,并且将老的 MemStore 添加到 Flush 队列,由单独的线程 Flush 到磁盘上,成为一个 StoreFile。当一个 Store 中的 StoreFile 达到一定阈值后,就会进行一次合并操作,将对同一个 key 的修改合并到一起,形成一个大的 StoreFile。当 StoreFile 的大小达到一定阈值后,又会对 StoreFile 进行切分操作,等分为两个 StoreFile。

9.1.4 HBase 读写流程

正如 HDFS 和 MapReduce 由客户端、数据节点和主节点组成一样,HBase 也是采用相同的主从结构模型,它由一个 Master 节点协调管理一个或多个 RegionServer 节点。

HBase 主节点负责启动整个 HBase 集群,通过"心跳机制"得到 RegionServer 节点工作状态,并管理 Region 数据的分发。当一个 RegionServer 节点发生故障或者宕机后,Master 节点中的 HMaster 进程将把该节点标记为故障,并协调其他负载较轻的 RegionServer 节点,将故障 RegionServer 节点中的数据复制到自己的节点中,以保证数据的完整性。

RegionServer 节点主要负责响应客户端的读写请求和数据的存储,定期地通过"心跳机制"向 HMaster 节点反馈自己的健康状态和数据位置。

HBase 读写流程如图 9.7 所示。

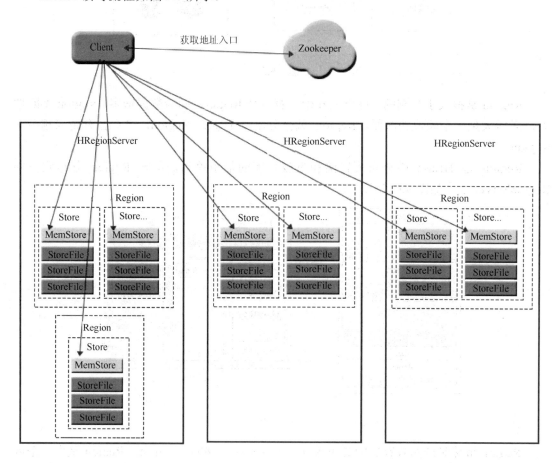

图 9.7

1. 写操作流程

(1) Client 通过 Zookeeper 的调度,向 HRegionServer 发出写数据请求,在 HRegion 中写数据。

(2) 数据被写入 HRegion 的 MemStore,直到 MemStore 达到预设阈值。

(3) MemStore 中的数据被 Flush 成一个 StoreFile。

(4) 随着 StoreFile 文件的不断增多,当其数量增长到一定阈值后,触发 Compact 合并操作,将多个 StoreFile 合并成一个 StoreFile,同时进行版本合并和数据删除。

（5）StoreFiles 通过不断的 Compact 合并操作，逐步形成越来越大的 StoreFile。

（6）单个 StoreFile 大小超过一定阈值后，触发 Split 操作，把当前 HRegion Split 成两个新的 HRegion。父 HRegion 会下线，新 Split 出的两个子 HRegion 会被 HMaster 分配到相应的 HRegionServer 上，使得原先一个 HRegion 的压力得以分流到两个 HRegion 上。

2. 读操作流程

（1）Client 访问 Zookeeper，查找-ROOT-表，获取.META.表信息。

（2）从.META.表查找，获取存放目标数据的 HRegion 信息，从而找到对应的 HRegionServer。

（3）通过 HRegionServer 获取需要查找的数据。

（4）HRegionserver 的内存分为 MemStore 和 BlockCache 两部分，MemStore 主要用于写数据，BlockCache 主要用于读数据。读请求将先到 MemStore 中查数据，如果无法查询到就会转到 BlockCache，再查询不到就会到 StoreFile，并把读的结果放入 BlockCache。

9.2　HBase 完全分布式

关于 HBase 安装的讲解视频可扫描二维码观看。

9.2.1　安装前的准备

（1）在官网下载 hbase-1.3.1-bin.tar.gz 安装包。

（2）复制安装包到 software 目录。

（3）解压 hbase-1.3.1-bin.tar.gz 安装包，解压后删除原包。

9.2.2　配置文件

（1）进入 HBase 的 conf 目录。

（2）修改 hbase-env.sh 文件。

插入：

```
export HBASE_HOME=/home/hadoop/software/hbase-1.3.1
export JAVA_HOME=/home/hadoop/software/jdk1.8.0_131
export HADOOP_HOME=/home/hadoop/software/hadoop-2.6.5
export HBASE_LOG_DIR=$HBASE_HOME/logs
export HBASE_PID_DIR=$HBASE_HOME/pids
export HBASE_MANAGES_ZK=false
```

（3）修改 hbase-site.xml 文件。

插入：

```
<!--设置HRegionServers共享目录,mycluster是在Hadoop中设置的名字空间-->
<property>
    <name>hbase.rootdir</name>
    <value>hdfs://mycluster/hbase</value>
</property>
```

```xml
<!--设置HMaster的rpc端口-->
<property>
    <name>hbase.master.port</name>
    <value>16000</value>
</property>
<!--设置HMaster的http端口-->
<property>
    <name>hbase.master.info.port</name>
    <value>16010</value>
</property>
<!--指定缓存文件存储的路径-->
<property>
    <name>hbase.tmp.dir</name>
    <value>/home/hadoop/software/hbase-1.3.1/tmp</value>
</property>
<!--开启分布模式-->
<property>
    <name>hbase.cluster.distributed</name>
    <value>true</value>
</property>
<!--指定Zookeeper集群位置-->
<property>
    <name>hbase.zookeeper.quorum</name>
    <value>Slave001,Slave002,Slave003</value>
</property>
<!--指定Zookeeper集群端口-->
<property>
    <name>hbase.zookeeper.property.clientPort</name>
    <value>2181</value>
</property>
<!--指定Zookeeper数据目录,需要与Zookeeper集群中的dataDir配置相一致-->
<property>
    <name>hbase.zookeeper.property.dataDir</name>
    <value>/home/hadoop/software/zookeeper-3.4.10/tmp/zookeeper</value>
</property>
```

(4) 配置 regionservers 文件。

插入:

```
Slave001
Slave002
Slave003
```

(5) 新建 backup-masters 文件,并配置。

插入:

```
Master002
```

（6）在 HBase 安装目录下创建 tmp（缓存文件）、logs（日志文件）、pid（pid 文件）的目录。

```
mkdir tmp logs pid
```

（7）将 HBase 配置好的安装文件同步到集群其他节点。

```
scp -r hbasexx Master002:~/software/
scp -r hbasexx Slave001:~/software/
scp -r hbasexx Slave002:~/software/
scp -r hbasexx Slave003:~/software/
```

（8）在集群的各个节点上配置环境变量（在 root 用户下操作）。

```
vi /etc/profile
```

插入：

```
export HBASE_HOME=/home/hadoop/software/hbaseXX
export PATH=$PAT:$HBASE_HOME/bin
```

（9）使 profile 文件生效。

```
source /etc/profile
```

9.2.3 集群启动

（1）启动 Zookeeper 集群（分别在 Slave001、Slave002、Slave003 上执行）。

```
./zkServer.sh start
```

备注：此命令分别在 Slave001、Slave002、Slave003 节点上启动了 QuorumPeerMain。

（2）启动 JournalNode 节点（在 Zookeerper 节点上执行：Slave001、Slave002、Slave003）。

```
hadoop-daemon.sh start journalnode
```

备注：此命令分别在 Slave001、Slave002、Slave003 节点上启动了 JournalNode。

（3）启动 HDFS（在 Msater001 上执行）。

```
start-dfs.sh
```

备注：此命令分别在 Master001、Master002 节点上启动了 NameNode 和 DFSZKFailover-Controller。分别在 Slave001、Slave002、Slave003 节点上启动了 DataNode。

（4）启动 YARN（在 Master001 上执行）。

```
start-yarn.sh
```

备注：此命令在 Master001 节点上启动了 ResourceManager，分别在 Slave001、Slave002、

Slave003 节点上启动了 NodeManager。

（5）启动 HBase（在 Master001 执行）。

```
start-hbase.sh
```

备注：此命令分别在 Master001、Master002 节点启动了 HMaster，分别在 Slave001、Slave002、Slave003 节点启动了 HRegionServer。

（6）检验。

在网页地址输入"192.168.xx.xx:16010"，如果能进入网页页面，说明配置成功。

9.3 HBase Shell

关于 HBase Shell 的使用的讲解视频可扫描二维码观看。

9.3.1 DDL 操作

DDL（Data Definition Language）是数据库模式定义语言，用于描述分布式数据库中要存储的现实世界实体。本节内容将执行关于 HBase 的 DDL 操作，包括数据库表的建立、查看所有表、查表结构、删除列族、删除表等操作。

创建一个 User 表，其结构如图 9.8 所示。

Row key	address			info	
	country	city	detailed_address	age	sex
zhangsan	China	Chengdu		20	female
wangqiang	China			25	male
liuyu	China	Chengdu		22	
zhanglun	China			27	
zhoujin	China	Chengdu		28	female
yauyu	China			19	female

图 9.8

（1）创建一个表名字为 User，并插入 address、info 和 member_id 三个列族。

```
create 'User','address','info', 'member_id';
```

（2）列出所有表。

```
hbase(main):012:0>list
TABLE
member
```

1 row(s) in 0.0160seconds

（3）显示 User 表的详细信息，查询结果如图 9.9 所示。

```
hbase(main):006:0>describe 'member'
```

```
hbase(main):006:0>describe 'user'
DESCRIPTION                                                    ENABLED
{NAME => 'user', FAMILIES => [{NAME=> 'address', BLOOMFILTER => 'NONE', REPLICATION_SCOPE => '0', true
VERSIONS => '3', COMPRESSION => 'NONE',TTL => '2147483647', BLOCKSIZE => '65536', IN_MEMORY => 'fa
lse', BLOCKCACHE => 'true'}, {NAME =>'info', BLOOMFILTER => 'NONE', REPLICATION_SCOPE => '0', VERSI
ONS => '3', COMPRESSION => 'NONE', TTL=> '2147483647', BLOCKSIZE => '65536', IN_MEMORY => 'false',
BLOCKCACHE => 'true'}]}
1 row(s) in 0.0230seconds
```

图 9.9

（4）删除一个列族：alter（修改表）、disable（禁用表）、enable（启用表）。

之前已经建立了 3 个列族，但是 member_id 这个列族是多余的，所以要将其删除。

```
hbase(main):003:0>alter 'User',{NAME=>'member_id',METHOD=>'delete'}
ERROR: Table memberis enabled. Disable it first before altering.
```

执行上面删除信息时会报错。因为删除列族时必须先禁用表，才能执行删除操作。可以用 disable 关键字标上一张表，例如：要删除 User 表，必先执行 disable 'User'后才能执行 alter 'User', {NAME=>'member_id', METHOD=>'delete'}操作。

```
hbase(main):004:0>disable 'User'
0 row(s) in 2.0390seconds
hbase(main):005:0>alter 'User',{NAME=>'member_id',METHOD=>'delete'}
0 row(s) in 0.0560seconds
```

通过 describe 命令查看到 member_id 字段已不存在，表示 member_id 字段已被删除，如图 9.10 所示。

```
hbase(main):006:0>describe 'User'
DESCRIPTION                                                    ENABLED
{NAME => 'user', FAMILIES => [{NAME=> 'address', BLOOMFILTER => 'NONE', REPLICATION_SCOPE => '0',false
VERSIONS => '3', COMPRESSION => 'NONE', TTL => '2147483647', BLOCKSIZE => '65536', IN_MEMORY => 'fa
lse', BLOCKCACHE => 'true'}, {NAME =>'info', BLOOMFILTER => 'NONE', REPLICATION_SCOPE => '0', VERSI
ONS => '3', COMPRESSION => 'NONE', TTL=> '2147483647', BLOCKSIZE => '65536', IN_MEMORY => 'false',
BLOCKCACHE => 'true'}]}
1 row(s) in 0.0230seconds
```

图 9.10

当需要删除的列族被删除后，要把 User 表重新启用。若不启用 User 表，User 将不能正常使用。

```
hbase(main):008:0> enable 'User'
0 row(s) in 2.0420seconds
```

判断表是否 enable 和 disable。

```
hbase(main):009:0> is_enable 'User'
```

```
true
0 row(s) in 2.0420seconds
hbase(main):009:0> is_disable 'User'
false
0 row(s) in 2.0420seconds
```

（5）可以使用 drop 命令删除一张表，需要两步：先把表禁止，再执行删除命令。
禁止 User 表：

```
hbase(main):029:0>disable 'User'
    0 row(s) in 2.0590seconds
```

删除 User 表：

```
hbase(main):030:0>drop 'User'
0 row(s) in 1.1070seconds
```

验证表是否存在可通过 list 列出所有的表。但如果表比较多，使用 list 就不是最佳方法了，这时可通过 exists 命令直接查看该表是否存在。

```
hbase(main):031:0>exists 'User'
Table User does exist
0 row(s) in 1.1070seconds
```

9.3.2 DML 操作

DML 命令是数据操作语言命令，包含的命令非常丰富，用于数据的写入、删除、修改、查询、清空等操作。本节将详细讲解 DML 操作的常用命令。

1. 插入几条记录

```
put 'User','zhangsan','address:country','China';
put 'User','zhangsan','address:city','Chengdu';
put 'User','zhangsan','info:age','20';
put 'User','zhangsan','info:sex','female';
put 'User','wangqiang','address:country','China';
put 'User','wangqiang','info:age','25';
put 'User','wangqiang','info:sex','male';
put 'User','liuyu','address:country','China';
put 'User','liuyu','address:city','Chengdu';
put 'User','liuyu','info:age','25';
put 'User','zhanglun','address:country','China';
put 'User','zhanglun','info:age','27';
put 'User','zhoujin','address:country','China';
put 'User','zhoujin','address:city','Chengdu';
put 'User','zhoujin','info:age','28';
put 'User','zhoujin','info:sex','female';
put 'User','yauyu','address:country','China';
```

```
put 'User','yauyu','info:age','19';
put 'User','yauyu','info:sex','female';
```

2. 获取一条数据

获取名字为'zhoujin'的所有数据：

```
hbase(main):001:0>get 'User','jiangxiaohui'
COLUMN              CELL
address:country     timestamp=1508804839610, value=China
address:city        timestamp=1508804837402, value=Chengdu
info:age            timestamp=1508804837475, value=28
info:sex            timestamp=1508804812019, value=female
4 row(s) in 0.1400 seconds
```

获取名字为'zhoujin'，并且列族为 info 的所有数据：

```
hbase(main):002:0>get 'User','zhoujin','info'
info:age            timestamp=1508804837475, value=28
info:sex            timestamp=1508804812019, value=female
2 row(s) in 0.1400 seconds
```

获取名字为'zhoujin'的年龄：

```
hbase(main):002:0>get 'User','zhoujin','info:age'
COLUMN              CELL
info:age            timestamp=1508804837475, value=28
1 row(s) in 0.0320seconds
```

3. 更新一条记录

若 zhoujin 的年龄录入错误要进行修改，但在 HBase 中没有特定的修改命令，可以通过新插入一条数据的方式覆盖以前的数据。

```
hbase(main):004:0>put 'User','zhoujin','info:age' ,'19'
0 row(s) in 0.0210seconds
```

4. 查看所有的年龄

```
hbase(main):047:0> scan 'User',{COLUMNS=>'info:age'}
ROW                 COLUMN+CELL
zhangsan            column=info:age, timestamp=1508804812019, value=20
wangqiang           column=info:age, timestamp=1508804812019, value=25
liuyu               column=info:age, timestamp=1508804812019, value=25
zhanglun            column=info:age, timestamp=1508804812019, value=27
zhoujin             column=info:age, timestamp=1508804837499, value=19
yauyu               column=info:age, timestamp=1508804812019, value=19
6 row(s) in 0.0740 seconds
```

5. 通过 timestamp 来获取 zhoujin 年龄修改之前的年龄

修改前的年龄：

```
hbase(main):010:0>get 'User', 'zhoujin', {COLUMN=>'info:age',TIMESTAMP=>
1508804837475}
COLUMN                  CELL
info:age                timestamp=1508804837475, value=28
1 row(s) in 0.0140seconds
```

修改后的年龄：

```
hbase(main):011:0>get 'User', 'zhoujin', {COLUMN=>'info:age',TIMESTAMP=>
1508804837499}
COLUMN                  CELL
info:age                timestamp=1508804837499, value=19
1 row(s) in 0.0180seconds
```

6. 全表扫描

```
hbase(main):013:0>scan 'User'

ROW             COLUMN+CELL
zhangsan        column=address:country, timestamp=1321586240244, value=China
zhangsan        column=address:city, timestamp=1321586239126, value=Chengdu
zhangsan        column=info:age, timestamp=1321586239197, value=20
zhangsan        column=info:sex, timestamp=1321586571843, value=male
wangqiang       column=address:country, timestamp=1321586239015, value=China
wangqiang       column=info:age, timestamp=1321586239071, value=25
wangqiang       column=info:sex,    timestamp=1321586248400,    value=jieyang
liuyu           column=address:country, timestamp=1321586248316, value=China
liuyu           column=address:city, timestamp=1321586248355, value=Chengdu
liuyu           column=info:age, timestamp=1321586249564, value=25
zhanglun        column=address:country, timestamp=1321586248202, value=China
zhanglun        column=info:age, timestamp=1321586248277, value=27
zhoujin         column=address:country, timestamp=1321586248241, value=China
zhoujin         column=address:city, timestamp=1321586240244, value=Chengdu
zhoujin         column=info:age, timestamp=1321586239126, value=28
zhoujin         column=info:sex, timestamp=1321586248355, value=female
yauyu           column=address:country, timestamp=1321586239197, value=China
yauyu           column=info:age,timestamp=1321586571843, value=19
yauyu           column=info:sex, timestamp=1321586248400, value=female 19
row(s) in 0.0570seconds
```

7. 删除某人的全部信息

假如 zhoujin 离职要删除所有信息，可以使用 deleteall 命令：

```
hbase(main):001:0>deleteall 'User','zhoujin'
0 row(s) in 0.3990seconds
```

8. 查询表中总共有多少行

```
hbase(main):019:0>count 'User'
15 row(s) in 0.0160seconds
```

9. 用 truncate 命令清空一张表

上述提到清空一张表必须经过两步：先禁用表，再删除表。现在使用 truncate 命令可直接清空一张表。观察执行过程可发现 truncat 命令同样执行禁用表和删除表两个步骤：

```
hbase(main):035:0>truncate 'User'
 Truncating 'User'table (it may take a while):
  - Disabling table...
  - Dropping table...
  - Creating table...
0 row(s) in 4.3430seconds
```

9.4 习题与思考

1. 用 HBase 存储社交网站站内短信信息，要求记录发送者、接收者、时间、内容。有关查询包括发送者和接收者均可分别列出其发出的所有信息列表。尝试进行数据建模创建 info_table 表。

2. 尝试利用 HBase 将表 9.1 中的数据插入到 info_table 表中。

表 9.1

发送者	时间	内容	接收者
张三	201801231450	HBase 是一款很好用的框架	李四
李四	201801231451	是吗？	张三
张三	201801231451	我们可以用它做许多结构化数据库做不到的事。	李四
李四	201801231452	哦！	张三
张三	201801231452	我们可以通过面列的方式快速统计出我们想要的信息，这是 Hive 和 HDFS 办不到的。	李四
张三	201801231458	还在吗！！！	李四

第 10 章　Sqoop 工 具

Sqoop 是一款 Hadoop 和关系型数据库之间进行数据导入导出的工具。借助这个工具，可通过 Sqoop 把数据从数据库（比如 MySQL，Oracle）导入到 HDFS 中，也可把数据从 HDFS 中导出到关系型数据库中。Sqoop 通过 Hadoop 的 MapReduce 导入和导出，提供了很高的并行性能以及良好的容错性。

Sqoop 适合以下人群使用：
- 系统和应用开发者；
- 系统管理员；
- 数据库管理员；
- 数据分析师；
- 数据工程师。

使用这款工具所必需的背景如下：
- 了解基本的计算机知识；
- 熟悉类似 Bash 的命令行（Sqoop 基本通过命令行进行操作）；
- 熟悉关系型数据库系统的管理（数据需从数据库导出）；
- 熟悉 Hadoop 基本操作（了解基本的 HDFS 操作和 MapReduce 的原理帮助理解 Sqoop 的过程）。

使用 Sqoop 之前需要先安装 Hadoop，文档基于 Linux 环境。若是在 Windows 环境中使用，需要安装 cygwin，物理机配置如图 10.1 所示。

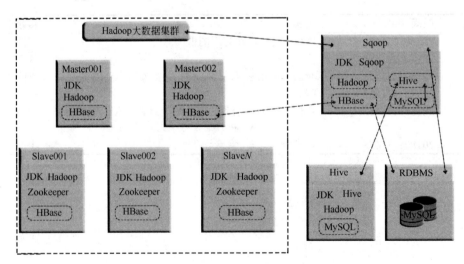

图　10.1

图 10.1 显示 Sqoop 独立于 Hadoop 大数据集群存在，因此要从 HDFS 中导入、导出数据必须依赖 Hadoop，在 Sqoop 客户中安装 Hadoop 的配置文件。Hive、HBase 同理。若与 Hive、HBase 或其他工具进行数据交互，需在 Sqoop 物理机下安装其配置文件。关系型数据库可安装在 Sqoop 物理机或其他物理机下，通过指定特定地址进行数据交互。

10.1　Sqoop 安 装

在官网 http://sqoop.apache.org 下载 Sqoop 安装文件（示例使用 sqoop-1.4.6-cdh5.5.2 版本）。

（1）解压安装文件：tar -zxf sqoop-1.4.6-cdh5.5.2.tar.gz。

（2）配置环境变量：vi /etc/profile。

添加：

```
#sqoop
export SQOOP_HOME=/home/hadoop/soft/sqoop-1.4.6-cdh5.5.2
export PATH=$PATH:$SQOOP_HOME/bin
```

（3）使环境变量生效：source /etc/profile。

（4）进入/home/hadoop/soft/sqoop-1.4.6-cdh5.5.2/conf 目录，复制 sqoop-env-template.sh 并重命名为 sqoop-env.sh。

```
cp sqoop-env-template.sh sqoop-env.sh
```

（5）编辑 sqoop-env.sh 并添加相关配置。

```
vi sqoop-env.sh
```

添加：

```
export HADOOP_COMMON_HOME=/home/hadoop/soft/hadoop-2.6.5
export HADOOP_MAPRED_HOME=/home/hadoop/soft/hadoop-2.6.5
```

（6）若对 MySQL 数据库进行交互，需将 MySQL 的驱动包 mysql-connector-java-5.1.41-bin.jar 上传至 Sqoop 安装目录的 lib 目录下。若对 Oracle 数据库进行交互导入 Oracel 驱动包即可。

（7）测试输入 Sqoop Version。如果出现以下内容说明安装成功（警告可忽略）。

```
Warning: /home/hadoop/soft/sqoop-1.4.6-cdh5.5.2/../hbase does not exist!
HBase imports will fail.
Please set $HBASE_HOME to the root of your HBase installation.
Please set $ZOOKEEPER_HOME to the root of your Zookeeper installation.
17/08/31 17:06:03 INFO sqoop.Sqoop: Running Sqoop version: 1.4.6-cdh5.5.2
Sqoop 1.4.6-cdh5.5.2
git commit id 8e266e052e423af592871e2dfe09d54c03f6a0e8
Compiled by jenkins on Mon Jan 25 16:10:15 PST 2016
```

10.2　Sqoop 的使用

10.2.1　MySQL 的导入导出

1. 查询 Sqoop 命令

```
sqoop help
```

2. 从 MySQL 导数据到 HDFS

```
sqoop import --connect jdbc:mysql://192.168.153.205:3306/bunfly --username root --password 123456 --table houses_number --fields-terminated-by '\t' -m 1
```

解释：

sqoop	sqoop命令
import	表示导入
--connect jdbc:mysql://192.168.153.205:3306/bunfly	用JDBC方式，连接MySQL数据库。数据库的IP是192.168.153.205，数据库端口是3306，数据库名是bunfly
--username root	数据库账户是root。
--password 12345	数据库密码是123456。
--table houses_number	导出数据库中的houses_number表。
--fields-terminated-by '\t'	文件分隔格式为'\t'。
-m 1	使用过程中用一个map作业。
--hive-import	把数据复制到Hive空间中。若不用这个选项则直接复制到HDFS中。数据在HDFS的/user/目录下。

注：如果要导入到 Hive 中，Hive 与 Sqoop 需装在同一台服务器上。

3. 导出

```
sqoop export --connect jdbc:mysql://192.168.153.205:3306/bunfly?characterEncoding=UTF-8 --username root -password 123456 --table houses_number --export-dir '/user/hadoop/houses_number/p*' --fields-terminated-by '\t'
```

解释：

--export-dir 'user/hadoop/houses_number/p*'	HDFS中的地址。
sqoop	Sqoop命令。
export	表示数据从Hive复制到MySQL中。
--connect jdbc:mysql://ip:3306/bunfly	用JDBC连接MySQL数据库，IP是MySQL数据所在的IP地址，MySQL数据库的端口号为3306，数据库名为bunfly。characterEncoding=UTF-8，令中文编码集为UTF-8（设置中文编码集后导入到MySQL的中文数据不会出现乱码）。
--username root	连接bunfly数据库的用户名。
--password admin	连接bunfly数据库的密码。
--table mysql2	mysql2是bunfly数据库的表，即将被导入的表名称。
--export-dir '/user/root/warehouse/mysql1'	Hive中被导出的文件目录。

```
--fields-terminated-by '\t'    ##Hive中被导出的文件字段的分隔符。
?characterEncoding=UTF-8       中文编码集。
```

📢注意：导出的数据表必须事先存在。

10.2.2 Oracle 的导入导出

从 Oracle 源数据库导入数据如下：

sqoop import --connect jdbc:oracle:thin:@192.168.1.200:1521:ORCL --username mfkddd --password amwvke --table I_ZHIB_YUANB -m 1

详细说明：

sqoop	Sqoop命令。
import	导入命令。
--connect jdbc:oracle:thin@	表示源数据库为Oracle，采用JDBC方法导入。
192.168.1.200	Oracle数据库服务器地址。
1521	Oracle数据库的端口号。
ORCL	Oracle数据库名。
--username mfkddd	ORCL数据库的用户名。
--password amwvke	ORCL数据库的密码。
--table I_SHIB_YUANB	ORCL数据库下的I_SHIB_YUANB表。
-m 1	启动的map数量，这里为1个。

示例如下，指定文件在 HDFS 中的存放地址：

```
sqoop import
--connect jdbc:oracle:thin:@192.168.1.200:1521:ORCL
--username mfkddd
--password amwvke
--table I_ZHIB_YUANB
--m 1
--target-dir /tmp/i_zhib_yuanb
```

说明：--target-dir 指定文件在 HDFS 中的存放地址。若要新指定一个地址，必须在该地址下新建一个目录存放 part-m-00000 文件，否则报错：xx already exists（xx 文件已经存在）。

验证：hadoop fs -cat /tmp/i_zhib_yuanb/part*出现导入内容，导入成功。

表数据导入子集通过 where 子句完成，相当于 SQL 中的 where 查询条件。

```
sqoop import
--connect jdbc:oracle:thin:@192.168.1.200:1521:ORCL
--username mfkddd
--password amwvke
--table O_ORG
--m 1
```

```
--where "org_type='03'"
--target-dir /tmp/o_org_2
```

说明：

```
--where "org_type='03'"
--where                      条件查询关键字。
org_tpye                     关系型数据库中的字段名称。
'03'                         需要过滤的内容。
```

从 HDFS 导出数据到 Oracle 数据库，示例如下：

```
sqoop export --connect jdbc:oracle:thin:@192.168.1.200:1521:ORCL
--username root -password 123456 --table houses_number --export-dir
'/user/hadoop/houses_number/p*' --fields-terminated-by '\t'
```

解释：

```
sqoop                                                          Sqoop命令
export                                                         导出
--connect jdbc:oracle:thin:@192.168.1.200:1521:ORCL            Oracle的url地址
--username root                                                Oracle数据库名
--password admin                                               Oracle数据库密码
--table houses_number                                          Oracle数据库表名
--export-dir '/user/hadoop/houses_number/p*'                   HDFS的文件地址
--fields-terminated-by '\t'                                    导出的文件字段分隔符
```

注意：导出的数据表必须是事先存在的。

10.3　习题与思考

1. 尝试配置 Sqoop 工具，从 Hive 中导出数据到 MySQL 中。
2. 尝试在 HDFS 中导出中文数据到 MySQL 中。

第 11 章　Flume 日志收集

　　Flume 是 Cloudera 提供的高可用、高可靠、分布式，进行海量日志采集、聚合和传输系统。Flume 支持在日志系统中定制各类数据发送方，用于收集数据。同时，Flume 可对数据进行简单处理，并写到各种数据接收方（可定制）。

　　Flume 是由 Cloudera 软件公司推出的可分布式日志收集系统。2009 年被 Apache 软件基金会捐赠，成为 Hadoop 相关组件之一。随着 Flume 的不断完善，特别是 Flume-ng 升级版本的推出以及内部组件的不断丰富，提高了用户在开发过程中的便利性，现已成为 Apache Top 项目之一。

　　Flume 的运行环境为必须安装 JDK 6.0 以上版本。Flume 目前只有 Linux 系统的启动脚本，没有 Windows 环境的启动脚本。

　　Flume 具有高可用、分布式配置工具，其设计原理基于将如日志数据的数据流从各种网站服务器上汇集起来存储到 HDFS、HBase 等集中存储器中，Flume 的结构如图 11.1 所示。

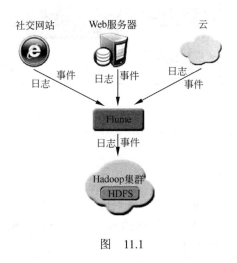

图　11.1

　　Flume 可以在电子商务网站做精准营销。例如，通过消费用户的访问点以及特定的节点区域来分析消费者的行为或者购买意图，便可更快地推送消费者想获取的信息。要实现这一点，需将获取的访问页面以及单击的产品数据等日志数据信息收集起来，并移交 Hadoop 平台进行分析。京东"还没有逛够"模块就是使用 Flume 和 Hadoop 技术。日志收集工具有很多，除了 Flume，还有 Facebook 的 Scribe、淘宝的 Time Tunnel 等。

Flume 的优点如下：

(1) Flume 可将应用产生的数据存储到任何集中存储器中，如 HDFS。

(2) 当收集数据的速度超过写入数据的速度，即收集信息遇到峰值时，收集的信息非常多，甚至超过系统写入数据的能力。此时 Flume 会在数据生产者和数据收容器间做出调整，保证其能够在两者之间提供一平稳的数据。

(3) Flume 的管道基于事务，以保证数据传送和接收的一致性。

(4) Flume 可靠性、容错性高，可升级，易管理，可定制。

(5) 可被水平扩展。

11.1 体系架构

11.1.1 Flume 内部结构

如图 11.2 所示，数据发生器产生的数据被单个运行在数据发生器所在服务器上的 Agent 收集，之后数据收容器从各个 Agent 上汇集数据，并将采集到的数据存入 HDFS。

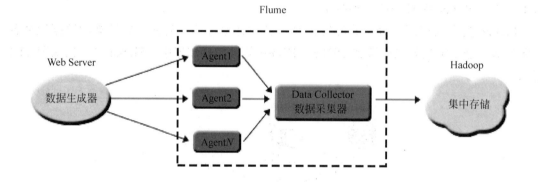

图 11.2

11.1.2 Flume 事件

Flume 的数据流由基本数据单位事件（Event）贯穿始终，并携带日志数据（字节数组形式）和头信息。这些 Event 由 Agent 外部的 Source 生成，当 Source 捕获事件后会进行特定的格式化，然后 Source 会把事件推入（单个或多个）Channel 中。将 Channel 看作缓冲区，它保存事件直到 Sink 将其处理完。Sink 负责持久化日志或把事件推向另一个 Source。

Flume 以 Agent 为最小的独立运行单位，一个 Agent 就是一个 JVM。单 Agent 由 Source、Sink 和 Channel 三大组件构成，数据模型如图 11.3 所示。

Agent：使用 JVM 运行 Flume 时，每台机器运行一个 Agent。但在一个 Agent 中包含多个 Source 和 Sinks。

Source：从客户端收集数据，并传递给 Channel。

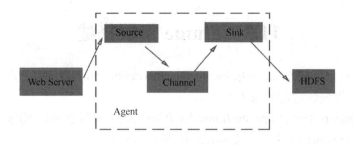

图 11.3

Channel：缓冲区，将 Source 传输的数据暂时存放。
Sink：从 Channel 收集数据，并写入到指定地址。
Event：日志文件、avro 对象等源文件。

假如 Agent 是一个池子，一端是进水口，另一端是出水口。进水口可配置各种水管，从多个地方进水；出水口也配置各种水管，但只有一个地方出水，指定水排出的地方。通过进水口将水输入到池里，然后出水口将池里的水抽到指定的地方。其中的水就是 Event，它包含了需要的源文件。即由 Source 将源文件收集起来并放到 Channel（缓冲池）中，而 Sink 将收集到的数据最终落到实处。

11.2 Flume 的特点

如图 11.4 所示，Flume 有两种角色的节点：代理节点（Agent）、收集节点（Collector）。

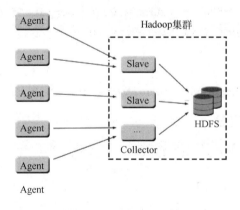

图 11.4

Agent 从各个数据源收集日志数据，并集中到 Collector，然后由收集节点汇总存入 HDFS。在 Flume 中，最重要的抽象是 dataflow（数据流）。dataflow 描述了数据从产生、传输、处理到写入目标的一条路径。Agent 数据流配置即把数据发送到得到数据的 Collector。Collector 接收 Agent 发过来的数据，并把数据发送到指定的目标机器。

Flume 框架在 jar 包上依赖 Hadoop 和 Zookeeper，并不要求 Flume 启动时必须启动 Hadoop 和 Zookeeper 服务。

11.3 Flume 集群搭建

(1) Flume 官方下载地址:http://www.apache.org/dist/flume/1.6.0/。其中 apache-flume-1.6.0-bin.tar.gz 版本是较稳定的版本。

(2) 通过 hadoop 用户将 apache-flume-1.6.0-bin.tar.gz 安装包通过 Xftp 工具上传到需要安装的节点的~/software 目录下(以 Slave001 节点为例)。

(3) 解压 apache-flume-1.6.0-bin.tar.gz 安装包。

```
[hadoop@Slave001 ~]$ cd software
[hadoop@Slave001 software]$ ls
hadoop-2.6.5  jdk1.8.0_131  zookeeper-3.4.10  apache-flume-1.6.0-bin.tar.gz
[hadoop@Slave001 software]$ tar -zxf apache-flume-1.6.0-bin.tar.gz
[hadoop@Slave001 software]$ ls
hadoop-2.6.5  jdk1.8.0_131  zookeeper-3.4.10  apache-flume-1.6.0-bin.tar.gz
apache-flume-1.6.0-bin
[hadoop@Slave001 software]$ cd apache-flume-1.6.0-bin
[hadoop@Slave001 apache-flume-1.6.0-bin]$ pwd
/home/hadoop/software/apache-flume-1.6.0-bin     #复制
```

(4) 修改配置文件。

```
[hadoop@Slave001 apache-flume-1.6.0-bin]$ cd conf
[hadoop@Slave001 conf]$ cp flume-env.sh.template flume-env.sh
[hadoop@Slave001 conf]$ vi flume-env.sh
```

改写:

```
export JAVA_HOME=/home/hadoop/software/jdk1.8.0_131     #自己的路径
```

(5) 配置 Flume 环境变量。

```
[hadoop@Slave001 conf]$ su -l root
```

密码:

```
[root@Slave001 ~]# vi /etc/profile
```

插入:

```
#flume
export FLUME_HOME=/home/hadoop/software/apache-flume-1.6.0-bin
export FLUME_CONF_DIR=$FLUME_HOME/conf
export PATH=$PATH:$FLUME_HOME/bin
[root@Slave001 ~]# reboot
```

说明：FLUME_HOME 路径是上面第（3）步中的复制路径，在文本编辑状态可通过鼠标快捷粘贴。

配置全局环境变量后使用 source /etc/profile 命令只针对 root 用户生效，若对全部用户都生效需使用 reboot 命令重启虚拟机。

（6）验证安装是否成功。

出现 Flume 版本号等信息说明安装成功，否则安装失败。

```
[hadoop@Slave001 ~]$ flume-ng version
Flume 1.6.0
Source code repository: https://git-wip-us.apache.org/repos/asf/flume.git
Revision: 2561a23240a71ba20bf288c7c2cda88f443c2080
Compiled by hshreedharan on Mon May 11 11:15:44 PDT 2015
From source with checksum b29e416802ce9ece3269d34233baf43f
```

（7）分发到各节点。

通过 scp 命令将 Flume 配置和环境变量配置发送到其他节点。为了教学方便，本书以 Slave001、Slave002、Slave003 三台节点作为 Flume 集群。

```
[hadoop@Slave001 ~]$ scp -r ~/software/apache-flume-1.6.0-bin/ Slave002:~/software/
[hadoop@Slave001 ~]$ scp -r ~/software/apache-flume-1.6.0-bin/ Slave003:~/software/
[hadoop@Slave001 ~]$ su -l root
[hadoop@Slave001 ~]$ scp  ~/etc/profile Slave002:/etc
[hadoop@Slave001 ~]$ scp  ~/etc/profile Slave003:/etc
```

分别打开 Slave002、Slave003 节点终端，进入到 root 用户，使用 reboot 命令重启虚拟机。重启后分别登录 Slave002、Slave003 节点，使用 flume-ng version 命令验证 Flume 安装是否成功。

11.4 Flume 实例

11.4.1 实例 1：实时测试客户端传输的数据

在 Slave001 节点/home/hadoop 目录创建 netcat.conf 文件，将下列内容插入到 netcat.conf，启动 netcat.conf 程序用于监听某个端口，并捕获传输的数据，在其他节点（是 Slave002 节点）使用 Telnet 协议发送数据。整个过程如同使用 QQ 聊天软件向另一个客户端发送消息，但对方仅接收不能回复。

（1）在 Slave001 中创建 netcat.conf 文件。Flume 可通过 Avro 监听端口并捕获传输的数据。具体示例如下：

```
#source+channels+sinks名字定义为agent
agent.sources = seqGenSrc
```

```
agent.channels = memoryChannel
agent.sinks = loggerSink
#描述source
agent.sources.seqGenSrc.type = netcat
agent.sources.seqGenSrc.bind = Slave001
agent.sources.seqGenSrc.port = 44444
#描述sink
agent.sinks.loggerSink.type = logger
#使用缓冲池
agent.channels.memoryChannel.type = memory
agent.channels.memoryChannel.capacity = 1000
agent.channels.memoryChannel.transactionCapacity = 100
#将source和sink绑定到channel（缓冲池）
agent.sources.seqGenSrc.channels = memoryChannel
agent.sinks.loggerSink.channel = memoryChannel
```

（2）启动 netcat.conf。

```
[hadoop@Slave001 ~] flume-ng agent -n agent -c conf -f /home/hadoop/netcat.conf
flume-ng agent：启动Flume的命令。
-n agent：Agent的名字，与netcat.conf配置文件的Agent名字一致。
-c conf：传输配置文件。
-f /home/hadoop/netcat.conf：文件路径。
```

（3）在 Slave002 节点中传输数据。

在 Slave002 中使用 Telnet 远程协议工具传输数据，使用 Telnet 工具须先安装 Telnet。

```
[hadoop@Slave002 ~]$ su -l root
```

密码：

```
[root@Slave002 ~]# yum list | grep telnet
telnet.x86_64           1:0.17-48.el6           @base
telnet-server.x86_64    1:0.17-48.el6           @base
[root@Slave002 ~]# yum install -y telnet.x86_64
[root@Slave002 ~]# yum install -y telnet-server.x86_64
[root@Slave002 ~]# su -l hadoop
[hadoop@Slave002 ~]$ telnet Slave001 44444
```

11.4.2 实例 2：监控本地文件夹并写入到 HDFS 中

监控本地文件夹：当文件夹的文件改变时，把改变内容加载到指定的 HDFS 文件夹。

```
[hadoop@Slave001 ~]$ cd ~/software/apache-flume-1.6.0-bin/example/
[hadoop@Slave001 example]$ vi monitor.conf
```

插入：

```
#agent name
agent.sources=source1
agent.sinks=sink1
agent.channels=channel1
#Spooling Directory
#set source1
agent.sources.source1.type=spooldir
agent.sources.source1.spoolDir=/home/hadoop/tmpfile/logdfs
agent.sources.source1.channels=channel1
agent.sources.source1.fileHeader = false
agent.sources.source1.interceptors = i1
agent.sources.source1.interceptors.i1.type = timestamp
agent.sources.source1.deletePolicy = immediate

#set sink1
agent.sinks.sink1.type=hdfs
agent.sinks.sink1.hdfs.path=/input
agent.sinks.sink1.hdfs.fileType=DataStream
agent.sinks.sink1.hdfs.writeFormat=TEXT
agent.sinks.sink1.hdfs.rollInterval=1
agent.sinks.sink1.channel=channel1
agent.sinks.sink1.hdfs.filePrefix=%Y-%m-%d
#set channel1
agent.channels.channel1.type=file
agent.channels.channel1.checkpointDir=/home/hadoop/tmpfile/logdfstmp/point
agent.channels.channel1.dataDirs=/home/hadoop/tmpfile/logdfstmp

[hadoop@Slave001 example]$ mkdir /home/hadoop/tmpfile/logdfs

[hadoop@Slave001 example]$ flume-ng agent -n agent -c conf -f ~/software/apache-flume-1.6.0-bin/example/monitor.conf
```

验证：

（1）当monitor.con程序启动后，可打开另一终端，进入/home/hadoop/tmpfile/logdfs目录。在该目录下通过vi工具创建一个新的文件夹，任意写入一些内容后保存退出。

（2）在大数据集群任意一台节点上输入下面语句查询导入结果。若结果与之前相同说明监控成功。

```
[hadoop@Slave002 ~]$ hadoop fs -cat /input/*
```

11.5 习题与思考

1．叙述 Flume 的功能、工作流程。

2．尝试从本地文件收集日志传送到 HDFS 中。

3．尝试从 Slave001 节点收集日志传送到 Slave002 节点，再由 Slave002 节点收集 Slave001 传送的日志，并将其传送至 HDFS。

第 3 篇　Spark 技术

第3章 Spark 基本

第 12 章　Spark 概 述

　　Spark 是基于内存的分布式、迭代式计算框架，可以基于内存也可以基于磁盘做迭代计算。根据 Apache Spark 开源组织官方描述，Spark 基于磁盘做迭代计算比基于磁盘迭代的 MapReduce 框架快 10 余倍；而基于内存的迭代计算则比 MapReduce 迭代快 100 倍以上。同时由于 Spark 是迭代式的计算框架，它更适合于将多步骤作业通过应用层面向过程的流水化操作，将其转换成底层的多个作业串联操作。在第一代计算框架 MapReduce 中需要程序员手动分解多个步骤操作到各个作业中去，因此如果基于 MapReduce 来编写数据分析应用会耗费大量代码。但如果基于 Spark 的 API 来构建应用会非常快速和高效。Spark 生态圈如图 12.1 所示。

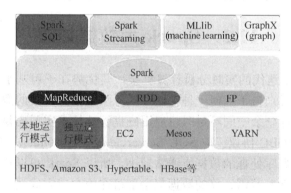

图 12.1

12.1　Spark 框架原理

　　Spark 是基于内存的一种迭代式计算框架，所处理的数据可以来自于任何一种存储介质，如关系数据库、本地文件系统、分布式存储、网络 Socket 字节流等。Spark 从数据源存储介质中装载需要处理的数据至内存，并将这些数据集抽象为 RDD（弹性分布式数据集）对象。然后采用一系列算子（封装计算逻辑的 API）处理 RDD，并将处理好的结果以 RDD 的形式输出到内存，或以数据流的方式持久化写到其他存储介质中。

　　Spark 框架拥有一系列用于迭代计算的算子库，通过调用算子可完成数据集在内存中的实时计算和处理。Spark 提交一个应用之后会启动一个 Driver 进程，该 Driver 进程负责与 Spark 集群中 Master 节点进行数据交互，同时跟踪和收集 Master 节点反馈的数据处理情况。Spark 集群中 Master 节点接收到 Driver 进程申请应用的请求后，会基于 Spark 框架内部机制分析应用，并将提交的应用拆解为可以直接执行的各个作业。这些作业之间根据设计的业务逻辑流程并发执行或串行执行，均由 Spark 的分析器来决策。每个作业的内部细分为各个任务，每个任务对应一个具体的执行进程。Spark 根据分析器拆分的作业以及任务的类型和数量将其分配到各个工作节点，每个工作节点都有一个 Worker 进程来响应 Master 节点的调度处理。分配的具体任务都由 Spark 在各个工作节点上生成的 Executor 进程来予以处理。在任务调度过程中，从一个任务到另一个任务的过渡或在一个 Executor 进程内部完成数据转移、或在不同的 Executor 进程之间完成数据转移。如果数据在不同的 Executor 之间完成数据转移则称为 Stage（迭代步骤）。计算过程中数据在不同的 Stage 之间迁移和交互，涉及数据在处理进程之间的重新分配过程。通常数据的重新分配会产生网络 I/O 操作，其将为分布式计算带来附加的集群负载。这是所有分布式计算均具备的共同特征，同时也是分布式计算仅适用于处理大数据集的一个根源。

每个节点上的任务操作结束的计算结果均反馈至 Worker 进程。Worker 进程负责将任务计算完成之后的状态汇报给 Master 进程，Master 进程负责数据输入、计算、调度和数据输出的整个迭代过程。每个任务结束之后 Master 进程会将其注销。同时根据输入数据集的分配启动新的任务继续处理剩余的数据。待指定的数据集全部处理结束完毕，Master 节点将数据集的处理结果反馈给 Spark 客户端，同时关闭与 Spark 客户端的连接。

12.2 Spark 大数据处理

Spark 是基于内存迭代的实时分析计算框架，其优势在于可基于内存做实时计算和分析处理。处理的数据可来自于任何一种存储介质。从数据源收集的数据优先被装载到内存中实现迭代处理，内存剩余空间不足时将启用磁盘缓冲技术实现磁盘迭代分析处理。

Spark 可以通过 JDBC 协议读取 SQL Server、MySQL、Oracle、DB2 等关系数据库中的数据。由于关系数据库处理的数据集通常较小，所以 Spark 可以在内存中完成这些数据的迭代计算和分析处理。若需计算的数据集全部可在内存中完成迭代，则分析效率是相当高的。计算分析结束之后的数据也可输出到关系数据库，以存储起来供前端系统使用。若前端系统对检索性能和效率要求高可将其分析结果输出到分布式缓存。

Spark 可通过分布式传输协议读取 HDFS、S3、Swift 等分布式存储系统的数据。由于分布式存储系统中数据集非常庞大，因此 Spark 处理这类数据时会优先将需要处理的数据集装载到内存，基于内存实现迭代处理。当迭代的数据集超过 Spark 的分布式缓存时，会将数据溢出到磁盘进行迭代处理。处理结束的数据仍可通过分布式协议写回分布式存储系统中用于二次分析的输入，若处理之后的结果数据集非常小且需要用于前端系统，可将其写入关系数据库或分布式缓存中去。

除静态数据集的分析和处理之外，Spark 还可使用生态圈中的 Streaming 子系统来完成实时流的分析和处理。在这种生产场景下其数据源自于网络 Socket 通道、本地 Event 流、Flume 推送、KafKa 中间件等，Spark 与这些数据源组件进行交互时需充分考虑数据在 I/O 通道上的传输过程，否则易在 I/O 通道的一端产生 OOM（OutOfMemery，泛指发生内存泄漏）异常。Spark 事件流的分析和处理广泛应用于电商领域推荐系统。在事件流的处理过程中，其处理的流化数据集是业务数据集的一个片段，这种数据集被称为离散数据集。离散数据集的输入和输出几乎实时进行，同时几乎都来自于前端系统的在线数据集。

12.3　RDD 数据集

RDD（Resilient Distributed Datasets）被译为弹性分布式数据集，它是 Spark 运行时（Spark Runtime）的内核实现。弹性分布式数据集具备如下特征。

1. 迭代模式的自动切换

处理的数据优先在内存中进行迭代，内存剩余空间不足时再将溢出部分数据缓冲到磁盘进行迭代处理。

2. 执行步骤的可恢复性

当一个应用存在多个操作步骤时，若在作业执行过程中某个步骤出错，只需在出错步骤的前一步骤后重新执行进行恢复，而无须从第一个步骤开始恢复。

3. 故障作业的高可靠性

若某个作业出错或失败，则会自动进行特定次数（默认 3 次）的重试。该重试包括作业本身的重试和基于作业操作底层节点任务计算失败的重试（默认 4 次）。

4. 故障数据的高度容错

若某个作业失败导致数据计算结果不完整，则作业会重新提交或重试，这时仅收集计算失败的数据分片进行重新计算，这意味着 Spark 可从细粒度上控制容错计算的数据量。

12.4　Spark 子系统

Spark 生态圈中包含诸多子系统，这些子系统是基于 Spark 核心 API 的应用框架。Spark 核心 API 以 Spark RDD 数据集为基础，同时 Spark RDD 又是 Spark 生态圈中其他子系统予以扩展的基础架构。Spark 生态圈中有如下常用子系统。

1. Spark SQL

该子系统可将传递给 Spark 内核的类 SQL 语句翻译成基于 Scala API 的 Spark 代码，然后将其包装成 Spark 应用提交给 Spark 集群处理，其数据源可以来自于数据库、文件系统等各种存储介质，且以指定的数据格式从这些存储系统中提取所需处理的数据。

2. Spark Streaming

该套子系统可基于离散数据流完成实时分析处理，处理的数据通常来自本地事件流、Socket 网络数据流、KafKa 中间件的缓冲数据流等。分析数据的特征是准实时计算和处理，数据的输出端通常指向业务平台的前端系统。

3. Spark MLlib

该套子系统为 Spark 框架提供机器学习算法库。机器学习俗称人工智能，它是指将一系列的样本输入数据，通过维度设计、数据回归、曲线拟合等方式得出一套经验公式或数理结论，然后将其应用于未来某个时空点产生的数据维度并预测可能发生的业务数据。MLlib 机器学习算法库基于 Spark 平台下的内存迭代方式进行数据分析。

4. 其他子系统

除了以上常用的子系统外，还有其他子系统，如基于图计算的 Spark GraphX、基于可视化表现的 Spark R 等，读者可自行参考相关书籍。

第 13 章　Scala 语言

Scala 是一种基于 JVM（Java 虚拟机）的跨平台编程语言，Scala 编译器可将 Scala 源码编译成符合 JVM 虚拟机规范的中间字节码文件，在 JVM 平台上解释和运行；Scala 语言是对 Java 语言的一种补充和扩展，其 API 可无缝兼容 Java 的 API，可认为 Java 的 API 是其 Scala API 的子集，Scala 是完全并彻底面向对象的一种编程语言。编程模式提供面向过程化和面向对象化两种编码设计模式；运行模式则提供编译和解释两种操作模式。

13.1　Scala 语法基础

13.1.1　变量、常量与赋值

关于声明变量、常量的视频讲解可扫描二维码观看。

1. 声明变量

变量的值可以被重新指定。类型可依赖于值自动推断。

```
var VariableName[:DataType][=Initial Value]
```

示例如下：

```
var ab:Int=30
ab: Int = 30
ab=50  提示：此处ab变量被重新赋值
ab: Int = 50
var ab=30
ab: Int = 30    提示：此处变量ab的类型被自动推断为Int类型
```

2. 声明常量

常量的值不可以被重新指定。因此在声明常量时必须指定值，类型也可依赖于值自动推断。

```
val VariableName[:DataType]=Initial Value
```

示例如下：

```
val ab:Int=30
ab: Int = 30
ab=50  提示：此处常量ab不能被重新赋值，如果重新赋值将发生以下错误
<console>:12: error: reassignment to val
```

```
       ab=50
       ^
val ab=30
ab: Int = 30   提示：此处常量ab的类型被自动推断为Int类型
```

13.1.2 运算符与表达式

Scala 中常用基础运算符如下：
- 算术运算符：+、-、*、/、%。
- 关系运算符：>、<、=、!=、>=、<=。
- 逻辑运算符：&&、||、!。
- 值变运算符：++、--。
- 成员运算符：·、[]。
- 条件运算符：?:。
- 赋值运算符：=。

将一系列的操作数通过运算符连接起来即变成表达式。最简单的表达式是一个常量或变量，表达式间的运算存在逻辑先后顺序。当不明确表达式之间的具体先后顺序时，可使用小括号将需优先计算的表达式括起来。

运算符与表达式示例如下：

```
var a:Int=(3+2-5)*0/5
a: Int = 0
```

提示：上述表达式中会先计算括号内的 3+2-5 部分然后再计算后面的乘除部分，因此最终得到的计算结果是 0。

13.1.3 条件分支控制

关于条件语句的视频讲解可扫描二维码观看。
if 条件分支语法（else 分支是可选的分支）：

```
if(条件表达式){
  ...............条件表达式返回true时执行................
}[else{
  ...............条件表达式返回false时执行................
}]
```

示例如下（比较两个数的大小）：

```
var a=100;
a: Int = 100
var b=200;
b: Int = 200
if(a>b){
  println(a);
}else{
```

```
println(b);
}
200
```

提示：a 的值小于 b 将导致 if 中的条件不成立，所以最终打印出 b 的值。

13.1.4 循环流程控制

关于循环语句的视频讲解可扫描二维码观看。

1. for 循环控制语句

```
for(迭代变量<-初始值 to 终点值){
..................循环体语句..................
}
```

示例如下（累加 1 到 100 的和）：

```
var i=1;
var sum=0;
for(i<-1 to 100){
  sum+=i;
}
println(sum);
```

2. while 循环控制语句

```
while（循环条件表达式）{
..................循环体语句..................
}
```

示例如下（累加 1 到 100 的和）：

```
var i:Int=1;
var sum=0;
while(i<=100){
  sum+=i;
  i=i+1;
}
println(sum);
```

13.1.5 Scala 数据类型

1. Scala 基本数据类型

Scala 的数据类型包括 Byte、Short、Int、Long、Float、Double、Char、String、Boolean、Unit、Null、Nothing、AnyVal、AnyRef、Any 等。前 9 种称为基本数据类型，其中 Unit 指代没有任何类型，Null 指代空类型，Nothing 指代所有类型的子类型，AnyVal 指代所有的值类型，AnyRef 指代所有的引用类型，Any 指代所有类型的超类型。

示例如下:

```
var sum:Any=3+2-5*0/5
sum: Any = 5
```

提示:计算出来的结果是一个值类型,该值类型可以自动上传到它的超类型——Any类型。

2. 集合数据类型

集合数据类型包括 Array、List、Set、Tuple、Map 等。其中 Array 称为数组类型,List 称为列表类型,Set 称为表列类型,Tuple 称为元组类型,Map 称为字典类型。其中元组类型根据元组中元素的数量和类型又可以分为多种元组子类型。

示例如下:

```
var arr:Array[String]=new Array[String](3);
arr: Array[String] = Array(null, null, null)
arr(0)="liyongfu"
arr(1)="lixiangyu"
arr(2)="limingzhu"
for(ele<-arr)println(ele)
liyongfu
lixiangyu
limingzhu
```

提示:方括号中 String 代表数组中的元素类型,小括号中的 3 代表数组中所能容纳的元素数量。

13.2　Scala 运算与函数

关于运算与函数的视频讲解可扫描二维码观看。

Scala 支持面向对象的编程风格和函数式的编程风格,函数式的编程风格是 Scala 不同于 Java 的一个亮点所在。在 Scala 中可以通过如下语法创建一个函数:

```
def funName(...参数列表...)[:type]={
    ....................函数体语句....................
}
```

函数定义示例如下:

```
def max(x: Int, y: Int) :Int={
  if (x > y){
    return x;
  }else{
    return y;
  }
}
max2: (x: Int, y: Int)Int
```

函数调用示例如下:

```
var maxData:Int=max(3,5)
res4: Int = 5
```

提示: 如果函数体中只有一条语句, 则函数体两端的大括号可省略。

13.3 Scala 闭包

关于闭包的视频讲解可扫描二维码观看。

闭包是在函数体内部可以访问函数体外部变量的过程。类似于 Java 方法体中访问成员变量的操作过程。

示例如下:

```
object Test {
    def main(args: Array[String]) {
      println( "muliplier(1) value = " + multiplier(1) )
      println( "muliplier(2) value = " + multiplier(2) )
    }
    var factor = 3
    val multiplier = (i:Int) => i * factor
}
```

提示:
(1) object 用于定义一个单态类, 在 Scala 中没有静态方法的概念。Scala 中的静态方法通过单态类来实现, 在 object 中定义的所有方法默认都是静态的。
(2) 上述函数体内部引用的 multiplier 变量是在函数体外部定义的。

13.4 Scala 数组与字符串

所有的开发语言中, 字符串是应用最为广泛的一种数据类型。同时字符串的底层是字符数组予以存储的, 因此可结合数组来学习字符串的操作。

13.4.1 Scala 数组

关于数组的视频讲解可扫描二维码观看。

数组是用于存储固定长度的一系列某种特定类型的元素集合。数组中的元素按线性方式排列并存储在内存中。

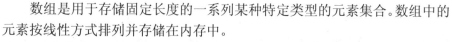

1. 定义 Scala 数组时必须指定数组元素的泛型, 否则将默认为 Nothing 类型, 拒绝将任何元素添加到数组集合中去。

定义 Scala 数组的示例 1:

```
var arr:Array=new Array(2);
arr: Array[Nothing] = Array(null, null)
arr(0)="liyongfu"
```

```
<console>:13: error: type mismatch;
 found   : String("liyongfu")
 required: Nothing
       arr(0)="liyongfu"
```

提示：由于定义数组时没有指定数组泛型导致候选添加元素发生错误。

2. 简洁方式创建数组

这种方式创建数组无须指定数组的长度和泛型。因为数组元素的个数及其类型可通过创建数组时枚举的值来判定。

简洁方式创建数组的示例如下：

```
val arr=Array("li","yong","fu")
arr: Array[String] = Array(li, yong, fu)
scala> arr.apply(0)
res30: String = li
arr.apply(1)
res31: String = yong
```

提示：Array 之后的小括号将数组中的所有元素直接枚举出来，无须指定数组元素的类型和个数。

13.4.2 Scala 字符串

1. 字符串常量

Scala 中的字符串常量有三种形式。

（1）单引号：表示单个字符，如'A'。
（2）双引号：表示字符串，如"ABC"。
（3）三双引号：表示具备可换行的多行字符串。如"""可换行的字串内容"""。

2. 字符串变量

关于字符串的视频讲解可扫描二维码观看。

Scala 中的 String 类型就是 java.lang.String 类型。Scala 没有自己的 String 类型。由于 Scala 中的字符串由 Java 的 API 来定义，因此延承了 Java 中字符串的全部特性，比如字符串的不可变性、不可扩展性等。字符串常用的 length 方法可测试字符串的长度，若需要连接两个字符串可以直接使用重载的"+"操作符。

字符串操作示例如下：

```
var str="liyongfu";
str: String = liyongfu
var len=str.length();
len: Int = 8
var str02:String=str+":36"
str02: String = liyongfu:36
```

13.5 Scala 迭代器

Scala 中的迭代器不是一种集合容器，而是用于访问集合内元素的一种方法。

关于迭代器的视频讲解可扫描二维码观看。

迭代器示例如下:

```
val itc = Iterator("Baidu", "Google", "Runoob", "Taobao")
itc: Iterator[String] = non-empty iterator
scala> itc.size
res4: Int = 4
val itc = Iterator("Baidu", "Google", "Runoob", "Taobao")
itc: Iterator[String] = non-empty iterator
scala> itc.length
res5: Int = 4
val itc = Iterator("Baidu", "Google", "Runoob", "Taobao")
itc: Iterator[String] = non-empty iterator
while (itc.hasNext){
    println(itc.next())
}
Baidu
Google
Runoob
Taobao
```

提示：size 和 length 属性均返回迭代器中元素的个数。上述三次测试均重新实例化的原因是调用过 size、length 或 next 属性后导致迭代器的指针已指向迭代器末尾。

13.6 Scala 类和对象

关于类和对象的视频讲解可扫描二维码观看。

与其他开发语言类似，Scala 语言也是面向对象的一种开发语言。类是具有相同属性和功能行为的一组对象，而对象是类的具体实例。使用具有一定特征的对象之前，必须先定义有此类对象共性特征的类，然后使用该类实例化一个需要使用的具体对象。

Scala 类定义的示例如下:

```
class Point(xc: Int, yc: Int) {
    var x: Int = xc
    var y: Int = yc
    def move(dx: Int, dy: Int) {
        x = x + dx
        y = y + dy
        println ("x 的坐标点: " + x);
        println ("y 的坐标点: " + y);
    }
}
```

提示：

（1）定义类使用关键字 class。

（2）上述定义了一个坐标类 Point，该类的构造方法中定义了两个整型参数，分别为 xc、yc。这两个构造参数用于初始化类的成员变量 x 横坐标和 y 纵坐标，上述类中还定义了一个方法 move，该方法没有返回值，其作用在于接收两个用于说明移动坐标横向增量和纵向增量的整型参数来移动坐标点。

Scala 类实例化的示例如下：

```
object Test {
  def main(args: Array[String]) {
    val pt = new Point(10, 20);
    pt.move(10, 10);
  }
}
```

提示：使用关键字 new 调用类的构造方法（同时传递构造方法所需的两个坐标参数）即可实例化坐标类为一个坐标对象，就可使用该坐标对象调用 move 方法（同时传递横纵两个方向上的移动增量参数）实现坐标点的移动行为。

13.7 习题与思考

1．编写一个 Conversions 对象，加入 inchesToCentimeters、gallonsToLiters 和 milesToKilometers。

2．提供一个通用的超类 UnitConversion 并定义扩展该超类的 InchesToCentimeters、GallonsToLiters 和 MilesToKilometers 对象。

3．定义一个 Point 类和一个伴生对象，使得可以不用 new 而直接用 Point(3,4)来构成 Point 实例。

4．尝试编写一个 Scala 应用程序，以反序打印命令行参数，用空格隔开。例如：Hello Bunfly 应该打印出 Bunfly Hello。

第 14 章　Spark 高可用环境

14.1　环境搭建

关于环境搭建的视频讲解可扫描二维码观看。

14.1.1　准备工作

在构建 Spark 集群之前需先构建 Hadoop 集群。因为在生产环境下，Spark 数据源的多数应用场景取自 HDFS 分布式存储。Hadoop 集群构建可参考第 5 章 5.2 节 Hadoop 环境搭建。

Hadoop 环境构建结束之后关闭各台虚拟机，并将每个虚拟机的物理内存调整到 2GB 以上，主机分配方案如下：

```
192.168.37.154 master01
192.168.37.155 master02
192.168.37.156 slave01
192.168.37.157 slave02
192.168.37.158 slave03
```

提示：master01 与 master02 是两台高可用的 Master 节点，其余三台 Slave 节点是 Worker 工作节点。

14.1.2　下载并安装 Spark

Spark 的安装包可以到官网的镜像站点（http://www-eu.apache.org）下载。本书以 Spark 2.1.1 版本进行讲解。

1. 安装 Spark

```
mkdir -p /software
tar -zxvf spark-2.1.1-bin-hadoop2.7.tgz -C /software/
mv /software/spark-2.1.1-bin-hadoop2.7/ spark-2.1.1
```

2. 配置 Spark 环境

1）配置 Spark 系统变量

```
vi /etc/profile
    JAVA_HOME=/software/jdk1.7.0_79
    SCALA_HOME=/software/scala-2.11.8
```

```
HADOOP_HOME=/software/hadoop-2.7.3
SPARK_HOME=/software/spark-2.1.1
PATH=$PATH:$JAVA_HOME/bin:$JAVA_HOME/lib:$SCALA_HOME/bin:$HADOOP_
HOME/bin:$HADOOP_HOME/sbin:$SPARK_HOME/bin:$HBASE_HOME/bin
export PATH JAVA_HOME SCALA_HOME HADOOP_HOME SPARK_HOME
```

提示：上述只有 SPARK_HOME 部分属于新增的环境变量，其余部分都是在安装 Hadoop 时已经设定好的环境变量。

2）配置 Spark 环境变量

```
vi /software/spark-2.1.1/conf/spark-env.sh
    export JAVA_HOME=/software/jdk1.7.0_79
    export SCALA_HOME=/software/scala-2.11.8
    export HADOOP_HOME=/software/hadoop-2.7.3
    export HADOOP_CONF_DIR=/software/hadoop-2.7.3/etc/hadoop
    export SPARK_MASTER_IP=master01,master02
    export SPARK_WORKER_MEMORY=1700
```

3）配置 Spark 工作节点

```
cp /software/spark-2.1.1/conf/slaves.template slaves /software/spark-2.1.1/conf/slaves
slave01
slave02
slave03
```

提示：安装和配置必须在 Spark 集群的每个节点上执行。

3. 启动 Spark 集群

1）在三个 Slave 节点上启动 ZK 集群

```
[hadoop@slave01 software]$ cd zookeeper-3.4.10/bin/ && ./zkServer.sh stop && jps
[hadoop@slave02 software]$ cd zookeeper-3.4.10/bin/ && ./zkServer.sh stop && jps
[hadoop@slave03 software]$ cd zookeeper-3.4.10/bin/ && ./zkServer.sh stop && jps
```

2）在 master01 节点上启动 HDFS 集群

```
[hadoop@master01 software]$ start-dfs.sh
    Starting namenodes on [master01 master02]
    master01: starting namenode, logging to /software/hadoop-2.7.3/logs/
    hadoop-hadoop-namenode-master01.out
    master02: ssh: connect to host master02 port 22: No route to host
    slave03: starting datanode, logging to /software/hadoop-2.7.3/logs/
    hadoop-hadoop-datanode-slave03.out
    slave01: starting datanode, logging to /software/hadoop-2.7.3/logs/
```

```
hadoop-hadoop-datanode-slave01.out
slave02: starting datanode, logging to /software/hadoop-2.7.3/logs/
hadoop-hadoop-datanode-slave02.out
Starting journal nodes [slave01 slave02 slave03]
slave03: starting journalnode, logging to /software/hadoop-2.7.3/logs/
hadoop-hadoop-journalnode-slave03.out
slave01: starting journalnode, logging to /software/hadoop-2.7.3/
logs/hadoop-hadoop-journalnode-slave01.out
slave02: starting journalnode, logging to /software/hadoop-2.7.3/logs/
hadoop-hadoop-journalnode-slave02.out
```

3）在 master01 节点上启动 Spark 集群的 Master 节点

```
[hadoop@master01 software]$ /software/spark-2.1.1/sbin/start-master.sh && jps
starting org.apache.spark.deploy.master.Master, logging to /software/
spark-2.1.1/logs/spark-hadoop-org.apache.spark.deploy.master.
Master-1-master01.out
3867 Master
3020 NameNode
3908 Jps
```

4）在 master02 节点上启动 Spark 集群的 Master 节点

```
[hadoop@master02 sbin]$ /software/spark-2.1.1/sbin/start-master.sh && jps
starting org.apache.spark.deploy.master.Master, logging to /software/
spark-2.1.1/logs/spark-hadoop-org.apache.spark.deploy.master.
Master-1-master02.out
3869 Master
3021 NameNode
3909 Jps
```

5）高可用集群测试
使用浏览器访问：

```
http://master01:8080/    提示：界面上显示Status:ALIVE
http://master02:8080/    提示：界面上显示Status: STANDBY
```

14.2 常见问题汇总

在 Spark 的高可用集群环境搭建完成之后，启动 Spark 可能出现一些问题。现将出现的常见问题和相应的解决方案收集如下。

1. WARN TaskSchedulerImpl: Initial job has not accepted any resources; check your cluster uito ensure that workers are registered and have sufficient memory

原因分析：集群的可用资源不足，导致 Spark 集群无法为新提交的应用分配可用资源。
解决方案：暂停低优先级的应用，让更多的可用资源优先执行给高优先级的应用。

2. Application isn't using all of the Cores: How to set the Cores used by a Spark App

原因分析：安装的 Spark 版本没有充分利用处理器的资源，导致 Spark 框架给出上述提示建议。

解决方案：修改 Spark 配置文件 spark-env.sh，修改参数 spark.deploy.defaultCores 为每台服务器节点的 CPU 内核数量。

3. Spark Executor OOM: How to set Memory Parameters on Spark

原因分析：Spark 集群在运行应用时可用内存不足导致内存泄漏。如果在启动 Spark 集群初期发生此种异常表示 Spark 初始化内存不足。

解决方案：考虑将大数据集分批量进行处理以减少大数据量对内存的冲击，有条件也可直接增加物理主机的内存。

4. Class Not Found: Classpath Issues

原因分析：相关类无法找到。这种异常可能发生在应用运行期间，也可能发生在 Spark 集群启动初期。若发生在 Spark 集群启动初期则表示 Spark 的安装包不完整或被损坏；若发生在应用运行期间则表示打包的 Job 中缺少依赖的 JAR 包。

解决方案：观察错误输出日志，分析发生 JAR 缺少的原因，并根据情况补充相关的依赖 JAR 包。

第 15 章　RDD 技 术

15.1　RDD 的实现

RDD（Resilient Distributed Datasets）称为弹性分布式数据集，是 Spark 的运行时（Spark Runtime）内核实现，一个 RDD 对象中可以包含多个分区（partition），Spark 应用运行时其每个 Partition 都对应着一个具体的任务；Spark 基于 RDD 的内核实现涵盖以下两个部分。

15.1.1　数据源

Spark 处理的数据源可以是 HDFS、Hive、HBase、S3、MySQL、Oracle 等；处理之后的数据可被输出到 HDFS、Hive、HBase、S3、MySQL、Oracle 及其当前的控制台终端。

15.1.2　调度器

YARN：是 Apache 组织基于 MapReduce 算法的一套调度框架。
Mesos：是 Apache Spark 默认的一套基于资源管理算法的调度框架。
AWS：是亚马逊旗下的一套分布式系统资源调度框架。

RDD 从逻辑上看是一个抽象分布式数据集的概念，其底层数据存储于集群中不同节点上的磁盘文件系统中，存储是按照分区（partition）方式进行的；所有 Spark 操作都可看成是一系列对 RDD 对象的操作，而 RDD 是数据集合的抽象，它可以使用 SparkContext（Spark 上下文）来创建，SparkContext 是 Spark 集群操作的入口。若是在 spark-shell 下操作，则 Spark 会自动创建一个基于已有配置的默认 SparkContext 对象；若是编写作业 Jar，则须自己手动创建（与 Hadoop 中的 FileSystem 相同，可通过 Configuration 配置参数来构建，或基于 classpath 中的配置文件）。

从 RDD 的实现逻辑来看，Spark 操作 RDD 的过程类似于 MapReduce 操作 HDFS 的过程，但 Spark 在操作过程中会缓存中间步骤的数据。Spark 的 Stage 有以下几点缓存特征：

（1）若 Spark 应用中的某个计算步骤非常耗时，则会缓存该步骤的计算结果。
（2）若 Spark 应用的计算步骤链很长，则会增加对缓存步骤计算结果的频次。
（3）shuffle 到其他节点上的数据会被目标节点缓存一次，以减少不必要的 I/O 次数。
（4）从一个作业切换到下一个作业时将发生一次 checkpoint 操作。在此，checkpoint 操作之前会缓存上一个作业的中间结果，checkpoint 操作会将数据放置于磁盘文件系统中以保证数据不发生丢失。

15.2 RDD 编程接口

RDD 的编程接口分为如下几类。

1. 数据源自于集合的接口

（1）parallelize：将内存中的集合对象包装成 Spark 中的 RDD 对象。

（2）createDataFrame：将内存中的集合对象包装成 Spark 中的 DataFrame 对象。

2. 数据源自于 RDD 的接口

（1）collect：将 RDD 对象转换成 Java 集合类型 List 对象。

（2）createDataFrame：将 RDD 对象转换成 DataFrame 对象。

3. 数据源自于 DataFrame 的接口

（1）javaRDD：将 DataFrame 对象转换成 Java API 中的 RDD 对象。

（2）collectAsList：将 DataFrame 对象转换成 Java API 中的 List 集合对象。

Spark 的编程入口根据不同的子系统由不同的接口实现。

（1）SparkContext：基于 RDD 内核的原生接口实现，该接口是 Spark 其他入口的实现基础。

（2）SQLContext：基于 SparkContext 的一种 SQL 子系统实现，主要用于 Spark SQL 子系统中。

（3）StreamingContext：基于 SparkContext 的一种流式子系统实现，主要用于 SparkContext 子系统中。

（4）HiveContext：基于 SQLContext 的一种 HQL 子系统实现，主要用于底层存储为 Hive 的子系统中。

15.3 RDD 操 作

关于 RDD 操作的视频讲解可扫描二维码观看。

15.3.1 Spark 基于命令行的操作

Spark 框架可借助 spark-shell 完成命令行操作。具体操作如下：

```
/software/spark-2.1.1/bin/spark-shell --master spark://master01:7077
scala> val mapRdd=sc.textFile("/spark/input").flatMap(_.split(" ")).
map(word=>(word,1)).reduceByKey(_+_).map(entry=>(entry._2,entry._1))
    mapRdd: org.apache.spark.rdd.RDD[(Int, String)] = MapPartitionsRDD[5]
    at map at<console>:24
scala> val sortRdd=mapRdd.sortByKey(false,1)
    sortRdd: org.apache.spark.rdd.RDD[(Int, String)] = ShuffledRDD[6] at
    sortByKey at <console>:26
scala> val mapRdd2=sortRdd.map(entry=>(entry._2,entry._1))
    mapRdd2: org.apache.spark.rdd.RDD[(String, Int)] = MapPartitionsRDD[7]
    at map at <console>:28
```

```
scala> mapRdd2.saveAsTextFile("/spark/output")
scala> :quit
```

提示：

（1）上述运行 spark-shell 的提示中显示 Scala 2.11.8 与 Spark 2.x 是兼容的，推荐使用 JDK 1.8+的版本。

（2）若在配置文件中指定 export SPARK_MASTER_IP=master01 参数配置，则在运行 spark-shell 或 spark-submit 时无须指定--master 参数；若在命令行中指定了--master 参数则会覆盖配置文件中配置的值。

命令行程序解析如下。

（1）Spark 默认以 Spark on Mesos 模式运行 spark-shell，指定 master 参数为：

```
spark://master01:7077
```

（2）Spark 命令行工具启动时将会先后启动 Spark 上下文（SparkContext）和 SQL 上下文（SQLContext），这两个对象分别被赋值给上下文中的 sc 变量和 SQLContext 变量，因此可在命令行中直接使用这两个变量。

（3）在 Spark 命令中多次使用到 Spark 的各种算子函数，这些函数的功能和作用介绍如下。

① textFile 函数　该函数将返回一个 org.apache.spark.rdd.RDD[String]类型的对象（RDD 在 Spark 的架构源码中被 Scala 定义为一个接口（trait）类型；方括号中的 String 表示 RDD 集合中的元素类型为字符串类型；RDD 的具体实现是 MapPartitionRDD（即字典分区 RDD）；textFile 操作仅执行数据抽象，并不立即执行数据的读取（read）操作。数据的读取操作会延迟到执行 action 动作时才发生；因此 textFile 函数属于一种 transformation 动作，transformation 动作是 lazy 级别的操作。

② count 函数　返回数据集中的记录数量，会启动一个 action（textFile 函数是一种转换（transformation），不会启动 action）；action 是指 Spark 会在底层启动作业（Job）或任务（Task）的计算操作；而转换（transformation）仅实现数据的封装和抽象。只涉及数据在节点本地上的抽象读（注意：没有真正意义上的读，因为没有产生任何 I/O 操作），但数据的写操作属于 action 操作，因为它涉及 I/O 操作过程，比如 saveAsTextFile 函数。

③ map 函数　该函数将集合中的各个元素映射为指定的对象，然后返回一个新的 RDD 集合对象。

④ flatMap 函数　该函数处理 DataSet 集合中的每一行，将每一行得到的集合元素释放到最外层集合中。此函数完成调用后将产生一个新的 RDD 对象结果集，该函数会深层次递归处理子级元素并将其提取到最外层的集合对象中去。

⑤ reduceByKey 函数　该函数将数据集中元组的第一个元素作为 Key 进行分组，同时递归迭代和归并函数中的两个参数。

⑥ sortByKey 函数　该函数根据数据集中元组的第一个参数进行排序处理，通过第二个参数指定并行度。

⑦ saveAsTextFile 函数　该函数将数据集作为文本格式存储到参数指定的目录下。

15.3.2 Spark 基于应用作业的操作

Spark 框架可借助 spark-submit 来提交一个编写好的 Job 应用到集群,从而完成 Spark 应用的分析和处理,这通常是生产场景中使用的一种操作方式。

```
/software/spark-2.1.1/bin/spark-submit --class org.apache.spark.
examples.JavaSparkPi --master spark://
master01:7077 ../examples/jars/spark-examples_2.11-2.1.1.jar 1
    17/08/10 00:43:14 INFO spark.SparkContext: Running Spark version 2.1.1
    17/08/10 00:43:14 WARN spark.SparkContext: Support for Java 7 is
    deprecated as of Spark 2.0.0
    17/08/10 00:43:15 WARN util.NativeCodeLoader: Unable to load
    native-hadoop library for your platform... using builtin-java classes
    where applicable
    17/08/10 00:43:16 INFO spark.SecurityManager: Changing view acls to:
    hadoop
    ...
    17/08/10 00:43:17 INFO spark.SparkEnv: Registering
    OutputCommitCoordinator
    17/08/10 00:43:17 INFO util.log: Logging initialized @5486ms
    17/08/10 00:43:18 INFO server.Server: jetty-9.2.z-SNAPSHOT
```

提示:

(1) 为节省篇幅,上述部分日志打印已被笔者省略。

(2) --class 参数用于指定提交 JAR 包中的运行主类。

(3) 最后的数字 1 代表运行的 slices 数量(即并行度),每一个 slice 都将启动一个 Task 来运行,每一个 Task 任务对应一个 JVM 进程。

(4) /software/spark-2.1.1/examples/jars/spark-examples_2.11-2.1.1.jar 是提交到 Spark 集群的 Job 作业打包 JAR,这与 Hadoop 提交作业的方式是相同的。

15.3.3 Spark 操作的基础命令与开发工具介绍

(1) 在 Spark 的安装目录下的 bin 目录中有以下两个运行命令。

① spark-submit:该命令用于提交 Spark 的 Job 应用,通常用于测试。

② spark-shell:Spark 交互式命令行工具,通常用于生产环境。

(2) 开发 Spark 应用可使用 Intellij IDEA 或 Eclipse,目前大多数企业中在生产上使用的是 Eclipse 工具,如果使用基于 Eclipse 来开发 Spark 应用,则考虑如下两点。

① 若基于 Scala 语言来开发 Spark 应用,推荐使用 Scala IDE For Eclipse 4.4LUNA 版本。

② 若基于 Java 语言来开发 Spark 应用,推荐使用 MyEclipse 2014 或 JavaEE For Eclipse 4.5 Mars 版本。

15.3.4 Spark 基于 YARN 的调度模式

(1) 由于 YARN 模式下不需要 Mesos 模式下的所有 Master 进程和所有 Worker 进程,

需首先关闭这些进程。

```
[hadoop@master01 sbin]$ ./stop-slaves.sh
    slave01: starting org.apache.spark.deploy.worker.Worker, logging to /
software/spark-2.1.1/logs/spark-hadoop-org.apache.spark.deploy.worker.
Worker-1-slave01.out
    slave02: starting org.apache.spark.deploy.worker.Worker, logging to /
software/spark-2.1.1/logs/spark-hadoop-org.apache.spark.deploy.worker.
Worker-1-slave02.out
    slave03: starting org.apache.spark.deploy.worker.Worker, logging to /
software/spark-2.1.1/logs/spark-hadoop-org.apache.spark.deploy.worker.
Worker-1-slave03.out
[hadoop@master01 sbin]$ ./stop-master.sh
    starting org.apache.spark.deploy.master.Master, logging to /software/
    spark-2.1.1/logs/spark-hadoop-org.apache.spark.deploy.master.Master
    -1-master01.out
```

注意：启动 Spark 高可用模式则高可用组下的所有 Master 进程停止。

```
[hadoop@master02 sbin]$ ./stop-master.sh
```

（2）启动 YARN 集群：

```
[hadoop@master01 software]$ start-yarn.sh
    starting yarn daemons
    starting resourcemanager, logging to /software/hadoop-2.7.3/
    logs/yarn-hadoop-resourcemanager-master01.out
    slave01: starting nodemanager, logging to /software/hadoop-
    2.7.3/logs/yarn-hadoop-nodemanager-slave01.out
    slave03: starting nodemanager, logging to /software/hadoop-
    2.7.3/logs/yarn-hadoop-nodemanager-slave03.out
    slave02: starting nodemanager, logging to /software/hadoop-
    2.7.3/logs/yarn-hadoop-nodemanager-slave02.out
```

注意：启动 YARN 高可用模式则应将高可用组下的其他 RM 进程开启。

```
[hadoop@master02 software]$ yarn-daemon.sh start resourcemanager
```

（3）在 YARN 模式下运行 spark-shell 需要指定 --master yarn 参数。

```
[hadoop@CloudDeskTop bin]$ hdfs dfs -rm -r /spark/output
    17/08/13 23:16:51 INFO fs.TrashPolicyDefault: Namenode trash
    configuration: Deletion interval = 0 minutes, Emptier interval = 0 minutes.
    Deleted /spark/output
[hadoop@CloudDeskTop bin]$ ./spark-shell --master yarn
    Setting default log level to "WARN".To adjust logging level
    use sc.setLogLevel(newLevel). For SparkR, use setLogLevel(newLevel).
    17/08/13 23:13:33 WARN spark.SparkContext: Support for Java 7 is
```

```
        deprecated as of Spark 2.0.0
        17/08/13 23:13:35 WARN util.NativeCodeLoader: Unable to load native-
        hadoop library for your platform... using builtin-java classes where
        applicable
        17/08/13 23:13:44 WARN yarn.Client: Neither spark.yarn.jars nor spark.
        yarn.archive is set, falling back to uploading libraries under
        SPARK_HOME.
        17/08/13 23:15:40 WARN metastore.ObjectStore: Failed to get database
        global_temp, returning NoSuchObjectException
        Spark context Web UI available at http://192.168.37.153:4040
        Spark context available as 'sc' (master = yarn, app id = application_
        1502633061749_0001).
        Spark session available as 'spark'.
        Welcome to
              ____              __
             / __/__  ___ _____/ /__
            _\ \/ _ \/ _ `/ __/  '_/
           /___/ .__/\_,_/_/ /_/\_\   version 2.1.1
              /_/

        Using Scala version 2.11.8 (Java HotSpot(TM) 64-Bit Server VM, Java
        1.7.0_79)
        Type in expressions to have them evaluated.
        Type :help for more information.
scala> sc.textFile("/spark/input").flatMap(_.split(" ")).map(word=>(word,
1)).reduceByKey(_+_).map(entry=>(entry._2,entry._1)).sortByKey(false,1).
map(entry=>(entry._2,entry._1)).saveAsTextFile("/spark/output")
scala> :q
```

提示:查看 HDFS 中的运行输出结果。

```
[hadoop@CloudDeskTop bin]$ hdfs dfs -cat /spark/output/part-00000
(is,4)
(my,4)
(118,1)
(1.67,1)
(35,1)
(weight,1)
(name,1)
(height,1)
(age,1)
(liyongfu,1)
```

(4) 在 YARN 模式下运行 spark-submit 需要指定 --master yarn。

```
[hadoop@CloudDeskTop bin]$ ./spark-submit --class org.apache.spark.
examples.JavaSparkPi --master yarn ../examples/jars/spark-examples_
```

```
     2.11-2.1.1.jar 1
     17/08/13 23:23:55 INFO spark.SparkContext: Running Spark version 2.1.1
     17/08/13 23:23:55 WARN spark.SparkContext: Support for Java 7 is
     deprecated as of Spark 2.0.0
     17/08/13 23:23:56 WARN util.NativeCodeLoader: Unable to load native-
     hadoop library for your platform... using builtin-java classes where
     applicable
     ...
     17/08/13 23:24:00 INFO util.log: Logging initialized @6433ms
     17/08/13 23:24:00 INFO server.Server: jetty-9.2.z-SNAPSHOT
```

提示：

① 在 YARN 模式下运行 spark-submit 须指定--master yarn --deploy-mode 参数为 cluster，即以集群方式运行。

② 上述操作均基于 master01 节点上执行。使用 Mesos 模式和 YARN 模式均须启动 ZK 集群和 HDFS 集群。在 YARN 模式下运行时，由于未使用 Master 进程与 Worker 进程，应将其关闭。一旦关闭 Master 进程，则无法使用浏览器访问 http://master01:8080 地址观察 Job 运行情况，因为此时资源调度组件实际上是 YARN，浏览器应通过 http://master01:8088 地址访问 Job 运行情况。

③ 由于 Spark 运行于 Mesos 模式下的技术相对 YARN 模式更加成熟，同时企业实践中以 Mesos 为主导模式，因此本书的讲解均是基于 Spark on Mesos 模式的。

15.3.5 Spark 基于 Scala 语言的本地应用开发

（1）在 Scala 4.4 的 Eclipse 版本中新建一个 Scala 工程和 Spark2.1.1-All 的用户库，将 ~/Spark2.1.1-All 目录中的所有 JAR 包均导入 Spark2.1.1-All 用户库，并将此用户库添加到当前的 Scala 工程。

（2）在 Scala 工程中创建一个带 main 方法的入口类。

执行 File → New → Scala Object 命令，在弹出的对话框的 Name 栏输入"com.bunfly.bigdata.spark.rdd.local.WordCount"，其中 WordCount 是类名，com.bunfly.bigdata.spark.rdd.local 是包名，下面是仿照 Hadoop 中的 WordCount 代码使用 Spark 来实现的。完整代码如下所示：

```
package com.bunfly.bigdata.spark.rdd.local
import org.apache.spark.SparkConf
import org.apache.spark.SparkContext
import org.apache.spark.rdd.RDD
object WordCount {
  def main(args: Array[String]): Unit = {
    val conf:SparkConf=new SparkConf();
    conf.setAppName("Local Scala Spark RDD");
    conf.setMaster("local");
    val sc:SparkContext=new SparkContext(conf);
    val fp:RDD[String]=sc.textFile("/project/scala/SparkRDD/input", 1);
```

```
    val wordList:RDD[String]=fp.flatMap((line:String)=>line.split(" "));
    val tupleWordList:RDD[Tuple2[String,Int]]=wordList.map(word=>
      (word,1));
    val tupleWordGroupList:RDD[Tuple2[String,Int]]=tupleWordList.
    reduceByKey(
    (preValue:Int,nextValue:Int)=>preValue+nextValue);
    val f:File=new File("/project/scala/SparkRDD/output");
    if(f.exists()&&f.isDirectory())f.delete();
    tupleWordGroupList.saveAsTextFile("/project/scala/SparkRDD/output");
  }
}
```

提示：若 master 主机设置为 local 则表示本地运行。本地运行时 Spark 会启动一个集群模拟器运行 Job 作业，本地运行仅在 Eclipse 或直接使用 Java 命令等运行，不可使用 spark-submit 提交运行，因为该命令装载 Spark 环境配置并连接到 Spark 集群，而本地模式下无须连接集群，仅适合在本地开发环境测试代码。

（3）使用 scala 命令运行本地模式

切换到工程目录下的 bin 目录，将上述编译后的代码打包后直接使用 scala 命令运行。

```
jar -cvfe WordCount.jar com.bunfly.bigdata.spark.rdd.local.WordCount.jar com/
scala WordCount.jar
```

15.3.6 Spark 基于 Scala 语言的集群应用开发

编写集群提交的应用代码如下：

```
package com.bunfly.bigdata.spark.rdd.cluster
import org.apache.spark.SparkConf
import org.apache.spark.SparkContext
import org.apache.spark.rdd.RDD
import java.io.File
import org.apache.hadoop.fs.FileSystem
import org.apache.hadoop.conf.Configuration
import java.net.URI
import org.apache.hadoop.fs.Path
object WordCount {
  var fs:FileSystem=null;
  {
    val hconf:Configuration=new Configuration();
    fs=FileSystem.get(new URI("hdfs://ns1/"), hconf, "hadoop");
  }
  def main(args: Array[String]): Unit = {
    val conf:SparkConf=new SparkConf();
    conf.setAppName("Cluster Scala Spark RDD");
    val sc:SparkContext=new SparkContext(conf);
```

```
        val fp:RDD[String]=sc.textFile("/spark/input", 1);
        val wordList:RDD[String]=fp.flatMap((line:String)=>line.split(" "));
        val tupleWordList:RDD[Tuple2[String,Int]]=wordList.
        map(word=>(word,1));
        val tupleWordGroupList:RDD[Tuple2[String,Int]]=tupleWordList.
        reduceByKey(
        (preValue:Int,nextValue:Int)=>preValue+nextValue);
        val dst:Path=new Path("/spark/output");
        if(fs.exists(dst)&&fs.isDirectory(dst))fs.delete(dst, true);
        tupleWordGroupList.saveAsTextFile("/spark/output");
        sc.stop();
    }
}
```

切换到工程目录下的 bin 目录，将上述编译后的代码打包后使用如下 spark-submit 命令运行：jar -cvfe WordCount.jar com.bunfly.bigdata.spark.rdd.cluster.WordCount.jar com/spark-submit--master spark://master01:7077 /project/scala/SparkRDD/clusterdist/wordcountcluster.jar 1。

15.3.7 Spark 基于 Java 语言的应用开发

1. Java 在 Spark 中的应用

Spark 自身使用 Scala 程序开发，Scala 语言是具备函数式编程和指令式编程的一种语言。Spark 源码基于 Scala 函数式编程进行设计，官方推荐 Spark 开发人员基于 Scala 的函数式编程实现 Spark 的 Job 开发，但目前仍以 Java 作为 Spark 在生产上的主流开发，其原因主要有以下几点。

（1）Java 已成为行业内的主流语言，社区相当活跃，相比 Scala 有更多的资料和文档供项目开发参考。

（2）Spark 项目会与其他已有的 Java 项目进行集成，即使用 Java 开发 Spark 项目将可以更好地实现与已有的各基于 Java 的平台进行对接和整合。

（3）行业内的 Scala 程序员目前相当匮乏，Spark 提供了基于 Scala 与 Java 的两套 API 实现。由于 Scala 的学习成本较之 Java 更高，因此很多公司和企业更倾向于使用 Java。

2. Spark 中 Java API 的特征

Java 的 API 是基于 Scala 的 API 的来对应设计的。Scala 的 API 基于函数式，而函数式编程的一个重要特征就是函数本身可作为函数的参数进行传递（即实现高阶函数调用）。Java 的编程方式是指令式，指令式编程中函数的参数类型可以是基本类型和对象类型而无法直接是函数类型。Spark 为与 Scala 的 API 设计一致，采用函数参数为对象类型的传递方式。由于受 Spark API 的限定，Spark 提供的各函数参数类型均固定，Scala 语法实现的 API 其函数的参数类型是函数类型，参数函数的类型依赖于参数函数的名称、参数函数的参数列表、参数函数的返回类型。

Java 语法实现的 Spark API 的函数参数类型是对象类型，由于参数对象类型受 Spark API 限定，因此 Spark 提供了专门针对函数参数对象类型的接口。在调用方法时通过传递接口类型的实例即可回调与函数式编程中相似的参数函数对象，即函数式编程中的参数函数相当于参数接口中定义的方法，参数接口可单独实现也可使用局部内部类来实现，由于参数接口回调大多是临时的而不是通用的，因此在生产上采用局部内部类来实现参数接口的情况更普遍。

值得注意的是，使用 Java 开发 Spark 项目时，其参数接口通常被 Spark 官方泛型定义，Spark 定义泛型化的参数接口是在编译期检查接口中回调函数的参数列表和返回类型。

提示：目前企业中以使用 Java 开发 Spark 居多，因此这里通过 Java 来讲解 Spark 的开发过程。

15.3.8　Spark 基于 Java 语言的本地应用开发

（1）使用 Eclipse 4.5（mars 版本）创建一个 Java 工程 SparkRDD，并在此工程下创建一个用于本地输入的目录 input。

```
[hadoop@CloudDeskTop software]$ mkdir -p /project/SparkRDD/input
```

（2）在 Eclipse 4.5 中创建 Spark2.1.1-All 用户库，将~/Spark2.1.1-All 目录中的所有 JAR 包均导入 Spark2.1.1-All 用户库中，并将此用户库添加到当前的 Java 工程中。

注意：由于更换 Eclipse 版本，之前 Scala-eclipse 中的用户库无法使用，因此针对 Eclipse 4.5 版本的 Spark 用户库需重新创建。

（3）将输入文件复制至创建的 input 目录：

```
[hadoop@CloudDeskTop project]$ cp -a /project/scala/SparkRDD/input/
{first.txt,second.txt} /project/SparkRDD/input/
```

（4）将上述 Scala 代码版本翻译为 Java 代码如下：

```java
package com.bunfly.bigdata.spark.rdd.local;
import java.io.File;
import java.util.Arrays;
import java.util.Iterator;
import org.apache.spark.SparkConf;
import org.apache.spark.api.java.JavaPairRDD;
import org.apache.spark.api.java.JavaRDD;
import org.apache.spark.api.java.JavaSparkContext;
import org.apache.spark.api.java.function.FlatMapFunction;
import org.apache.spark.api.java.function.Function2;
import org.apache.spark.api.java.function.PairFunction;
import scala.Tuple2;
public class WordCount {
    private static void deleteDir(File f){
        if(f.isFile()||f.listFiles().length==0){
            f.delete();
            return;
        }
```

```java
            File[] files=f.listFiles();
            for(File fp:files)deleteDir(fp);
            f.delete();
        }
        public static void main(String[] args) {
            SparkConf conf=new SparkConf();
            conf.setAppName("");
            conf.setMaster("local");
            JavaSparkContext sc=new JavaSparkContext(conf);
            JavaRDD<String> lineRdd=sc.textFile("/project/SparkRDD/input", 1);
            JavaRDD<String> wordRdd=lineRdd.flatMap(new FlatMapFunction<String,
        String>(){
                @Override
                public Iterator<String> call(String line) throws Exception {
                    String[] wordArr=line.split(" ");
                    return Arrays.asList(wordArr).iterator();
                }
            });
            JavaPairRDD<String, Integer> wordTupleList=wordRdd.mapToPair(new
        PairFunction<String,String,Integer>(){
                @Override
                public Tuple2<String, Integer> call(String word) throws Exception {
                    return new Tuple2<String,Integer>(word,1);
                }
            });
            JavaPairRDD<String, Integer> wordGroupList=wordTupleList.
        reduceByKey(new
        Function2<Integer,Integer,Integer>(){
                @Override
                public Integer call(Integer prev, Integer next) throws Exception {
                    return prev+next;
                }
            });
            File fp=new File("/project/SparkRDD/output");
            if(fp.exists())deleteDir(fp);
            wordGroupList.saveAsTextFile("/project/SparkRDD/output");
        }
    }
```

（5）本地测试：基于本地（local）模式的 Spark 应用可直接在 Eclipse 中测试。

15.3.9 Spark 基于 Java 语言的集群应用开发

（1）在工程目录 SparkRDD 下创建一个用于存储打包文件的目录 clusterdist。

```
[hadoop@CloudDeskTop software]$ /project/SparkRDD/clusterdist
```

（2）翻译 Scala 集群开发的源代码为 Java 代码。代码如下：

```java
package com.bunfly.bigdata.spark.rdd.cluster;
import java.io.IOException;
import java.net.URI;
import java.net.URISyntaxException;
import java.util.Arrays;
import java.util.Iterator;
import org.apache.hadoop.conf.Configuration;
import org.apache.hadoop.fs.FileSystem;
import org.apache.hadoop.fs.Path;
import org.apache.spark.SparkConf;
import org.apache.spark.api.java.JavaPairRDD;
import org.apache.spark.api.java.JavaRDD;
import org.apache.spark.api.java.JavaSparkContext;
import org.apache.spark.api.java.function.FlatMapFunction;
import org.apache.spark.api.java.function.Function2;
import org.apache.spark.api.java.function.PairFunction;
import scala.Tuple2;
public class WordCount {
    private static FileSystem fs;
    static{
        Configuration conf=new Configuration();
        try {
            fs=FileSystem.get(new URI("hdfs://ns1/"), conf, "hadoop");
        } catch (IOException e) {
            e.printStackTrace();
        } catch (InterruptedException e) {
            e.printStackTrace();
        } catch (URISyntaxException e) {
            e.printStackTrace();
        }
    }
    public static void main(String[] args) throws IOException {
        SparkConf conf=new SparkConf();
        conf.setAppName("");
        JavaSparkContext sc=new JavaSparkContext(conf);
        JavaRDD<String> lineRdd=sc.textFile("/spark/input", 1);
        JavaRDD<String> wordRdd=lineRdd.flatMap(new FlatMapFunction
            <String, String>(){
                @Override
                public Iterator<String> call(String line) throws Exception {
                    String[] wordArr=line.split(" ");
                    return Arrays.asList(wordArr).iterator();
                }
        });
```

```java
            JavaPairRDD<String, Integer> wordTupleList=wordRdd.mapToPair(new
            PairFunction<String,String,Integer>(){
                @Override
                public Tuple2<String, Integer> call(String word) throws Exception {
                    return new Tuple2<String,Integer>(word,1);
                }
            });
            JavaPairRDD<String, Integer> wordGroupList=wordTupleList.
            reduceByKey(new
            Function2<Integer,Integer,Integer>(){
                @Override
                public Integer call(Integer prev, Integer next) throws Exception {
                    return prev+next;
                }
            });
            Path dist=new Path("/spark/output");
            if(fs.exists(dist))fs.delete(dist, true);
            wordGroupList.saveAsTextFile("/spark/output");
        }
    }
```

（3）打包 Spark 应用至 clusterdist 目录。

删除之前的输出目录：

```
[hadoop@CloudDeskTop bin]$ hdfs dfs -rm -r /spark/output
```

切换至 Spark 工程目录的 bin 目录下，将 com 文件夹打包至工程目录下的 clusterdist 目录下：

```
[hadoop@CloudDeskTop software]$ cd /project/SparkRDD/bin/
[hadoop@CloudDeskTop bin]$ jar -cvf /project/SparkRDD/clusterdist/wordcountcluster.jar com/
```

（4）提交 Job 到 Spark 集群。

```
[hadoop@CloudDeskTop software]$ cd /software/spark-2.1.1/bin/
[hadoop@CloudDeskTop bin]$ ./spark-submit --master spark://master01:7077 --class com.bunfly.bigdata.spark.rdd.cluster.WordCount /project/SparkRDD/clusterdist/wordcountcluster.jar 1
    17/08/16 00:38:40 WARN util.NativeCodeLoader: Unable to load
    native-hadoop library for your platform... using builtin-java classes
    where applicable
    17/08/16 00:38:42 INFO spark.SparkContext: Running Spark version 2.1.1
    17/08/16 00:38:42 WARN spark.SparkContext: Support for Java 7 is
    deprecated as of Spark 2.0.0
    ...
    17/08/16 00:38:44 INFO spark.SparkEnv: Registering
```

```
            OutputCommitCoordinator
        17/08/16 00:38:44 INFO util.log: Logging initialized @8137ms
        17/08/16 00:38:44 INFO server.Server: jetty-9.2.z-SNAPSHOT
```

（5）查看输出目录下是否有数据生成。

```
[hadoop@CloudDeskTop bin]$ hdfs dfs -ls /spark
Found 2 items
drwxr-xr-x   - hadoop supergroup          0 2017-08-09 09:13 /spark/input
drwxr-xr-x   - hadoop supergroup          0 2017-08-16 00:39 /spark/output
[hadoop@CloudDeskTop bin]$ hdfs dfs -ls /spark/output
Found 2 items
-rw-r--r--   3 hadoop supergroup          0 2017-08-16 00:39 /spark/output/_SUCCESS
-rw-r--r--   3 hadoop supergroup         90 2017-08-16 00:39 /spark/output/
part-00000
[hadoop@CloudDeskTop bin]$ hdfs dfs -cat /spark/output/part-00000
(118,1)
(is,4)
(name,1)
(1.67,1)
(35,1)
(height,1)
(age,1)
(liyongfu,1)
(my,4)
(weight,1)
```

提示：

（1）将 Job 提交到集群时，请勿直接在 Eclipse 工程中测试。这种操作可预测性小，易出现异常。若需直接在 Eclipse 中测试，可以设置提交的 master 节点。如下：

```
SparkConf conf=new SparkConf();
conf.setAppName("Cluster Java Spark RDD");
conf.setMaster("spark://master01:7077");
JavaSparkContext sc=new JavaSparkContext(conf);
```

（2）由于 Job 涉及 HDFS 的文件操作，需连接到 HDFS 完成，因此需将 Hadoop 的配置文件复制至工程的根目录。

```
[hadoop@CloudDeskTop software]$ cd hadoop-2.7.3/etc/hadoop/
[hadoop@CloudDeskTop hadoop]$ cp -a core-site.xml hdfs-site.xml /project/
SparkRDD/src/
```

15.4 习题与思考

1. RDD 是什么？为什么会产生 RDD？
2. 尝试利用 Spark RDD 技术做词频统计。

第 16 章　Spark SQL

16.1　Spark SQL 架构原理

16.1.1　Hive 的两种功能

（1）作为数据仓库提供存储功能（Hive 的元数据，如库、表等结构信息，均由 Hive 自身来维护，但数据本身存储在 HDFS 集群中）。

（2）作为查询引擎提供检索查询功能（如 Hive-SQL，简称为 HQL）。Spark SQL 仅代替 Hive 的查询引擎，而存储本身需借助于其他存储介质，其中 Hive 仓库就是 Spark SQL 常用的一种存储介质。在企业实际生产环境下，Hive 存储+Spark SQL 查询引擎属最佳的实践组合。

查询引擎的作用是：将应用层提供的 SQL 语句翻译为计算框架的 Job 代码，并将其 Job 提交到集群执行的一种查询机制。

Hive 查询引擎的功能如下。

（1）将 HQL 翻译成 MR 代码，产生一到多个 Job 作业（一个 HQL 有可能被翻译成一连串的 Job 作业）。

（2）将 Job 打成 JAR 包并发布到 Hadoop 集群中运行。

Hive 就相当于 Hadoop 的一个自动化的客户端，其查询引擎存在较大延迟，因此才有 Spark SQL 查询引擎。Spark SQL 的查询引擎较之 Hive 的 HQL 查询引擎快 6 倍左右（100 万条测试数据的情况下）。Spark SQL 底层使用 Spark Core 将数据从 HDFS 载入内存，然后在内存中执行 RDD 的迭代计算；而 HQL 底层则是使用 MapReduce 执行基于磁盘的迭代计算。

16.1.2　Spark SQL 的重要功能

（1）Spark SQL 可操作 Hive、HBase、MySQL、Oracle、DB2 等中的数据（可自定义以支持更多类型的数据来源）。

（2）提升了数据仓库的计算能力和计算复杂度。

（3）基于 Spark SQL 推出的 DataFrame 可实现数据仓库直接使用机器学习、图计算等复杂算法库深度数据挖掘数据仓库。

（4）Spark SQL（DataFrame、DataSet、Tungsten）是数据仓库、数据挖掘及其科学计算和分析引擎工具。

传统关系型数据库主要用于实时事务性分析计算。关系数据库的缺点是无法进行分布式计算和分布式存储，除实时性事务分析计算均可使用以下的流行架构组合来构建分析架构：

Hive（仓库存储）+Spark SQL（高速计算）+DataFrame（复杂度分析和挖掘）+[DataSet]

16.1.3 Spark SQL 的 DataFrame 特征

Spark SQL 是基于 DataFrame 实现的列式查询引擎，这是 DataFrame 区别于 RDD 的一个重要特征。R 及 Python 语言中都有 DataFrame。Spark 中的 DataFrame 作为一种分布式表，描述的二维表中详细地指定了每个列的名称和类型。因此 DataFrame 可优化到每条记录的列级别。而 RDD 的基本单位不是列级别而是行级别，因此 RDD 仅优化到行级别。数据处理与分析在 IT 领域中的阶段如下：

第一个阶段：C/C+++ 文件存储。
第二个阶段：J2EE+ 数据库。
第三个阶段：HiveQL + Hadoop。
第四个阶段：Spark SQL + Hive。
第五个阶段：Spark SQL + DataFrame + Hive。
第六个阶段：Spark SQL + DataSet + DataFrame + Hive。

16.2 Spark SQL 操作 Hive

关于 Spark SQL 操作的视频讲解可扫描以下两个二维码观看。

16.2.1 添加配置文件，便于 Spark SQL 访问 Hive 仓库

```
vi /software/spark-1.6.2-bin-hadoop2.4/conf/hive-site.xml
    <configuration>
        <property>
            <name>hive.metastore.uris</name>
            <value>thrift://YunHive:9083</value>
        </property>
    </configuration>
```

📢 注意：上述 thrift 协议的 uris 指向 Hive 服务端（Hive 的安装节点）。

16.2.2 安装 JDBC 驱动

将 MySQL 的 JDBC 驱动包复制到 Spark 安装目录下的 lib 子目录下（便于连接 MySQL 数据库）。复制驱动包的操作可不做，因为 Spark 通过 Hive 提供的 9083 的 thrift 协议服务获取 MySQL 数据库中的元数据，所以只需 Hive 服务端能够连接 MySQL 数据库即可。

16.2.3 启动 MySQL 服务及其 Hive 的元数据服务

1. 启动 MySQL 服务

```
service mysqld start
```

或

```
/etc/init.d/mysqld start
```

2. 启动 Hive 元数据服务

```
nohup hive --service metastore &>metastore.log &
```

16.2.4 启动 HDFS 集群和 Spark 集群

```
start-dfs.sh
start-all.sh
```

16.2.5 启动 spark-shell 并测试

```
/software/spark-2.1.1/bin/spark-shell --master spark://YunMaster01:7077
```

(1) 创建 HiveContext 对象，参数 sc 是 SparkContext 类型的对象。

```
val hiveContext=new org.apache.spark.sql.hive.HiveContext(sc)
```

(2) 切换数据库到 Hive。

```
hiveContext.sql("use hive")
#显示Hive库中的所有表
hiveContext.sql("show tables").collect.foreach(println)
```

(3) 执行数据查询并显示结果。

```
hiveContext.sql("select count(*) from t_user").collect.foreach(println)
hiveContext.sql("select count(*) from t_user where userAge>30").collect.foreach(println)
hiveContext.sql("select count(*) from t_user where userAge>30 and userName like '%li%'").collect.foreach(println)
```

◁))注意：

(1) Spark SQL 和 Spark Streaming 均基于 SparkCore 内核完成 RDD 计算。

(2) SQLContext 是 HiveContext 的超接口，HiveContext 是 SQLContext 接口的 Hive 实现（即底层的数据仓库可使用非 Hive 的其他存储来实现）。

16.3 Spark SQL 操作 HDFS

Spark SQL 可直接读取关系数据库中的数据并将其包装为 DataFrame，然后基于 DataFrame 注册一个临时表，根据此临时表完成 SQL 的分析处理。

16.3.1 操作代码

Spark SQL 操作 HDFS 的代码如下：

```java
        package com.bunfly.bigdata.spark.sql.cluster;
        import java.io.IOException;
        import java.net.URI;
        import java.net.URISyntaxException;
        import java.sql.Date;
        import org.apache.hadoop.conf.Configuration;
        import org.apache.hadoop.fs.FileSystem;
        import org.apache.hadoop.fs.Path;
        import org.apache.spark.SparkConf;
        import org.apache.spark.api.java.JavaRDD;
        import org.apache.spark.api.java.JavaSparkContext;
        import org.apache.spark.api.java.function.Function;
        import org.apache.spark.sql.Dataset;
        import org.apache.spark.sql.Row;
        import org.apache.spark.sql.SQLContext;
        import org.apache.spark.sql.SparkSession;
import com.bunfly.bigdata.spark.sql.po.User;
@SuppressWarnings("resource")
public class RDDToHdfs {
    private static FileSystem fs;
    static{
        Configuration conf=new Configuration();
        try {
            fs=FileSystem.get(new URI("hdfs://ns1/"), conf, "hadoop");
        } catch (IOException | InterruptedException | URISyntaxException e) {
            e.printStackTrace();
        }
    }
    public static void main(String[] args) throws IOException {
        SparkConf conf=new SparkConf();
        conf.setAppName("RDD To HDFS Cluster");
        JavaSparkContext sc=new JavaSparkContext(conf);
        SQLContext sqlContext=new SQLContext(SparkSession.builder().
        sparkContext(sc.sc())
        .getOrCreate());
        JavaRDD<String> lineRDD=sc.textFile("/spark/input");
        JavaRDD<User> userRDD=lineRDD.map(new Function<String,User>(){
            @Override
            public User call(String line) throws Exception {
                String[] fields=line.split("\t");
                Long userId=Long.parseLong(fields[0]);
                String userName=fields[1];
                Integer weight=Integer.parseInt(fields[2]);
                Date birthday=Date.valueOf(fields[3]);
                return new User(userId,userName,weight,birthday);
```

```
            }
        });
        Dataset<Row> userDataSet=sqlContext.createDataFrame(userRDD,
        User.class);
        userDataSet.createOrReplaceTempView("t_user");
        Dataset<Row> queryDataSet=sqlContext.sql("select * from t_user
        where weight>100");
        JavaRDD<Row> rowRDD=queryDataSet.toJavaRDD();
        JavaRDD<User> userRDDResult=rowRDD.map(new Function<Row,User>(){
            @Override
            public User call(Row row) throws Exception {
                Long userId=row.getAs("userId");
                String userName=row.getAs("userName");
                Integer weight=row.getAs("weight");
                Date birthday=row.getAs("birthday");
                return new User(userId,userName,weight,birthday);
            }
        });
        Path dist=new Path("/spark/output");
        if(fs.exists(dist))fs.delete(dist,true);
        userRDDResult.saveAsTextFile("/spark/output");
    }
}
```

16.3.2 工程文件

将工程打包成 JAR 文件放入工程目录下的 dist 目录下。

```
[hadoop@CloudDeskTop bin]$ pwd
    /project/SparkSQL-JAVA/bin
[hadoop@CloudDeskTop bin]$ jar -cvfe ../dist/RDDToHdfs.jar
    com.bnyw.bigdata.spark.sql.cluster.RDDToHdfs com/
[hadoop@CloudDeskTop bin]$ cd ../dist/
[hadoop@CloudDeskTop dist]$ pwd
    /project/SparkSQL-JAVA/dist
[hadoop@CloudDeskTop dist]$ ls
    RDDToHdfs.jar
```

16.3.3 创建测试数据

```
[hadoop@CloudDeskTop input]$ pwd
    /project/SparkSQL-JAVA/input
[hadoop@CloudDeskTop input]$ ls
    users
[hadoop@CloudDeskTop input]$ cat users
    1    ligang    118  1982-12-28
    2    zhanghua  120  1983-08-04
```

```
    3   zhaoyu      106 1978-04-08
    4   huanghua    98  1987-08-07
    5   chenglan    88  1992-07-08
    6   huanghao    115 1996-06-05
[hadoop@CloudDeskTop input]$ hdfs dfs -put users /spark/input/
[hadoop@CloudDeskTop input]$ hdfs dfs -ls /spark/input
    Found 1 items
    -rw-r--r--   3 hadoop supergroup   150 2017-08-17 11:20 /spark/
    input/users
```

16.3.4 运行 Job 并提交到集群

```
[hadoop@CloudDeskTop bin]$ pwd
    /software/spark-2.1.1/bin
[hadoop@CloudDeskTop bin]$ ./spark-submit --master spark://master01:7077
--class com.bnyw.bigdata.spark.sql.cluster.RDDToHdfs /project/
SparkSQL-JAVA/dist/RDDToHdfs.jar 1
17/08/17 12:53:23 WARN util.NativeCodeLoader: Unable to load native-hadoop
library for your platform... using builtin-java classes where applicable
17/08/17 12:53:25 INFO spark.SparkContext: Running Spark version 2.1.1
17/08/17 12:53:25 WARN spark.SparkContext: Support for Java 7 is deprecated
as of Spark 2.0.0
...
17/08/17 12:53:27 INFO spark.SparkEnv: Registering OutputCommitCoordinator
17/08/17 12:53:27 INFO util.log: Logging initialized @8820ms
17/08/17 12:53:28 INFO server.Server: jetty-9.2.z-SNAPSHOT
```

16.3.5 查看运行结果

```
[hadoop@CloudDeskTop input]$ hdfs dfs -ls /spark
    Found 2 items
    drwxr-xr-x   - hadoop supergroup     0 2017-08-17 11:20 /spark/input
    drwxr-xr-x   - hadoop supergroup     0 2017-08-17 11:26 /spark/ output

[hadoop@CloudDeskTop input]$ hdfs dfs -ls /spark/output
    Found 3 items
    -rw-r--r--   3 hadoop supergroup     0 2017-08-17 11:26 /spark/
    output/_SUCCESS
    -rw-r--r--   3 hadoop supergroup    74 2017-08-17 11:26 /spark/
    output/part-00000
    -rw-r--r--   3 hadoop supergroup    26 2017-08-17 11:26 /spark/
    output/part-00001
[hadoop@CloudDeskTop input]$ hdfs dfs -cat /spark/output/part-00000
    1   ligang      118 1982-12-28
    2   zhanghua    120 1983-08-04
```

```
    3       zhaoyu    106 1978-04-08
[hadoop@CloudDeskTop input]$ hdfs dfs -cat /spark/output/part-00001
    6       huanghao  115 1996-06-05
```

提示：Spark 处理之后的数据写入到 HDFS 集群。若需将其转移到关系数据库中可启动一个定时任务并使用 Sqoop 工具完成。

16.4 Spark SQL 操作关系数据库

Spark SQL 可读取关系数据库中的数据进行分析处理，需借助于 JDBC 协议和相应的驱动来完成。以下业务场景以客户端本地的数据通过 Spark SQL 处理后的结果写入远端关系数据库中，供前端在线事务系统使用。

16.4.1 添加访问 MySQL 的驱动包

在 Eclipse 4.5 中建立工程 RDDToJDBC，并创建文件夹 lib 用于放置第三方驱动包。

```
[hadoop@CloudDeskTop software]$ cd /project/RDDToJDBC/
[hadoop@CloudDeskTop RDDToJDBC]$ mkdir -p dist
[hadoop@CloudDeskTop RDDToJDBC]$ ls
bin  dist  src
```

16.4.2 添加必要的开发环境

将 MySQL 的 JAR 包复制到工程目录 RDDToJDBC 下的 lib 目录下：

```
cp -a /software/hive-1.2.2/lib/mysql-connector-java-3.0.17-ga-bin.jar /project/RDDToJDBC/lib/
```

将 Spark 的开发库 spark 2.1.1 -all 追加到 RDDToJDBC 工程的 classpath 路径中去，可在 Eclipse 中通过添加用户库的方式解决。

16.4.3 使用 Spark SQL 操作关系数据库

源码如下：

```
package com.bnyw.bigdata.spark.rdd.local;
import java.sql.Connection;
import java.sql.DriverManager;
import java.sql.PreparedStatement;
import java.sql.SQLException;
import java.util.Arrays;
import java.util.Iterator;
import org.apache.spark.SparkConf;
import org.apache.spark.api.java.JavaPairRDD;
import org.apache.spark.api.java.JavaRDD;
import org.apache.spark.api.java.JavaSparkContext;
```

```java
    import org.apache.spark.api.java.function.FlatMapFunction;
    import org.apache.spark.api.java.function.Function2;
    import org.apache.spark.api.java.function.PairFunction;
    import org.apache.spark.api.java.function.VoidFunction;
import scala.Tuple2;
@SuppressWarnings("serial")
public class WordCount {
    private static Connection conn;
    private static PreparedStatement pstat;
    static{
        try {
            Class.forName("com.mysql.jdbc.Driver");
            conn=DriverManager.getConnection("jdbc:mysql:
            //192.168.37.143:3306/bnyw?
            characterEncoding=utf8", "root", "123456");
            pstat=conn.prepareStatement("insert into wordcount(word,
            count) values(?,?)");
        } catch (Exception e) {
            e.printStackTrace();
        }
    }
    private static int save(Tuple2<String, Integer> line) throws SQLException{
        pstat.setString(1, line._1);
        pstat.setInt(2, line._2);
        return pstat.executeUpdate();
    }
    private static void saveToDB(JavaPairRDD<String, Integer> pairRDD)
    throws Exception{
        pairRDD.foreach(new VoidFunction<Tuple2<String,Integer>>(){
            @Override
            public void call(Tuple2<String, Integer> line) throws Exception {
                save(line);
            }
        });
    }
    public static void main(String[] args) throws Exception {
        SparkConf conf=new SparkConf();
        conf.setAppName("Local Java Spark RDD");
        conf.setMaster("local");
        JavaSparkContext sc=new JavaSparkContext(conf);
        JavaRDD<String> lineRdd=sc.textFile("/home/hadoop/data/srcdata/
        wordcount/", 1);
        JavaRDD<String> wordRdd=lineRdd.flatMap(new FlatMapFunction
        <String, String>(){
```

```java
            @Override
            public Iterator<String> call(String line) throws Exception {
                String[] wordArr=line.split(" ");
                return Arrays.asList(wordArr).iterator();
            }
        });
        JavaPairRDD<String, Integer> wordTupleList=wordRdd.mapToPair(new PairFunction<
        String,String,Integer>(){
            @Override
            public Tuple2<String, Integer> call(String word) throws Exception {
                return new Tuple2<String,Integer>(word,1);
            }
        });
        JavaPairRDD<String, Integer> wordGroupList=wordTupleList.reduceByKey(new
        Function2<Integer,Integer,Integer>(){
            @Override
            public Integer call(Integer prev, Integer next) throws Exception {
                return prev+next;
            }
        });
        saveToDB(wordGroupList);
    }
}
```

16.4.4 初始化 MySQL 数据库服务

1. 启动 MySQL 数据库服务

```
[root@DB03 ~]# cd /software/mysql-5.5.32/multi-data/3306/
[root@DB03 3306]# ls
data  my.cnf  my.cnf.bak  mysqld
[root@DB03 3306]# ./mysqld start
Starting MySQL...
```

2. 建立 bunfly 库

```
[root@DB03 bin]# ./mysql -h192.168.37.143 -P3306 -uroot -p123456 -e "show databases;"
+--------------------+
| Database           |
+--------------------+
| information_schema |
| hive               |
| mydb               |
```

```
| mysql                   |
| performance_schema      |
| test                    |
| test2                   |
+-------------------------+
[root@DB03 bin]# ./mysql -h192.168.37.143 -P3306 -uroot -p123456 -e "create
database bunfly character set utf8;"
[root@DB03 bin]# ./mysql -h192.168.37.143 -P3306 -uroot -p123456 -e "show
databases;"
+-------------------------+
| Database                |
+-------------------------+
| information_schema      |
| bunfly                  |
| hive                    |
| mydb                    |
| mysql                   |
| performance_schema      |
| test                    |
| test2                   |
+-------------------------+
```

3. 建立 wordcount 表

```
[root@DB03 bin]# ./mysql -h192.168.37.143 -P3306 -uroot -p123456 -e "create
table if not exists bunfly.wordcount(wid int(11) auto_increment primary
key,word varchar(30),count int(3))engine=myisam charset=utf8;"
[root@DB03 bin]# ./mysql -h192.168.37.143 -P3306 -uroot -p123456 -e "desc
bunfly.wordcount;"
+-------+-------------+------+-----+---------+----------------+
| Field | Type        | Null | Key | Default | Extra          |
+-------+-------------+------+-----+---------+----------------+
| wid   | int(11)     | NO   | PRI | NULL    | auto_increment |
| word  | varchar(30) | YES  |     | NULL    |                |
| count | int(3)      | YES  |     | NULL    |                |
+-------+-------------+------+-----+---------+----------------+
```

目前数据库表中不存在数据。

```
[root@DB03 bin]# ./mysql -h192.168.37.143 -P3306 -uroot -p123456 -e "select
* from bnyw.wordcount;"
```

16.4.5 准备 Spark SQL 源数据

```
[hadoop@CloudDeskTop bin]$ ls ~/data/srcdata/wordcount/
    words
[hadoop@CloudDeskTop bin]$ cat ~/data/srcdata/wordcount/words
```

```
my name is lixiang
my age is 36
my height is 1.67
my weight is 120
```

16.4.6 运行 Spark 代码

在 Eclipse 4.5 中直接运行 Spark 代码，观察 Eclipse 控制台输出。

```
[root@DB03 bin]# ./mysql -h192.168.37.143 -P3306 -uroot -p123456 -e "select
 * from bunfly.wordcount;"
+-----+---------+-------+
| wid | word    | count |
+-----+---------+-------+
|   1 | is      |     4 |
|   2 | name    |     1 |
|   3 | 1.67    |     1 |
|   4 | 36      |     1 |
|   5 | 120     |     1 |
|   6 | lixiang |     1 |
|   7 | height  |     1 |
|   8 | age     |     1 |
|   9 | my      |     4 |
|  10 | weight  |     1 |
+-----+---------+-------+
```

提示：上述 Spark SQL 计算客户端本地数据，然后将计算结果输出到关系数据库 MySQL。接下来开始讲解 Spark SQL 处理 HDFS 分布式存储中的数据，并将其写出到远端关系数据库 MySQL 中。

16.4.7 创建 dist 文件夹

在 Eclipse 4.5 中的工程 RDDToJDBC 下创建一个 dist 文件夹，用于放置打包文件。

```
[hadoop@CloudDeskTop software]$ cd /project/RDDToJDBC/
[hadoop@CloudDeskTop RDDToJDBC]$ mkdir -p dist
[hadoop@CloudDeskTop RDDToJDBC]$ ls
bin  dist  src
```

16.4.8 安装数据库驱动

将关系数据库的驱动包放置到 Spark 安装目录下的 jars 目录下。

```
cp -a /software/hive-1.2.2/lib/mysql-connector-java-3.0.17-ga-bin.jar /
software/spark-2.1.1/jars/
scp -r /software/hive-1.2.2/lib/mysql-connector-java-3.0.17-ga-bin.jar
master01:/software/spark-2.1.1/jars/
scp -r /software/spark-2.1.1/jars/mysql-connector-java-3.0.17-ga-bin.jar
```

```
master02:/software/spark-2.1.1/jars/
scp -r /software/spark-2.1.1/jars/mysql-connector-java-3.0.17-ga-bin.jar
slave01:/software/spark-2.1.1/jars/
scp -r /software/spark-2.1.1/jars/mysql-connector-java-3.0.17-ga-bin.jar
slave02:/software/spark-2.1.1/jars/
scp -r /software/spark-2.1.1/jars/mysql-connector-java-3.0.17-ga-bin.jar
slave03:/software/spark-2.1.1/jars/
```

16.4.9 基于集群操作

基于集群操作的源码如下：

```java
package com.bunfly.bigdata.spark.rdd.cluster;
import java.sql.Connection;
import java.sql.DriverManager;
import java.sql.PreparedStatement;
import java.sql.SQLException;
import java.util.Arrays;
import java.util.Iterator;
import org.apache.spark.SparkConf;
import org.apache.spark.api.java.JavaPairRDD;
import org.apache.spark.api.java.JavaRDD;
import org.apache.spark.api.java.JavaSparkContext;
import org.apache.spark.api.java.function.FlatMapFunction;
import org.apache.spark.api.java.function.Function2;
import org.apache.spark.api.java.function.PairFunction;
import org.apache.spark.api.java.function.VoidFunction;
import scala.Tuple2;
@SuppressWarnings("serial")
public class WordCount {
    private static int count;
    private static Connection conn;
    private static PreparedStatement pstat;
    static{
        try {
            Class.forName("com.mysql.jdbc.Driver");
            conn=DriverManager.getConnection("jdbc:mysql://
            192.168.37.143:3306/bnyw? characterEncoding=utf8",
            "root", "123456");
            pstat=conn.prepareStatement("insert into wordcount(word,count)
            values(?,?)");
        } catch (Exception e) {
            e.printStackTrace();
        }
    }
```

```java
        private static void batchSave(Tuple2<String, Integer> line,boolean
    isOver) throws
SQLException{
        if(isOver){
            pstat.setString(1, line._1);
            pstat.setInt(2, line._2);
            pstat.addBatch();
            pstat.executeBatch();
            pstat.clearBatch();
            pstat.clearParameters();
        }else{
            pstat.setString(1, line._1);
            pstat.setInt(2, line._2);
            pstat.addBatch();
            if(++count%100==0){
                pstat.executeBatch();
                pstat.clearBatch();
                pstat.clearParameters();
                count=0;
            }
        }
    }
    private static void saveToDB(JavaPairRDD<String, Integer> pairRDD)
    throws Exception{
        try{
            final long rddLineNum=pairRDD.count();
            pairRDD.foreach(new VoidFunction<Tuple2<String,Integer>>(){
                private int count;
                @Override
                public void call(Tuple2<String, Integer> line) throws
                Exception {
                    if(++count>=rddLineNum){
                        batchSave(line,true);
                    }else{
                        batchSave(line,false);
                    }
                }
            });
        }finally{
            pstat.close();
            conn.close();
        }
    }
    public static void main(String[] args) throws Exception {
        SparkConf conf=new SparkConf();
```

```java
        conf.setAppName("Cluster Java Spark RDD");
        JavaSparkContext sc=new JavaSparkContext(conf);
        JavaRDD<String> lineRdd=sc.textFile("/spark/input", 1);
        JavaRDD<String> wordRdd=lineRdd.flatMap(new FlatMapFunction
        <String, String>(){
            @Override
            public Iterator<String> call(String line) throws Exception {
                String[] wordArr=line.split(" ");
                return Arrays.asList(wordArr).iterator();
            }
        });
        JavaPairRDD<String, Integer> wordTupleList=wordRdd.mapToPair(new
            PairFunction<String,String,Integer>(){
            @Override
            public Tuple2<String, Integer> call(String word) throws Exception {
                return new Tuple2<String,Integer>(word,1);
            }
        });
        JavaPairRDD<String, Integer> wordGroupList=wordTupleList.
            reduceByKey(new
    Function2<Integer,Integer,Integer>(){
            @Override
            public Integer call(Integer prev, Integer next) throws Exception {
                return prev+next;
            }
        });
        saveToDB(wordGroupList);
    }
}
```

在集群模式下，Spark 操作关系数据库通过启动一个 Job 来完成，而启动 Job 则通过 RDD 操作触发。因此在 Spark 集群模式下其关系数据库的所有操作必须位于 RDD 操作级别才有效，否则数据的操作无法影响关系数据库。而 RDD 级别之外的操作都属于 Spark Core 的客户端 Driver 级别（如 Spark SQL 和 Spark Streaming）。上述代码中，只有 RDD 对象被 foreachXXX 时才会进入到 SparkCore 级别的 Job 操作。RDD 之外的操作是 Driver 级别的操作，无法在基于 RDD 级别的 SparkCore 操作过程中启动 Job，其数据均被封装成 Job 提交到集群，并在集群的各个节点执行分配的 Task。数据在各 Task 节点之间传递需数据本身支持可序列化，因此在 Spark 应用中高频率出现的内部类对象（比如上述 VoidFunction）都需要支持可序列化，即在这些内部类对象中出现的成员需要可序列化，因此在内部类对象上下文中编写代码时须注意不能出现不可序列化的对象或引用（如不能出现基于瞬态的流化对象 Connection、Statement、Thread 等），即在内部类对象上下文中出现的对象引用须实现 java.io.Serializable 接口。

16.4.10 打包工程代码到 dist 目录下

```
[hadoop@CloudDeskTop bin]$ pwd
    /project/RDDToJDBC/bin
[hadoop@CloudDeskTop bin]$ ls
    com
[hadoop@CloudDeskTop bin]$ jar -cvfe /project/RDDToJDBC/dist/RDDToJDBC.jar com.bnyw.bigdata.spark.rdd.cluster.WordCount com/
[hadoop@CloudDeskTop bin]$ cd ../dist/
[hadoop@CloudDeskTop dist]$ ls
    RDDToJDBC.jar
```

16.4.11 启动集群并提交 Job 应用

1. 启动三台 Slave 节点的 Zookeeper 进程

```
[hadoop@slave01 software]$ cd /software/zookeeper-3.4.10/bin/ && ./zkServer.sh start &&
    cd - && jps
[hadoop@slave02 software]$ cd /software/zookeeper-3.4.10/bin/ && ./zkServer.sh start &&    cd - && jps
    [hadoop@slave03 software]$ cd /software/zookeeper-3.4.10/bin/ && ./zkServer.sh start &&    cd - && jps
```

2. 启动 HDFS 集群

```
[hadoop@master01 software]$ start-dfs.sh
Starting namenodes on [master01 master02]
master01: starting namenode, logging to /software/hadoop-2.7.3/logs/hadoop-hadoop-namenode-master01.out
master02: ssh: connect to host master02 port 22: No route to host
slave02: starting datanode, logging to /software/hadoop-2.7.3/logs/hadoop-hadoop-datanode-slave02.out
slave01: starting datanode, logging to /software/hadoop-2.7.3/logs/hadoop-hadoop-datanode-slave01.out
slave03: starting datanode, logging to /software/hadoop-2.7.3/logs/hadoop-hadoop-datanode-slave03.out
Starting journal nodes [slave01 slave02 slave03]
slave01: starting journalnode, logging to /software/hadoop-2.7.3/logs/hadoop-hadoop-journalnode-slave01.out
slave02: starting journalnode, logging to /software/hadoop-2.7.3/logs/hadoop-hadoop-journalnode-slave02.out
slave03: starting journalnode, logging to /software/hadoop-2.7.3/logs/hadoop-hadoop-journalnode-slave03.out
```

3. 启动 Spark 集群

```
[hadoop@master01 sbin]$ ./start-slaves.sh
```

```
slave02: starting org.apache.spark.deploy.worker.Worker, logging to /
software/spark-2.1.1/logs/spark-hadoop-org.apache.spark.deploy.worker.
Worker-1-slave02.out
slave03: starting org.apache.spark.deploy.worker.Worker, logging to /
software/spark-2.1.1/logs/spark-hadoop-org.apache.spark.deploy.worker.
Worker-1-slave03.out
slave01: starting org.apache.spark.deploy.worker.Worker, logging to /
software/spark-2.1.1/logs/spark-hadoop-org.apache.spark.deploy.worker.
Worker-1-slave01. out
slave02: failed to launch: nice -n 0 /software/spark-2.1.1/bin/spark-class
org.apache.spark.deploy.worker.Worker --webui-port 8081 spark://
master01:7077
slave02: full log in /software/spark-2.1.1/logs/spark-hadoop-org.apache.
spark.deploy.worker.Worker-1-slave02.out
slave01: failed to launch: nice -n 0 /software/spark-2.1.1/bin/spark-class
org.apache.spark.deploy.worker.Worker --webui-port 8081 spark://
master01:7077
slave01: full log in /software/spark-2.1.1/logs/spark-hadoop-org.apache.
spark.deploy.worker.Worker-1-slave01.out
```

4. 在 Spark SQL 客户端节点提交 Spark 应用

1）将数据库中的旧数据删除

```
[root@DB03 bin]# ./mysql -h192.168.37.143 -P3306 -uroot -p123456 -e
"truncate table bunfly.wordcount;"
[root@DB03 bin]# ./mysql -h192.168.37.143 -P3306 -uroot -p123456 -e "select
* from bunfly.wordcount;"
```

2）准备源数据

```
[hadoop@CloudDeskTop bin]$ ls ~/data/srcdata/wordcount/
words
[hadoop@CloudDeskTop bin]$ cat ~/data/srcdata/wordcount/words
    my name is lixiang
    my age is 36
    my height is 1.67
    my weight is 120
[hadoop@CloudDeskTop bin]$ hdfs dfs -put ~/data/srcdata/wordcount/words /
spark/input/
[hadoop@CloudDeskTop bin]$ hdfs dfs -ls /spark/input
    Found 1 items
    -rw-r--r--   3 hadoop supergroup    67 2017-08-20 06:46 /
spark/input/words
[hadoop@CloudDeskTop bin]$ hdfs dfs -cat /spark/input/words
    my name is lixiang
    my age is 36
```

```
my height is 1.67
my weight is 120
```

3）提交 Spark 应用

```
[hadoop@CloudDeskTop lib]$ cd /software/spark-2.1.1/bin/
[hadoop@CloudDeskTop bin]$ ./spark-submit --master spark://master01:7077
--class com.bnyw.bigdata.spark.rdd.cluster.WordCount /project/
RDDToJDBC/dist/RDDToJDBC.jar
17/08/20 17:20:22 INFO spark.SparkContext: Running Spark version 2.1.1
17/08/20 17:20:22 WARN spark.SparkContext: Support for Java 7 is deprecated
as of Spark2.0.0
17/08/20 17:20:23 WARN util.NativeCodeLoader: Unable to load native-hadoop
library for your platform... using builtin-java classes where applicable
...
17/08/20 17:20:26 INFO spark.SparkEnv: Registering OutputCommitCoordinator
17/08/20 17:20:26 INFO util.log: Logging initialized @5575ms
17/08/20 17:20:26 INFO server.Server: jetty-9.2.z-SNAPSHOT
```

16.4.12 检查关系数据库中是否已有数据

```
[root@DB03 bin]# ./mysql -h192.168.37.143 -P3306 -uroot -p123456 -e "select
* from bnyw.wordcount;"
+-----+---------+-------+
| wid | word    | count |
+-----+---------+-------+
|  1  | is      |   4   |
|  2  | name    |   1   |
|  3  | 1.67    |   1   |
|  4  | 36      |   1   |
|  5  | 120     |   1   |
|  6  | lixiang |   1   |
|  7  | height  |   1   |
```

16.5 习题与思考

1. 通过 Spark SQL 创建学生表、课程表、成绩表和老师表，学生表和成绩表的数据结构分别如表 16.1 和表 16.2 所示。并导入数据，数据内容如 16.3 表和表 16.4 所示。

表 16.1

字段名	数据类型	可否为空	含义
Sno	Varchar2(3)	否	学号
Sname	Varchar2(8)	否	姓名
Ssex	Varchar2(2)	否	性别
Sbirthday	Varchar2(8)	是	生日
SClass	Varchar2(5)	是	班级

表 16.2

属性名	数据类型	可否为空	含义
Sno	Varchar2(3)	否	学号（外键）
Cname	Varchar2(5)	否	课程
Degree	Number(4,1)	是	成绩

表 16.3

Sno	Sname	Ssex	Sbirthday	SClass
20173598	张三	女	19940808	计算机一班
20173820	李四	女	19941209	计算机一班
20173743	王五	男	19931230	计算机二班

表 16.4

Sno	Cname	Degree
20173598	计算机基础	58
20173598	英语	48
20173598	数学	69
20173820	计算机基础	33
20173820	英语	50
20173820	数学	59
20173743	计算机基础	47
20173743	英语	55
20173743	数学	40

2．查询 Student 表中的所有记录的 Sname、Ssex 和 SClass 列。

3．查询有成绩不及格的所有学生。

4．查询每门课程的平均成绩。

5．按 Sno 字段查询所有学科的总成绩。

第 17 章 Spark Streaming

17.1 架构与原理

关于 Spark Streaming 架构与原理的视频讲解可扫描二维码观看。

17.1.1 Spark Streaming 中的离散流特征

Spark Streaming 是从间隙（间隙时间可通过程序设定）拉取数据源（如 Flume、KafKa 等）中的数据来进行分析和处理的。每次拉取的数据都对应一个处理单位，处理单位即数据片段，又称为离散数据流（简称为离散流）。每次处理的数据片段会被 Spark Streaming 包装成不可变的 RDD 推送到 Spark 内核进行处理，编写的 Spark Streaming 应用代码基于每次处理的单个数据片段来执行 Job，即一个数据片段对应一个 RDD，而一个 RDD 则对应一个被 Spark Streaming 封装的 Job 实例，RDD 的模板类称为离散流类即 DStream。

实际上，Spark Streaming 是针对流式处理的特征对 Spark 内核的一种 RDD 封装，以时间为分片将拉取的离散数据流封装为一系列独立的 RDD 实例，封装的离散 RDD 之间彼此独立，相互之间无任何关系（离散数据流是一种无状态的流）。

包装离散 RDD 为 Job 是 Spark Streaming 框架根据设定的间隔时间产生的，即每隔一段时间（可程序设定）产生一个 RDD 和 Job。编写的代码是 Job 的模板，Spark Streaming 框架是根据编写的 Job 模板类产生 Job 实例，并运行 Job 实例处理 RDD 中的数据。

Streaming 推送到 Spark 内核的 RDD 计算完成后可将其存储到中间件、缓存、数据库或 HDFS 分布式存储系统中。

17.1.2 Spark Streaming 的应用场景

Spark Streaming 与 Spark SQL 相同，均属于 Spark 生态圈中的子系统，是建立在 Spark 内核基础之上的应用框架。Spark Streaming 将接收到的离散数据流转换为 RDD 实例，并将其封装为 Job 应用提交给底层的 Spark 内核进行处理。离散流中的每一个 RDD 实例均对应一次 Job 的提交和执行。对于具有离散流特征的数据处理过程适合于分析在线实时数据。数据在前端系统中一旦产生则立即将其推送至 Spark Streaming 处理，并将处理后的数据反馈给前端系统。如此前端系统便可对线上产生数据实时分析和反馈处理。目前，实时分析技术在电商领域中应用广泛。

17.2 KafKa 中间件

KafKa 是由 LinkedIn 开发的一个分布式消息系统，使用 Scala 编写，由于可水平扩展和高吞吐率被广泛使用。目前开源分布式处理系统如 Cloudera、Apache Storm、Spark 趋向于支持与 KafKa 的集成。

17.2.1 KafKa 的特点

（1）低延迟、稳定可靠的数据流传送。

（2）副本机制实现了集群的容错能力，同时使用稳定的 Zookeeper 集群管理 KafKa 集群的元数据。

（3）实现了主题、分区及其队列模式（消息队列 MQ(Message Queue)）以及生产者、消费者构架模式。

（4）持久化存储消息数据流至磁盘，同时启用基于操作系统内核的 ZeroCopy 技术，提升数据传输 I/O 效率。

17.2.2 ZeroCopy 技术

1. 传统 I/O 操作技术

协议引擎（基于磁盘 I/O 或网络 I/O 的协议）→操作系统内核→用户应用层→操作系统内核→协议引擎（基于磁盘 I/O 或网络 I/O 的协议）。

2. ZeroCopy 操作技术

协议引擎（基于磁盘 I/O 或网络 I/O 的协议）→操作系统内核→协议引擎（基于磁盘 I/O 或网络 I/O 的协议）。

用户态切换至内核态需消耗系统资源，使用 ZeroCopy 消除多余的缓冲副本将提升数据传输性能。实际上 NIO 技术中 FileChannel 的 transferTo 接口借助 ZeroCopy 技术提升性能（消除了缓冲区之间的数据副本，提升了 I/O 性能）。用户应用层的 transferTo 方法调用仅发送了一个内核指令来完成操作，此操作并不涉及数据从操作系统内核流入到用户应用层，因此避免了系统内核缓冲区与用户应用缓冲区之间的相互数据复制。但如果用户应用层涉及数据操作则无法避免数据流入到用户应用层，此时必定存在内核缓冲区到用户应用缓冲区之间的数据复制过程。

17.2.3 KafKa 的通信原理

生产者组件和消费者组件均可连接到 KafKa 集群，而 KafKa 被认为是组件通信之间所使用的一种消息中间件。Topic 表示分类主题，生产者将数据主动推送到 KafKa 集群中的某个主题类别之下，消费者组件主动到 KafKa 集群中拉取数据。Topic（主题）和 partition（分区）均为 KafKa 定义的用于抽象集群数据的一种逻辑存储结构。主题和分区的概念均为针对整个集群定义的一种抽象概念，而非针对某个具体的 KafKa 节点。KafKa 在生产环境上应用的最佳实践架构是 Flume+KafKa+Spark Streaming。

17.2.4 KafKa 的内部存储结构

KafKa 内部分为很多主题,每个主题又被分为很多分区,每个分区中的数据按队列模式进行编号存储。被编号的日志数据称为此日志数据块在队列中的偏移量(offset),偏移量越大的数据块越新,即越靠近当前时间。

17.2.5 KafKa 的下载

1. 准备工作

由于 KafKa 是运行在 JVM 平台上的一款中间件,因此安装 KafKa 之前须首先安装 JDK。由于 KafKa 使用 Scala 语言并发,在运行之前需借助 Scala 的编译器来完成源码编译,因此在安装 KafKa 之前还需安装 Scala。关于 JDK 与 Scala 的安装可参考 Hadoop 环境搭建相关章节。

2. Zookeeper 的安装

若 KafKa 需以集群方式运行,则 KafKa 需借助 Zookeeper 完成元数据管理。KafKa 中间件的集群属于无中心化的集群模式,管理 KafKa 的 Zookeeper 自身需实现高可靠能力。因此管理 KafKa 集群的 Zookeeper 须以集群方式运行,在安装 KafKa 之前需构建 Zookeeper 集群。关于 Zookeeper 集群的搭建可参考 Hadoop 相关章节。

3. 下载 KafKa 安装包

最新版本的 KafKa 安装包可在官网(http://kafka.apache.org/downloads.html)下载。最新版本的安装包须与当前系统中的 Scala 版本匹配,本书采用 kafka_2.10-0.9.0.1.tgz 版本。

17.2.6 KafKa 集群搭建

1. 在所有 Slave 节点上安装 KafKa

```
tar -zxvf kafka_2.10-0.9.0.1.tgz -C /software/
```

提示:为完成 KafKa 在后台执行,可下载 slf4j-nop 组件,并将 KafKa 的后台执行依赖包复制至 kafka_2.10-0.9.0.1/libs 目录。

2. 在所有的 Slave 节点上配置系统环境变量

```
vi /etc/profile
```

配置 KafKa 的安装目录:

```
export KAFKA_HOME=/software/kafka_2.10-0.9.0.1
PATH=$PATH:$KAFKA_HOME/bin
source /etc/profile
```

3. 配置 KafKa

```
vi /software/kafka_2.10-0.9.0.1/config/server.properties
```

配置节点标识:

```
broker.id=0
```

配置 Zookeeper 集群：

```
zookeeper.connect=YunSlave01:2181,YunSlave02:2181,YunSlave03:2181
```

提示：上述配置以 Slave01 节点为例讲解配置，其他两个 Slave 节点上的 brokerId 值必须取不同值。

17.2.7 启动并使用 KafKa 集群

1. 启动 Zookeeper 集群

```
cd /software/zookeeper-3.10/bin
./zkServer.sh start
jps
cd -
```

2. 启动 KafKa 集群

```
cd /software/kafka_2.10-0.9.0.1/bin
nohup ./kafka-server-start.sh ../config/server.properties &
jps
cd -
```

3. 创建主题

```
./kafka-topics.sh --create --zookeeper YunSlave01:2181,YunSlave02:2181,YunSlave03:2181    --replication-factor 3 --partitions 1 --topic HelloKafka
```

参数解释：
--zookeeper：用于指定创建的主题信息需存储到的 ZK 集群节点列表。
--replication-factor：用于指定主题目录的副本数量，其中副本包括主题下的所有数据。
--partitions：用于指定创建的主题下的分区数量。
--topic：用于指定创建的主题名称。

4. 查看 KafKa 主题

```
./kafka-topics.sh --describe --zookeeper YunSlave01:2181,YunSlave02:2181,YunSlave03:2181 --topic HelloKafka
```

提示：连接到 ZK 集群，并从中提取 KafKa 集群中指定的主题目录的描述信息。

5. 启动 KafKa 生产者

```
./kafka-console-producer.sh --broker-list YunSlave01:9092,YunSlave02:9092,YunSlave03:9092 --topic HelloKafka
this is first test!
this is second test02!
```

提示：最后两行代表生产者需将其发送到KafKa集群的数据。此处假设KafKa集群与Zookeeper集群均安装在YunSlave的三台节点上，其中：

--broker-list：用于指定KafKa集群节点列表。

--topic：用于指定数据存储到KafKa集群的主题目录。

6. 启动KafKa消费者

```
./kafka-console-consumer.sh --zookeeper YunSlave01:2181,YunSlave02:2181,YunSlave03:2181 --from-beginning --topic HelloKafka
```

提示：上述消费者命令启动后会立即从KafKa集群中取出数据并在控制台打印。

消费者通过ZK集群中提供的KafKa节点信息连接到KafKa集群，其中：

--topic：用于指定从KafKa集群提取数据的主题目录。

📢注意：

（1）kafka-console-producer.sh和kafka-console-consumer.sh均为用于测试KafKa的控制台工具。

（2）由于KafKa是一个消息中间件，具有数据存储和缓冲的作用，因此生产者和消费者之间可任意启动，即生产者和消费者之间并没有直接构成通信。

17.2.8 停止KafKa集群

停止KafKa集群需分别到每个KafKa节点执行以下命令来停止每个节点上的Kafka进程。

```
cd /software/kafka_2.10-0.9.0.1/bin
./kafka-server-stop.sh
```

停止之后使用jps命令查看，发现Kafka进程消失：

```
jps
```

17.2.9 KafKa集成Flume

（1）如需要在KafKa中集成Flume，可先搭建Flume集群。

（2）在KafKa集群中集成Flume须使用KafKa的Flume插件，此插件在对应的KafKa版本通常都有实现。Flume作为KafKa的前端生产者对接到KafKa集群中，在完成对接时只须配置Flume的Sink组件即可。

（3）KafKa集成Flume集群的Sink组件配置示例：

```
producer.sources.s.type = spooldir
producer.sources.s.spoolDir = /home/hadoop/dir/logdfs
producer.sinks.r.type = org.apache.flume.plugins.KafkaSink
producer.sinks.r.metadata.broker.list=YunSlave01:9092,YunSlave02:9092,YunSlave03:9092
producer.sinks.r.partition.key=0
```

```
producer.sinks.r.partitioner.class=org.apache.flume.plugins.
SinglePartition
producer.sinks.r.serializer.class=kafka.serializer.StringEncoder
producer.sinks.r.request.required.acks=0
producer.sinks.r.max.message.size=1000000
producer.sinks.r.producer.type=sync
producer.sinks.r.custom.encoding=UTF-8
producer.sinks.r.custom.topic.name=HelloKafka
```

提示：
① 关于 Flume 的相关安装和使用请自行参考 Hadoop 相关章节。
② 在集成 Flume 的 KafKa 集群中须同时启动 Flume 集群和 KafKa 集群。
③ 在大规模的数据处理和分析场景下，使用 Flume+KafKa 完成数据收集是最佳实践。

17.3　Socket 事件流操作

17.3.1　netcat 网络 Socket 控制台工具

netcat 是一款基于 Socket 的控制台网络通信工具（简称 nc），它同时具备绑定服务端和启动客户端的能力。服务端负责将控制台输入的数据通过 Socket 传送到连接此服务端的客户端控制台并将其显示出来，其通信原理如图 17.1 所示。

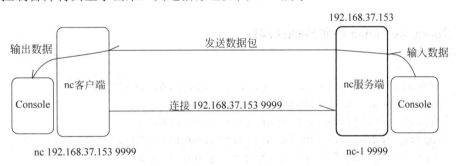

图　17.1

通常 Linux 系统安装完毕 nc 命令就已存在。需要在安装 Linux 系统时选中安装网络测试工具。

（1）启动 nc 服务端，并绑定当前主机的 9999 端口。

```
nc -l 9999
```

（2）启动 nc 客户端，并连接到端口为 9999 的 nc 服务端。

```
nc 192.168.37.153 9999
```

提示：如 nc 客户端主动断开连接，则 nc 服务端进程随之结束。在后续 Spark Streaming 测试中，其 Spark Streaming 端充当 nc 客户端的角色。

（3）运行 Spark 官方提供的 Spark Streaming 样例。

关闭 nc 客户端并启动 nc 服务端，同时在控制台输入如下命令运行样例测试。Spark Streaming 的运行过程如下：

```
./run-example org.apache.spark.examples.streaming.NetworkWordCount localhost 9999
```

运行命令解释：

上述命令表示 Spark Streaming 将通过主动连接 localhost 9999 主机服务进程来读取（拉取）需处理的数据。若无法连接给定的主机进程服务或已连接但无法读取到数据，将输出空的处理结果。未读取到数据是否需要启动 Job 来完成空处理取决于 Spark Streaming 程序。

Spark Streaming 与 Strom 相同，均为通过一定的时间间隔到给定的数据源。通常与之配合的数据源是 KafKa 集群，其数据来自于如 Flume、电商应用、终端设备等其他终端，KafKa 从这些终端主动拉取数据，然后执行统计分析。Spark Streaming 拉取一次数据便统计一次，这种统计模式取决于应用程序本身。

Spark Streaming 与 Strom 相同，一旦启动便不会停止，这是流式处理所具备的一般特征。

17.3.2 基于本地的 Spark Streaming 流式数据分析示例

1. 场景描述

通过间隙扫描 Socket 端口源数据，并将拉取到的数据放置于 Spark Streaming 控制台打印出来。

2. Spark Streaming 扫描 Socket 端口

源码实现如下：

```java
package com.bnyw.bigdata.spark.streaming.local;
import org.apache.spark.SparkConf;
import org.apache.spark.api.java.JavaRDD;
import org.apache.spark.api.java.function.VoidFunction;
import org.apache.spark.streaming.Durations;
import org.apache.spark.streaming.api.java.JavaReceiverInputDStream;
import org.apache.spark.streaming.api.java.JavaStreamingContext;
@SuppressWarnings("unchecked")
public class SocketToPrint {
    public static void main(String[] args) throws InterruptedException {
        SparkConf conf=new SparkConf();
        conf.setMaster("local[2]");
        conf.setAppName("WordCountOnLine");
        JavaStreamingContext jsc=new JavaStreamingContext(conf, Durations.seconds(5));
        JavaReceiverInputDStream<String> JobLines=jsc.socketTextStream("CloudDeskTop",9999);
        JobLines.foreachRDD(new VoidFunction<JavaRDD<String>>(){
            @Override
```

```java
            public void call(JavaRDD<String> javaRDD) throws Exception {
                long size=javaRDD.count();
                System.out.println("-----foreachRDD-call-collection-
                size:"+size+"-------");
                javaRDD.foreach(new VoidFunction<String>(){
                    @Override
                    public void call(String line) throws Exception {
                        System.out.println(line);
                    }
                });
            }
        });
        jsc.start();
        System.out.println("--------already start---------");
        jsc.awaitTermination();
        System.out.println("--------already await---------");
        
        jsc.close();
        System.out.println("--------already close----------");
    }
}
```

DStream、RDD 与 Partition 的基础单位均为行记录 Row，但 DStream 中可将多条记录包装成一个个 RDD，而在 RDD 中也可将多条记录包装成 Partition。从 RDD 级别以下的操作都属于 Spark Core 的 Job 操作，Job 操作均涉及集群节点的参与。而在 RDD 之外的所有操作都是客户端 Driver 级别的操作，不会涉及 Job 参与。javaRDD.count()返回的是 RDD 集合中行记录 row 的数量（类型为 long），javaRDD.getNumPartitions()返回的是 RDD 集合中包装的分区 partition 的数量（类型为 int）。由于 partition 是 row 的集合，即一个 partition 中包含多行记录 row，因此 row 的数量为 0 则 partition 的数量必定为 0。

3. 基于本地 Spark Streaming 的 Socket 扫描测试

（1）在 CloudDeskTop 节点上开启 Socket 监听服务。

```
[hadoop@CloudDeskTop software]$ nc -l 9999
```

（2）在 Eclipse 中运行上述代码。一旦代码开始运行，控制台将持续输出数据，但是均为输出。

```
System.out.println("-----foreachRDD-call-collection-size:0-------");
```

（3）返回至 CloudDeskTop 节点的控制台输入数据。

```
liyongfu
```

再次查看 Eclipse 的控制台打印，此时输出：

```
System.out.println("-----foreachRDD-call-collection-size:1-------");
```

```
liyongfu
```

（4）若返回至 CloudDeskTop 节点的控制台上，快速输入两行数据：

```
liyongfu
ligang
```

则 Eclipse 的控制台打印此时输出：

```
-----foreachRDD-call-collection-size:2-------
liyongfu
17/08/19 18:09:00 INFO Executor: Finished task 0.0 in stage 4.0 (TID 5).
925 bytes result sent to driver
17/08/19 18:09:00 INFO TaskSetManager: Starting task 1.0 in stage 4.0 (TID
6, localhost, executor driver, partition 1, ANY, 6261 bytes)
17/08/19 18:09:00 INFO Executor: Running task 1.0 in stage 4.0 (TID 6)
17/08/19 18:09:00 INFO TaskSetManager: Finished task 0.0 in stage 4.0 (TID
5) in 62 ms on localhost (executor driver) (1/2)
17/08/19 18:09:00 INFO BlockManager: Found block input-0-1503137339200
locally ligang
```

4. 应用代码分析

JobLines 是 DStream 类型，属于离散流集合。它的数据来自 Socket 端口的扫描获取，一个离散流的集合对象（如 JobLines）对应于一个特定的时间片段（如这里的 5 秒）收集的数据记录行。该数据记录以 RDD 为单位进行包装，并将其放置于离散流集合中，离散流集合对象中的每一个 RDD 集合均包含若干条数据记录。即便 Streaming 框架未从数据源获得任何数据，离散流集合对象 JobLines 中也至少存在一个尺寸为空的 RDD 集合对象（代码中的 javaRDD），因此数据源没有数据的情况下至少会打印如下输出：

```
-----foreachRDD-call-collection-size:0-------
```

若在特定时间片段上获取的数据记录数量较少，则离散流集合对象中可能仅存在一个 RDD 集合。一个离散流集合对象中存在多少个 RDD 集合对象取决于时间片段的长短和时间片段内的数据记录的数量大小，该过程由 Streaming 框架自动计算。

Spark Streaming 同样属于 Spark Core 的上层应用，而 Spark Core 基于 RDD 执行 Job 操作（一个 RDD 对应一个 Job）。因此 Spark Streaming 必须将离散流集合对象中的多条记录包装成一个个的 RDD 对象，以传递给底层的 Spark Core 进行处理。

JobLines.foreachRDD 及该代码逻辑之外的语句均属建立在 RDD 之上的 Spark Streaming 框架中的离散流操作语句，该操作语句相对 Spark Core 属于其 SparkCore 的客户端 Driver 执行的代码。因此该部分代码在本地运行模式或集群运行模式下，其标准输出均打印到控制台终端显示出来。而 javaRDD.foreach 及其包装体中的代码语句均基于 RDD 本身进行操作的 Spark Core 的代码，Spark Core 操作 RDD 集合时，将其包装成 Job 作业提交至 Spark 集群中运行（即在每一个 Worker 节点上运行这些代码）。因此基于 RDD 操作中的标准输出语句不会显示到终端控制台。

17.3.3 基于集群的 Spark Streaming 流式数据分析示例

（1）场景描述：通过间隙扫描 Socket 端口源的数据，将拉取到的数据存储到 HDFS 分布式文件系统。

（2）Spark Streaming 扫描 Socket 端口。源码实现如下：

```java
package com.bnyw.bigdata.spark.streaming.cluster;
import java.io.IOException;
import java.net.URI;
import java.net.URISyntaxException;
import org.apache.hadoop.conf.Configuration;
import org.apache.hadoop.fs.FileSystem;
import org.apache.hadoop.fs.Path;
import org.apache.spark.SparkConf;
import org.apache.spark.api.java.JavaRDD;
import org.apache.spark.api.java.function.VoidFunction;
import org.apache.spark.streaming.Durations;
import org.apache.spark.streaming.api.java.JavaReceiverInputDStream;
import org.apache.spark.streaming.api.java.JavaStreamingContext;
@SuppressWarnings("unchecked")
public class SocketToHDFS {
    private static FileSystem fs;
    static{
        Configuration conf=new Configuration();
        try {
            fs=FileSystem.get(new URI("hdfs://ns1/"), conf, "hadoop");
        } catch (IOException | InterruptedException | URISyntaxException e) {
            e.printStackTrace();
        }
    }
    public static void main(String[] args) throws InterruptedException, IOException {
        Path output=new Path("/spark/streaming/output");
        if(fs.exists(output))fs.delete(output, true);
        SparkConf conf=new SparkConf();
        conf.setAppName("WordCountOnLine");
        JavaStreamingContext jsc=new JavaStreamingContext(conf, Durations.seconds(5));
        JavaReceiverInputDStream<String> JobLines=jsc.socketTextStream("CloudDeskTop",9999);
        JobLines.foreachRDD(new VoidFunction<JavaRDD<String>>(){
            @Override
            public void call(JavaRDD<String> javaRDD) throws Exception {
                long size=javaRDD.count();
                System.out.println("-----foreachRDD-call-collection-size:
```

```
                "+size+"-------");
                if(0!=size)javaRDD.saveAsTextFile("/spark/streaming/
                output");
            }
        });
        jsc.start();
        jsc.awaitTermination();
        jsc.close();
    }
}
```

(3)切换至工程的 bin 目录,将 com 文件夹打包至工程目录下的 dist 目录。

```
cd /project/SparkStreaming-JAVA/bin
jar -cvfe /project/SparkStreaming-JAVA/dist/SocketToHDFS.jar
com.bnyw.bigdata.spark.streaming.cluster.SocketToHDFS com/
```

(4)在集群模式下运行 Spark Streaming 的 Socket 测试。
① 启动 Zookeeper 集群。

```
cd zookeeper-3.4.10/bin/
./zkServer.sh start
jps
cd -
```

② 启动 HDFS 集群。

```
cd /software/hadoop-2.7.3/bin
./start-dfs.sh
Starting namenodes on [master01 master02]
master01: starting namenode, logging to /software/hadoop-2.7.3/
logs/hadoop-hadoop-namenode-master01.out
master02: ssh: connect to host master02 port 22: No route to host
slave01: starting datanode, logging to /software/hadoop-2.7.3/logs/
hadoop-hadoop-datanode-slave01.out
slave02: starting datanode, logging to /software/hadoop-2.7.3/logs/
hadoop-hadoop-datanode-slave02.out
slave03: starting datanode, logging to /software/hadoop-2.7.3/logs/
hadoop-hadoop-datanode-slave03.out
Starting journal nodes [slave01 slave02 slave03]
slave02: starting journalnode, logging to /software/hadoop-2.7.3/logs/
hadoop-hadoop-journalnode-slave02.out
slave01: starting journalnode, logging to /software/hadoop-2.7.3/logs/
hadoop-hadoop-journalnode-slave01.out
slave03: starting journalnode, logging to /software/hadoop-2.7.3/logs/
hadoop-hadoop-journalnode-slave03.out
```

③ 启动 Spark 集群。

```
cd spark-2.1.1/sbin/
./start-master.sh
starting org.apache.spark.deploy.master.Master, logging to /
software/spark-2.1.1/logs/spark-hadoop-org.apache.spark.deploy.master.
Master-1-master01.out
[hadoop@master01 sbin]$ ./start-slaves.sh
slave01: starting org.apache.spark.deploy.worker.Worker, logging to /
software/spark-2.1.1/logs/spark-hadoop-org.apache.spark.deploy.worker.
Worker-1-slave01. out
slave03: starting org.apache.spark.deploy.worker.Worker, logging to /
software/spark-2.1.1/logs/spark-hadoop-org.apache.spark.deploy.worker.
Worker-1-slave03.out
slave02: starting org.apache.spark.deploy.worker.Worker, logging to /
software/spark-2.1.1/logs/spark-hadoop-org.apache.spark.deploy.worker.
Worker-1-slave02.out
slave01: failed to launch: nice -n 0 /software/spark-2.1.1/bin/spark-class
org.apache.spark.deploy.worker.Worker --webui-port 8081
spark://master01:7077
slave01: full log in /software/spark-2.1.1/logs/spark-hadoop-org.
apache.spark.deploy.worker.Worker-1-slave01.out
slave03: failed to launch: nice -n 0 /software/spark-2.1.1/bin/spark-class
org.apache.spark.deploy.worker.Worker --webui-port 8081
spark://master01:7077
slave03: full log in /software/spark-2.1.1/logs/spark-hadoop-org.apache.
spark.deploy.worker.Worker-1-slave03.out
slave02: failed to launch: nice -n 0 /software/spark-2.1.1/bin/spark-class
org.apache.spark.deploy.worker.Worker --webui-port 8081 spark://master01:
7077
slave02: full log in /software/spark-2.1.1/logs/spark-hadoop-org.
apache.spark.deploy.worker.Worker-1-slave02.out
```

（5）提交 Spark Streaming 应用到 Spark 集群。

① 检查输出目录是否存在。

```
[hadoop@CloudDeskTop dist]$ hdfs dfs -ls /spark
    Found 1 items
drwxr-xr-x   - hadoop supergroup          0 2017-08-17 11:20 /spark/input
```

② 启动 nc 监听服务。

```
[hadoop@CloudDeskTop dist]$ nc -l 9999
```

③ 再开启一个终端提交 Streaming 应用进行测试（此处无须指定读取数据的并行度）。

```
[hadoop@CloudDeskTop bin]$ pwd
    /software/spark-2.1.1/bin
```

```
[hadoop@CloudDeskTop bin]$ ./spark-submit --master spark://master01:7077
--class com.bnyw.bigdata.spark.streaming.cluster.SocketToHDFS /project/
SparkStreaming-JAVA /dist/SocketToHDFS.jar
```

④ 待控制台周期性地出现如下提示时,在 nc 服务端的控制台下输入数据(如 my name is liyongfu):

```
-----foreachRDD-call-collection-size:0-------
```

⑤ 再次检查输出目录是否已存在:

```
[hadoop@CloudDeskTop dist]$ hdfs dfs -ls /spark
Found 2 items
drwxr-xr-x   - hadoop supergroup    0 2017-08-17 11:20 /spark/input
drwxr-xr-x   - hadoop supergroup    0 2017-08-19 17:47 /spark/streaming
[hadoop@CloudDeskTop dist]$ hdfs dfs -ls /spark/streaming
Found 1 items
drwxr-xr-x   - hadoop supergroup    0 2017-08-19 17:48 /spark/streaming/output
[hadoop@CloudDeskTop dist]$ hdfs dfs -ls /spark/streaming/output
Found 3 items
-rw-r--r--   3 hadoop supergroup    0 2017-08-19 17:48 /spark/streaming/output/_SUCCESS
drwxr-xr-x   - hadoop supergroup    0 2017-08-19 17:48 /spark/streaming/output/_temporary
-rw-r--r--   3 hadoop supergroup   20 2017-08-19 17:48 /spark/streaming/output/part-00000
[hadoop@CloudDeskTop dist]$ hdfs dfs -cat /spark/streaming/output/part-00000
my name is liyongfu
```

Spark Streaming 并不适合使用 javaRDD.saveAsTextFile 方式存储流式数据到 HDFS 集群,因为在特定的时间片段下产生的离散流数据记录量相对较小,由于离散流中分为多个 RDD 集合,每一个 RDD 集合被推送到 Spark Core 上执行 saveAsTextFile 方法时都将产生一个新的 Job,而每一个新的 Job 都会在对应的输出目录(如此处的 output 目录)下创建类似于 part-0000X 的文件。在 DStream 的 foreachRDD 迭代中,前面的 RDD 在执行 saveAsTextFile 方法时产生的 part-0000X 文件会被后面的 RDD 覆盖,因此在 Spark Streaming 应用执行完成之后仅保留最后一个 RDD 对象的数据。为实现离散流分析后的数据不被覆盖,可使用文件流的追加方式来实现。

17.3.4 基于集群模式下的集群文件 I/O 流分析示例

上述集群操作显示,不同的 RDD 均创建相同的文件以产生文件覆盖。为收集所有数据不可使用 saveAsTextFile 方法,应在 Task 中使用 FileSystem 的 I/O 流进行操作。

1. 基于文件 I/O 流的源码开发

```java
package com.bnyw.bigdata.spark.streaming.cluster;
import java.io.IOException;
import java.net.URI;
import java.net.URISyntaxException;
import java.util.Iterator;
import org.apache.hadoop.conf.Configuration;
import org.apache.hadoop.fs.FSDataOutputStream;
import org.apache.hadoop.fs.FileSystem;
import org.apache.hadoop.fs.Path;
import org.apache.spark.SparkConf;
import org.apache.spark.api.java.JavaRDD;
import org.apache.spark.api.java.function.VoidFunction;
import org.apache.spark.streaming.Durations;
import org.apache.spark.streaming.api.java.JavaReceiverInputDStream;
import org.apache.spark.streaming.api.java.JavaStreamingContext;
@SuppressWarnings("unchecked")
public class SocketToHDFS2 {
    private static int count;
private static FileSystem fs;
private static final String outPath="/spark/streaming/output";
static{
    Configuration conf=new Configuration();
    try {
        fs=FileSystem.get(new URI("hdfs://ns1/"), conf, "hadoop");
    } catch (IOException | InterruptedException | URISyntaxException e) {
        e.printStackTrace();
    }
}
private static void saveLine(Iterator<String> its) throws IOException{
    Path outFile=new Path(outPath+"/part-"+count++);
    FSDataOutputStream dos=fs.create(outFile, true);
    try{
        while(its.hasNext()){
            String line=its.next();
            dos.writeUTF(line+"\n");
        }
        dos.flush();
    }finally{
        dos.close();
    }
}
public static void main(String[] args) throws InterruptedException, IOException {
```

```java
            Path output=new Path(outPath);
            if(fs.exists(output))fs.delete(output, true);
            boolean flag=fs.mkdirs(output);
            if(!flag)return;
            SparkConf conf=new SparkConf();
            conf.setAppName("WordCountOnLine");
            JavaStreamingContext jsc=new JavaStreamingContext(conf,Durations.seconds(5));
            JavaReceiverInputDStream<String> JobLines=jsc.socketTextStream("CloudDeskTop",9999);
            JobLines.foreachRDD(new VoidFunction<JavaRDD<String>>(){
                @Override
                public void call(JavaRDD<String> javaRDD) throws Exception {
                    long size=javaRDD.count();
                    System.out.println("-----foreachRDD-call-collection-size:"+size+"-------");
                    if(0==size)return;
                    javaRDD.foreachPartition(new VoidFunction<Iterator<String>>(){
                        private int count;
                        @Override
                        public void call(Iterator<String> its) throws Exception {
                            saveLine(its);
                        }
                    });
                }
            });
            jsc.start();
            jsc.awaitTermination();
            jsc.close();
        }
    }
```

2. 将编写的 Spark Streaming 代码打包至工程目录下的 dist 目录

```
[hadoop@CloudDeskTop bin]$ pwd
    /project/SparkStreaming-JAVA/bin
[hadoop@CloudDeskTop bin]$ jar -cvfe ../dist/SocketToHDFS2.jar com.bnyw.bigdata.spark.streaming.cluster.SocketToHDFS2 com/
```

3. 启动 nc 服务器

```
[hadoop@CloudDeskTop bin]$ nc -l 9999
```

4. 启动 Streaming 应用并测试

```
[hadoop@CloudDeskTop bin]$ pwd
```

```
            /software/spark-2.1.1/bin
[hadoop@CloudDeskTop bin]$ ./spark-submit --master spark://master01:7077
--class com.bnyw.bigdata.spark.streaming.cluster.SocketToHDFS2
/project/SparkStreaming-JAVA/dist/SocketToHDFS2.jar
```

5. 在 nc 服务端控制台粘贴/etc/sysconfig/network-scripts/ifcfg–eth0 文件中的内容并查看 HDFS 集群中是否写入数据

```
[hadoop@CloudDeskTop software]$ hdfs dfs -ls /spark/streaming/output
Found 2 items
-rw-r--r--   3 hadoop supergroup     173 2017-08-20 14:12 /spark/streaming/
output/part-0
-rw-r--r--   3 hadoop supergroup     15 2017-08-20 14:11
/spark/streaming/output/part-1
 [hadoop@CloudDeskTop software]$ hdfs dfs -cat /spark/streaming/output/
part-0
    DEVICE=eth0
    TYPE=Ethernet
    ONBOOT=yes
    NM_CONTROLLED=yes
    BOOTPROTO=static
    IPADDR=192.168.37.154
    NETMASK=255.255.255.0
    GATEWAY=192.168.37.2
    DNS1=192.168.37.2
[hadoop@CloudDeskTop software]$ hdfs dfs -cat /spark/streaming/
output/part-1
    DNS2=8.8.8.8
```

17.4　KafKa 事件流操作

17.4.1　基于 Receiver 模式的 KafKa 集成

（1）创建 Spark Streaming 上下文。

```
JavaStreamingContext jsc=JavaStreamingContext.getOrCreate
(checkpointDirectory,factory);
```

（2）消费组连接 KafKa 集群中对应主题的线程数量（连接某一主题所需线程数量，即为主题与线程数量的映射表）。一般情况下，主题中的一个分区（Partition）对应一个线程。

```
(Thread)Map<String,Integer> topicThreadNum=new HashMap<String,Integer>();
```

（3）此处指定两个拉取线程，实际操作 Streaming 需要至少启动 3 个线程，因为还需要另一个处理线程。该处指定的 Key 是 topic 主题名称，该主题名称需要在 KafKa 集群中事先创建完毕，代码如下：

```
topicThreadNum.put("HelloKafKa", 2);
```

（4）创建基于 KafKa 中间件的输出流。

```
JavaPairReceiverInputDStream<String,String> lineList=KafkaUtils.
createStream(jsc,"YunSlave01:2181,YunSlave02:2181,YunSlave03:2181",
"FirstConsumerGroup",topicThreadNum);
```

上述方法调用返回一个二元素元组（其中第一个元素代表 Key，此 Key 一般为空或无意义的数据，第二个元素代表 value，此 value 表示为行记录数据）。createStream 方法参数解释如下：

第一个参数：表示 StreamingContext。

第二个参数：表示 Zookeeper 集群节点列表，说明 Spark Streaming 通过连接 Zookeeper 集群获知 KafKa 集群信息，进而连接 KafKa 集群拉取数据。

第三个参数：表示 Spark Streaming 作为 KafKa 集群的消费组 ID 名称，该 ID 自定义并保持唯一。

第四个参数：表示 Spark Streaming 连接 KafKa 集群中指定主题名称的线程数量。

第五个参数：表示 Streaming 对读取到的数据的存储级别（默认值为 StorageLevel.MEMORY_AND_DISK_SER_2，表示存储至内存和磁盘）。

（5）得到 DStream 对象后其后续操作与 Socket 操作流程相同。

```
JavaDStream<String> wordList=lineList.flatMap(new FlatMapFunction<
Tuple2<String,String>,String>(){
    @Override
    public Iterable<String> call(Tuple2<String,String> tuple) throws
    Exception {
        String line=tuple._2;
        return Arrays.asList(line.split(" "));
    }
});
```

（6）启动集群。

① 启动 Zookeeper 集群：

```
cd /software/zookeeper-3.4.10/bin
./zkServer.sh start
jps
cd -
```

② 启动 KafKa 集群：

```
cd /software/kafka_2.10-0.9.0.1/bin
nohup ./kafka-server-start.sh ../config/server.properties &
```

③ 创建主题：

```
--topic参数指定Streaming客户端连接的主题名称
./kafka-topics.sh --create --zookeeper
YunSlave01:2181,YunSlave02:2181,YunSlave03:2181
--replication-factor 3
--partitions 1
--topic HelloKafka
```

④ 创建用于产生测试数据的 producer 并在 Shell 控制台输入两行测试数据，--topic 参数指定 Streaming 客户端连接的主题名称：

```
./kafka-console-producer.sh --broker-list
YunSlave01:9092,YunSlave02:9092,YunSlave03:9092 --topic HelloKafka
this is first test!
this is second test02!
```

⑤ 启动 HDFS 集群：

```
start-dfs.sh
Starting namenodes on [master01 master02]
master01: starting namenode, logging to /software/
hadoop-2.7.3/logs/hadoop-hadoop-namenode-master01.out
master02: ssh: connect to host master02 port 22: No route to host
slave02: starting datanode, logging to /software/hadoop-2.7.3/logs/
hadoop-hadoop-datanode-slave02.out
slave03: starting datanode, logging to /software/hadoop-2.7.3/logs/
hadoop-hadoop-datanode-slave03.out
slave01: starting datanode, logging to /software/hadoop-2.7.3/
logs/hadoop-hadoop-datanode-slave01.out
Starting journal nodes [slave01 slave02 slave03]
slave01: starting journalnode, logging to
/software/hadoop-2.7.3/logs/hadoop-hadoop-journalnode-slave01.out
slave02: starting journalnode, logging to
/software/hadoop-2.7.3/logs/hadoop-hadoop-journalnode-slave02.out
slave03: starting journalnode, logging to
/software/hadoop-2.7.3/logs/hadoop-hadoop-journalnode-slave03.out
```

⑥ 启动 Spark 集群：

```
start-master.sh
starting org.apache.spark.deploy.master.Master, logging to /software/
spark-2.1.1/logs/spark-hadoop-org.apache.spark.deploy.master.Master-1-
master01.out
[hadoop@master01 sbin]$ ./start-slaves.sh
slave01: starting org.apache.spark.deploy.worker.Worker, logging to /
software/spark-2.1.1/logs/spark-hadoop-org.apache.spark.deploy.worker.
Worker-1-slave01.out
```

```
slave03: starting org.apache.spark.deploy.worker.Worker, logging to /
software/spark-2.1.1/logs/spark-hadoop-org.apache.spark.deploy.worker.
Worker-1-slave03.out
slave02: starting org.apache.spark.deploy.worker.Worker, logging to /
software/spark-2.1.1/logs/spark-hadoop-org.apache.spark.deploy.worker.
Worker-1-slave02.out
```

⑦ 运行 Spark-Streaming 程序并观察其控制台的打印输出。

17.4.2 基于 Direct 模式的 KafKa 集成

KatKa using Receiver 模式下需借助于 ZK 集群的连接来进行操作,由于 Streaming 中的多个线程在并发访问 ZK 集群时可能导致 ZK 集群中保存的同一个状态(offset 偏移量)被多个线程同时读取,将导致 KafKa 集群中的数据被重复处理。同时可能因状态切换不一致导致某些数据未经处理而丢失(此处可能需开启日志复写模式(WAL))。但如使用 Direct 模式,可使 Streaming 作为消费者直接操作 KafKa 集群,以保证同一个 KafKa 集群中的数据仅被处理一次,保证事务一致性(同时无须开启 WAL 模式)。

Direct 模式下,Streaming 将使用 KafKa 原生的 API 直接操作 KafKa 集群(此时不再借助于 ZK 集群)。Direct 模式下属于本地数据操作模式(即 KafKa 集群与 Streaming 存储数据的节点处理数据的节点相同)。此时 Streaming 中 RDD 的 partition 与 KafKa 集群中主题 Topic 中的 partition 一致,如此 Topic 中的 partition 相当于 HDFS 集群中的一个 Block。

实际上,Direct 模式即将 Spark 构建于以 KafKa 集群作为底层文件系统的基础之上进行数据处理,Streaming 保证事务的一致性的原因在于 Streaming 启用 checkpoint 机制(若 Streaming 程序版本升级则可手动指定读取 checkpoint 文件实现恢复)持久化存储 offset 偏移量,直接将底层 KafKa 集群读取到的每一条消息均包装为一个转换(transfmations),当后续延迟处理数据时启动 action 操作。

1. 创建 Spark Streaming 上下文

```
JavaStreamingContext jsc=JavaStreamingContext.getOrCreate
(checkpointDirectory,factory);
```

2. 创建 KafKa 元数据参数字典

```
Map<String,String> kafkaParam=new HashMap<String,String>();
kafkaParam.put("metadata.broker.list","YunSlave01:9092,YunSlave02:9092,
YunSlave03:9092");
```

3. 创建主题列表

```
Set<String> topics=new HashSet<String>();
topics.add("SparkStreamingDirected");
```

4. 创建直接基于底层 KafKa 集群文件系统的 Spark Streaming 输入流

```
JavaPairInputDStream<String,String>
```

```
lineList=KafkaUtils.createDirectStream(jsc, String.class,
String.class, StringDecoder.class, StringDecoder.class, kafkaParam,
topics);
```

5. 获取到 DStream 对象

其后续的操作与 Socket 模式相同。

```
JavaDStream<String> wordList=lineList.flatMap(new FlatMapFunction<
Tuple2<String,String>,String>(){
    @Override
    public Iterable<String> call(Tuple2<String,String> tuple) throws
    Exception {
        String line=tuple._2;
        return Arrays.asList(line.split(" "));
    }
});
```

6. 启动集群

1）启动 Zookeeper 集群

```
cd /software/zookeeper-3.4.10/bin
./zkServer.sh start
jps
cd -
```

2）启动 KafKa 集群

```
cd /software/kafka_2.10-0.9.0.1/bin
nohup ./kafka-server-start.sh ../config/server.properties &
```

3）创建消息主题

```
--topic参数指定Streaming客户端连接的主题名称
./kafka-topics.sh --create --zookeeper
YunSlave01:2181,YunSlave02:2181,YunSlave03:2181
--replication-factor 3
--partitions 1
--topic SparkStreamingDirected
```

4）创建用于产生测试数据的 producer 并在 shell 控制台输入两行测试数据

--topic 参数指定 Streaming 客户端连接的主题名称。

```
./kafka-console-producer.sh --broker-list
YunSlave01:9092,YunSlave02:9092,YunSlave03:9092 --topic
SparkStreamingDirected
this is first test!
this is second test02!
```

5)启动 HDFS 集群

```
start-dfs.sh
Starting namenodes on [master01 master02]
master01: starting namenode, logging to /software/hadoop-2.7.3/logs/
hadoop-hadoop-namenode-master01.out
master02: ssh: connect to host master02 port 22: No route to host
slave02: starting datanode, logging to /software/hadoop-2.7.3/logs/
hadoop-hadoop-datanode-slave02.out
slave03: starting datanode, logging to /software/hadoop-2.7.3/logs/
hadoop-hadoop-datanode-slave03.out
slave01: starting datanode, logging to /software/hadoop-2.7.3/logs/
hadoop-hadoop-datanode-slave01.out
Starting journal nodes [slave01 slave02 slave03]
slave01: starting journalnode, logging to
/software/hadoop-2.7.3/logs/hadoop-hadoop-journalnode-slave01.out
slave02: starting journalnode, logging to
/software/hadoop-2.7.3/logs/hadoop-hadoop-journalnode-slave02.out
slave03: starting journalnode, logging to
/software/hadoop-2.7.3/logs/hadoop-hadoop-journalnode-slave03.out
```

6)启动 Spark 集群

```
start-master.sh
starting org.apache.spark.deploy.master.Master, logging to /software/
spark-2.1.1/logs/spark-hadoop-org.apache.spark.deploy.master.Master-1-
master01.out
[hadoop@master01 sbin]$ ./start-slaves.sh
slave01: starting org.apache.spark.deploy.worker.Worker, logging to
/software/spark-2.1.1/logs/spark-hadoop-org.apache.spark.deploy.worker.
Worker-1-slave01.out
slave03: starting org.apache.spark.deploy.worker.Worker, logging to
/software/spark-2.1.1/logs/spark-hadoop-org.apache.spark.deploy.worker.
Worker-1-slave03.out
slave02: starting org.apache.spark.deploy.worker.Worker, logging to
/software/spark-2.1.1/logs/spark-hadoop-org.apache.spark.deploy.worker.
Worker-1-slave02.out
```

7)运行 Spark-Streaming 程序并观察其控制台的打印输出

17.5 I/O 文件事件流操作

Spark Streaming 可直接扫描给定目录下的数据,然后将数据读入 Spark Streaming 的离散流中,将离散流包装为 RDD 并封装为 Job 应用提交到 Spark 内核进行分析和处理。

17.5.1 基于路径扫描的 Spark Streaming

源码如下：

```java
package com.bnyw.bigdata.spark.streaming.cluster;
import java.sql.Connection;
import java.sql.Date;
import java.sql.DriverManager;
import java.sql.PreparedStatement;
import java.sql.SQLException;
import java.util.Iterator;
import org.apache.spark.SparkConf;
import org.apache.spark.api.java.JavaRDD;
import org.apache.spark.api.java.function.Function;
import org.apache.spark.api.java.function.VoidFunction;
import org.apache.spark.streaming.Durations;
import org.apache.spark.streaming.api.java.JavaDStream;
import org.apache.spark.streaming.api.java.JavaStreamingContext;
import scala.Tuple4;
public class HDFSToDB {
    //批量计数器
    private static int count;
    //数据库连接对象
    private static Connection conn;
    //数据库操作语句对象
    private static PreparedStatement pstat;
    static{
        try {
            Class.forName("com.mysql.jdbc.Driver");
            conn=DriverManager.getConnection("jdbc:mysql://
            192.168.37.143:
            3306/bnyw?characterEncoding=utf8", "root", "123456");
pstat=conn.prepareStatement("insert into tuser(uno,uname,height,birthday) values(?,?,?,?)");
        } catch (Exception e) {
            e.printStackTrace();
        }
    }
    //批量存储RDD分区中的数据
    private static void batchSave(Iterator<Tuple4<Integer, String, Double, Date>>
    partition,boolean isOver) throws SQLException{
        if(isOver){
```

```java
                while(partition.hasNext()){
                    Tuple4<Integer, String, Double, Date> row=partition.
                    next();
                    pstat.setInt(1, row._1());
                    pstat.setString(2, row._2());
                    pstat.setDouble(3, row._3());
                    pstat.setDate(4, row._4());
                    pstat.addBatch();
                }
                pstat.executeBatch();
                pstat.clearBatch();
                pstat.clearParameters();
            }else{
                while(partition.hasNext()){
                    Tuple4<Integer, String, Double, Date> row=partition.
                    next();
                    pstat.setInt(1, row._1());
                    pstat.setString(2, row._2());
                    pstat.setDouble(3, row._3());
                    pstat.setDate(4, row._4());
                    pstat.addBatch();
                }
                if(++count%100==0){
                    pstat.executeBatch();
                    pstat.clearBatch();
                    pstat.clearParameters();
                    count=0;
                }
            }
        }
}
//存储RDD集合中的数据
private static void saveToDB(JavaRDD<Tuple4<Integer, String,
Double, Date>> javaRDD) throws Exception{
    try{
        final int partitionNum=javaRDD.getNumPartitions();
        javaRDD.foreachPartition(new VoidFunction<Iterator
        <Tuple4<Integer,String,Double,Date>>>(){
            private int count;
            @Override
            public void call(Iterator<Tuple4<Integer, String,
            Double, Date>> its) throws Exception {
                if(++count>=partitionNum){
                    batchSave(its,true);
                }else{
```

```java
                    batchSave(its,false);
                }
            }
        });
    }finally{
        pstat.close();
        conn.close();
    }
}
public static void main(String[] args) throws Exception{
    SparkConf conf=new SparkConf();
    conf.setAppName("Cluster HDFS To HDFS");
    JavaStreamingContext jsc=new JavaStreamingContext
    (conf,Durations.seconds(8));
    JavaDStream<String> dStreamLines=jsc.textFileStream
    ("/spark/streaming/input");
    dStreamLines.foreachRDD(new VoidFunction<JavaRDD<String>>(){
        @Override
        public void call(JavaRDD<String> javaRDD) throws Exception {
            int partitionNum=javaRDD.getNumPartitions();
            System.out.println("------foreachRDD-partitionNum:"
            +partitionNum+"--------------------");
            if(0==partitionNum)return;
            //将行记录映射为元组方便后续执行数据库操作
            JavaRDD<Tuple4<Integer, String, Double, Date>> mapRDD=javaRDD.
            map(new
                Function<String,Tuple4<Integer,String,Double,Date>>(){
                    @Override
                    public Tuple4<Integer, String, Double, Date> call
                    (String line) throws
                        Exception {
                        String[] fieldValues=line.split("\t");
                        Integer uid=Integer.parseInt(fieldValues[0]);
                        String uname=fieldValues[1];
                        Double height=Double.parseDouble(fieldValues
                        [2]);
                        Date birthday=Date.valueOf(fieldValues[3]);
                        return new Tuple4<Integer, String, Double
                        Date>(uid,uname,height,birthday);
                    }
            });
            //将RDD存储到关系数据库
            saveToDB(mapRDD);
        }
```

```
        });

        jsc.start();
        jsc.awaitTermination();
        jsc.close();
    }
}
```

17.5.2 打包至工程的 dist 目录

1. 切换至工程的编译路径

```
[hadoop@CloudDeskTop bin]$ pwd
    /project/SparkStreaming-JAVA/bin
```

2. 删除旧的 JAR 包

```
[hadoop@CloudDeskTop bin]$ rm -rf ../dist/*
```

3. 压缩新 JAR 包

```
[hadoop@CloudDeskTop bin]$ jar -cvfe ../dist/HDFSToDB.
    com.bnyw.bigdata.spark.streaming.cluster.HDFSToDB com/
[hadoop@CloudDeskTop bin]$ ls ../dist/
    HDFSToDB.jar
```

17.5.3 启动集群

1. 启动 Zookeeper 集群

```
cd zookeeper-3.4.10/bin/
./zkServer.sh start
cd -
jps
```

2. 启动 HDFS 集群

```
./start-dfs.sh
Starting namenodes on [master01 master02]
master01: starting namenode, logging to /software/hadoop-2.7.3/logs/
hadoop-hadoop-namenode-master01.out
master02: ssh: connect to host master02 port 22: No route to host
slave02: starting datanode, logging to /software/hadoop-2.7.3/logs/
hadoop-hadoop-datanode-slave02.out
slave03: starting datanode, logging to /software/hadoop-2.7.3/logs/
hadoop-hadoop-datanode-slave03.out
slave01: starting datanode, logging to /software/hadoop-2.7.3/logs/
```

```
hadoop-hadoop-datanode-slave01.out
Starting journal nodes [slave01 slave02 slave03]
slave01: starting journalnode, logging to /software/hadoop-2.7.3/logs/
hadoop-hadoop-journalnode-slave01.out
slave02: starting journalnode, logging to /software/hadoop-2.7.3/logs/
hadoop-hadoop-journalnode-slave02.out
slave03: starting journalnode, logging to /software/hadoop-2.7.3/logs/
hadoop-hadoop-journalnode-slave03.out
```

3. 启动 Spark 集群

```
./start-master.sh
starting org.apache.spark.deploy.master.Master, logging to
/software/spark-2.1.1/logs/spark-hadoop-org.apache.spark.deploy.master.
Master-1-master01.out
[hadoop@master01 sbin]$ ./start-slaves.sh
slave01: starting org.apache.spark.deploy.worker.Worker, logging to/
software/spark-2.1.1/logs/spark-hadoop-org.apache.spark.deploy.worker.
Worker-1-slave01.out
slave03: starting org.apache.spark.deploy.worker.Worker, logging
to/software/spark-2.1.1/logs/spark-hadoop-org.apache.spark.deploy.worker.
Worker-1-slave03.out
slave02: starting org.apache.spark.deploy.worker.Worker, logging to /
software/spark-2.1.1/logs/spark-hadoop-org.apache.spark.deploy.worker.
Worker-1-slave02.out
```

4. 准备数据

创建 Spark 的 HDFS 输入源路径如下：

```
[hadoop@master01 sbin]$ hdfs dfs -ls /spark/streaming
Found 1 items
drwxr-xr-x   - hadoop supergroup          0 2017-08-20 14:12 /spark/streaming/
output
[hadoop@master01 sbin]$ hdfs dfs -mkdir -p /spark/streaming/input
[hadoop@master01 sbin]$ hdfs dfs -ls /spark/streaming
Found 2 items
drwxr-xr-x   - hadoop supergroup          0 2017-08-21 00:27 /spark/streaming/
input
drwxr-xr-x   - hadoop supergroup          0 2017-08-20 14:12 /spark/streaming/
output
    [hadoop@CloudDeskTop install]$ cat user1
    100  ligang      1.67    1982-12-28
    200  zhanghua    1.72    1983-10-12
    300  chenqian    1.78    1981-11-18
    400  guanyu      1.56    1996-08-08
```

```
      500 zhangfei      1.64      1994-02-25
      600 zhaoyun       1.76      1978-06-05
```

5. 创建 MySQL 数据表

```
[root@DB03 ~]# cd /software/mysql-5.5.32/multi-data/3306/
[root@DB03 3306]# ./mysqld start
Starting MySQL...
[root@DB03 3306]# lsof -i:3306
COMMAND  PID  USER    FD    TYPE  DEVICE  SIZE/OFF  NODE  NAME
mysqld   1663 mysql   11u   IPv4  13983       0t0   TCP   *:mysql (LISTEN)
[root@DB03 3306]# cd /software/mysql-5.5.32/bin/
[root@DB03 bin]# ./mysql -h192.168.37.143 -P3306 -uroot -p123456 -Dbnyw
-e"show tables"
        +----------------+
        | Tables_in_bnyw |
        +----------------+
        | wordcount      |
        +----------------+
[root@DB03 bin]# ./mysql -h192.168.37.143 -P3306 -uroot -p123456 -Dbnyw
-e"createtable if not exists tuser(uid int(11) auto_increment primary
key,uno int(11),uname varchar(30),height double(4,2),birthday
date)engine=innodb charset=utf8"
[root@DB03 bin]# ./mysql -h192.168.37.143 -P3306 -uroot -p123456 -Dbnyw
-e"desc tuser"
     +----------+-------------+------+-----+---------+----------------+
     | Field    | Type        | Null | Key | Default | Extra          |
     +----------+-------------+------+-----+---------+----------------+
     | uid      | int(11)     | NO   | PRI | NULL    | auto_increment |
     | uno      | int(11)     | YES  |     | NULL    |                |
     | uname    | varchar(30) | YES  |     | NULL    |                |
     | height   | double(4,2) | YES  |     | NULL    |                |
     | birthday | date        | YES  |     | NULL    |                |
     +----------+-------------+------+-----+---------+----------------+
[root@DB03 bin]# ./mysql -h192.168.37.143 -P3306 -uroot -p123456 -Dbnyw
-e"select* from tuser"
```

6. 提交 Streaming 应用到 Spark 集群

（1）查看表中是否有数据。

```
[root@DB03 bin]# ./mysql -h192.168.37.143 -P3306 -uroot -p123456 -Dbnyw
-e"select* from tuser"
```

（2）提交 Streaming 应用。

```
[hadoop@CloudDeskTop bin]$ pwd
/software/spark-2.1.1/bin
```

```
[hadoop@CloudDeskTop bin]$ ./spark-submit --master spark://master01:7077
--class com.bnyw.bigdata.spark.streaming.cluster.HDFSToDB /project/
SparkStreaming-JAVA/dist/HDFSToDB.jar
```

（3）待控制台出现以下数据打印时，开始往 Streaming 扫描的 HDFS 集群路径下上传数据。

```
------foreachRDD-partitionNum:0---------------------
```

（4）上传本地数据至 Streaming 扫描的 HDFS 集群路径。

```
[hadoop@CloudDeskTop install]$ hdfs dfs -put user1 /spark/streaming/input/
```

（5）再次观察控制台打印的数据。

```
------foreachRDD-partitionNum:1---------------------
```

（6）观察关系数据库中是否已有数据。

```
[root@DB03 bin]# ./mysql -h192.168.37.143 -P3306 -uroot -p123456 -Dbnyw
-e"select* from tuser"
        +-----+------+----------+--------+------------+
        | uid | uno  | uname    | height | birthday   |
        +-----+------+----------+--------+------------+
        |  1  | 100  | ligang   |  1.67  | 1982-12-28 |
        |  2  | 200  | zhanghua |  1.72  | 1983-10-12 |
        |  3  | 300  | chenqian |  1.78  | 1981-11-18 |
        |  4  | 400  | guanyu   |  1.56  | 1996-08-08 |
        |  5  | 500  | zhangfei |  1.64  | 1994-02-25 |
        |  6  | 600  | zhaoyun  |  1.76  | 1978-06-05 |
        +-----+------+----------+--------+------------+
```

（7）复制一份本地文件，再次上传并观察数据库中的数据是否增长。

```
[hadoop@CloudDeskTop install]$ cp -a user1 user2
[hadoop@CloudDeskTop install]$ hdfs dfs -put user2 /spark/streaming/input/
[root@DB03 bin]# ./mysql -h192.168.37.143 -P3306 -uroot -p123456 -Dbnyw
-e"select * from tuser"
        +-----+------+----------+--------+------------+
        | uid | uno  | uname    | height | birthday   |
        +-----+------+----------+--------+------------+
        |  1  | 100  | ligang   |  1.67  | 1982-12-28 |
        |  2  | 200  | zhanghua |  1.72  | 1983-10-12 |
        |  3  | 300  | chenqian |  1.78  | 1981-11-18 |
        |  4  | 400  | guanyu   |  1.56  | 1996-08-08 |
        |  5  | 500  | zhangfei |  1.64  | 1994-02-25 |
        |  6  | 600  | zhaoyun  |  1.76  | 1978-06-05 |
```

```
|  7 | 100 | ligang   | 1.67 | 1982-12-28 |
|  8 | 200 | zhanghua | 1.72 | 1983-10-12 |
|  9 | 300 | chenqian | 1.78 | 1981-11-18 |
| 10 | 400 | guanyu   | 1.56 | 1996-08-08 |
| 11 | 500 | zhangfei | 1.64 | 1994-02-25 |
| 12 | 600 | zhaoyun  | 1.76 | 1978-06-05 |
+--+-----+--------+------+-------+----------+
```

实际上 Spark Streaming 是一种基于事件的路径扫描机制。在指定扫描的路径下若存在新增文件的动作（Action）则触发 Spark Streaming 框架的事件处理函数（EventHandler），Spark Streaming 框架将读取新增文件进行流化处理。

第 18 章　Spark 机器学习

18.1　机器学习原理

关于 Spark 机器学习的模型构建和模型训练的视频讲解可扫描二维码观看。

18.1.1　机器学习的概念

机器学习（俗称人工智能）是指根据业务领域内相关维度并基于已有样本数据集，通过数理上的回归分析、拟合迭代等算法得出一套经验公式或数理结论，而后将其应用到未来某个时空点上以预测可能产生或发生的数据。

Spark 的机器学习建立在 MLLIB 算法库的基础之上。从数理的角度来看，Spark 的机器学习构建在向量和矩阵之上，即建立在 RDD/DataFrame/DataSet 之上的函数算法库，算法库中的这些函数实现了机器学习的各种算法（如各种回归算法、拟合算法、分类算法等），该算法函数操作的数据单位是 RDD、DataFrame 或 DataSet。

Spark 数据来源的最底层封装是 RDD，其与 Spark 的版本无任何关系，版本的发展仅为提供如 DataFrame、DataSet 等更高层的 API；使用 DataFrame 和 DataSet 是为了提供一个更高层的统一优化层面。机器学习即将 RDD、DataFrame 或 DataSet 包装成矩阵或向量，并依赖于此矩阵或向量调用算法库中的函数来实现模型推导。

机器学习是依据统计学的算法和推导得出一个规律（即模型或数理公式），推导此规律的过程就是模型的训练过程。规律一旦得出（模型一旦被训练出）便可将当前的数据应用到此规律（模型）中得出结论，该结论即依赖于已有规律和当前数据的一个预测值。

规律的推导（即模型的训练）须依赖于统计大量基础数据而得出结论，数据量越大则统计的精确度越高，得出的规律（模型）越准确。因此，大数据机器学习是依赖于海量数据集的统计算法而得出的一种模型。依赖于海量数据集进行演绎和推导模型的过程称为模型训练，训练出的模型准确度取决于数据集的大小。

18.1.2　机器学习的分类

（1）监督学习：比如分类，即给出学习样本的同时指定样本类别。
（2）非监督学习：比如聚类，即给出学习样本时未指定样本类别。

18.1.3 Spark 机器学习的版本演变

RDD：自 Spark 产生以来的基础和核心 API。
DataFrame：自 Spark 1.3.x 开始，主要用于处理来自关系型数据库中的源数据。
DataSet：自 Spark 1.6.x 开始，至 Spark 2.x 成熟。
提示：生产环境下建议使用 Spark 2.1.1 成熟版本。

18.1.4 DataFrame 数据结构

该数据结构由新一代基于存储列信息的 Row 对象的语法解析框架（Catalyst）和执行引擎（Tungsten）完成计算，可在数据传递给 RDD 计算之前完成基于行列数据的过滤，实现一个 pushdown 的下推操作。此处 RDD 充当服务层，DataFrame 充当底层的执行引擎，数据源首先通过 DataFrame 过滤，然后上行到 RDD 计算。

18.1.5 DataSet 数据结构

在 DataFrame 基础上提供基于 Encoder 编译时的类型安全检查，将字段类型兼容性检查由运行时提前到编译期间，将减少发生类型转换失败的错误；无须在计算时做反序列化转换，直接基于 Encoder 的二进制计算将提升计算效率和性能；优化内存管理的 GC 机制（降低 GC 发生频率）将降低内存使用率和网络数据传送量；降低使用 Scala 与使用 Java 语言编程的差异性（由于 DataSet 需进行编译期类型检查，因此 DataSet 目前不支持 Python 和 R）；除此之外，DataSet 对流处理、SQL 及其 ML 的 API 编程进行了统一。

18.1.6 执行引擎的性能与效率

Spark 第二代 Tungsten 引擎（即 Spark 2.x 中 DataSet 使用的计算引擎）与第一代 Tungten 引擎（DataFrame 使用）相比，在过滤和分组方面有了极大的性能提升（这些操作包括 filter、join、group、distinct 等）。过滤性能上的提升主要体现在以下几个方面。

（1）将多个算子产生的多次集合过滤操作合并为一个或更少的算子产生的单次过滤或更少的过滤操作。

（2）极大地减少虚函数调用以及运行时具体实例的查找和绑定过程，从而降低 CPU 的操作负载。

（3）中间数据放置于 CPU 寄存器而不是放置于内存中，对于复杂类型的数据（如一行数据是一个复杂对象时）可以优化以基于列式迭代（有多少列就迭代多少次，每迭代一次就取出一整列数据）的方式来进行过滤操作，极大地减少了迭代次数。

18.1.7 Spark 2.x 的新特性

在 Spark 的发展史中到目前为止有两个稳定版本，分别是 Spark 1.6.2 和 Spark 2.1.1。其中，Spark 2.x 的优点是擅长密集型计算，缺点是没有对 I/O 操作进行优化。Spark 2.x 中的 Structured Streaming 可直接基于事件的扫描或读取数据流，并对读取的 Spark 数据流可直接执行交互式 SQL 查询，这是 Spark 1.x 中不具备的能力。

在 Spark 2.x 中处理的 DataFrame、DataSet 对应的表无下边界，因为数据流不断产生，这种机制称为"End To End Continues Application",这是 Spark 2.x 区别于以往版本的重要特征。

18.2 线性回归

线性回归算法(Linear Regression)是利用线性回归函数对一个或多个自变量和因变量之间的关系建模的一种回归分析方法。仅一个自变量的情况称为简单一元线性回归,多个自变量的回归分析被称为多元线性回归。

线性回归算法的原理如下:

根据已有的特征值集合(集合中包括多组特征值,每组特征值 feature 使用(x,y,z,...)来表示,会是一到多个自变量。如果一组特征值中仅一个自变量则表示为一元线性回归,如果存在多个自变量则表示为多元线性回归)和特征值集合中每组特征值对应的函数值(函数值 label 使用 f(x,y,z,...)表示)来进行模型训练(即执行回归迭代),模型训练的结果将产生一个类似于 f(x,y,z,...)=(x,y,z...)的经验公式。线性回归中的函数值 label 与特征值组内的每个自变量(x 或 y 或 z,..)均呈单调线性关系。线性回归分析算法是其他数据挖掘算法的基础。

18.2.1 线性回归分析过程

(1)构造预测函数(h 函数);
(2)构造 cost 函数(J(θ));
(3)利用梯度下降(上升)算法来计算 J(θ)的最小值;
(4)梯度下降(上升)算法的向量化(以减少循环层次)。

18.2.2 矩阵分析过程

Spark 输入流中的每一条记录均可切分为各个字段,多条记录的组合即构成一个分析样本的二维表结构,此二维表又称为输入矩阵。其行为输入的每一条记录,其列为输入记录中切分出来的各个字段。基于 RDD 和 DataFrame 的矩阵称为批量样本矩阵。基于 DataSet 的样本流没有结束点,因此这种样本矩阵称为动态输入样本矩阵,即随着时间的推移其矩阵的行记录将不断增长。

18.2.3 基于本地模式的线性回归分析

(1)建立 Scala 工程 SparkML,并在工程目录下创建 input 和 output 目录,sample 目录用于存放 Spark-ML 的样本数据,input 目录用于存放需要预测的数据,output 目录用于存放预测后的数据。

```
[hadoop@CloudDeskTop SparkML]$ pwd
    /project/scala/SparkML
[hadoop@CloudDeskTop SparkML]$ mkdir -p sample/lr input/lr output/lr
[hadoop@CloudDeskTop SparkML]$ ls
    bin  input  output  sample  src
```

（2）构建样本数据。

```
[hadoop@CloudDeskTop sample]$ pwd
    /project/scala/SparkML/sample
[hadoop@CloudDeskTop sample]$ cat data
[hadoop@CloudDeskTop sample]$ cat data
    1    7
    2    12
    3    17
    4    18
    5    25
    6    27
    7    34
    8    37
    9    42
    10   42
```

（3）本地线性回归的源码。

```scala
package com.bnyw.bigdata.spark.ml.local
import java.io.File
import org.apache.spark.SparkConf
import org.apache.spark.SparkContext
import org.apache.spark.api.java.JavaRDD
import org.apache.spark.ml.feature.VectorAssembler
import org.apache.spark.ml.linalg.Vector
import org.apache.spark.ml.linalg.DenseVector
import org.apache.spark.ml.regression.LinearRegression
import org.apache.spark.ml.regression.LinearRegressionModel
import org.apache.spark.rdd.RDD
import org.apache.spark.sql.Row
import org.apache.spark.sql.DataFrame
import org.apache.spark.sql.SQLContext
import org.apache.spark.sql.types.DataTypes
import org.apache.spark.sql.types.StructField
import org.apache.spark.sql.types.StructType
object LinearRegressionML {
  def delDir(file:File):Unit={
    if(file.isFile||file.listFiles().length==0){
      file.delete();
      return;
    }
    val fileList:Array[File]=file.listFiles();
    fileList.foreach(file=>delDir(file));
    file.delete();
  }
```

```scala
def main(args: Array[String]): Unit = {
  val conf:SparkConf=new SparkConf();
  conf.setAppName("Local Scala Spark LinearRegressionML");
  conf.setMaster("local");
  val sc:SparkContext=new SparkContext(conf);
  val lines:RDD[String]=sc.textFile("/project/scala/SparkML/sample/lr", 1);
  val featureField:StructField=DataTypes.createStructField("eno",
    DataTypes.DoubleType, false);
  val labelField:StructField=DataTypes.createStructField("score",
    DataTypes.DoubleType, false);
  val fieldList:java.util.List[StructField]=new java.util.ArrayList[StructField]();
  fieldList.add(featureField);
  fieldList.add(labelField);
  val structType:StructType=DataTypes.createStructType(fieldList);
  val rows:RDD[Row]=lines.map((line:String)=>{
    val fieldVals:Array[String]=line.split('\t');
    val d1:Double= fieldVals(0).toDouble;
    val d2:Double=fieldVals(1).toDouble;
    Row(d1,d2);
  });
  val sqlContext:SQLContext=new SQLContext(sc);
  val dflines:DataFrame=sqlContext.createDataFrame(rows, structType);
  val inputColumns:Array[String]=Array[String]("eno");
  val assembler:VectorAssembler = new VectorAssembler().setInputCols(inputColumns).
    setOutputCol("features");
  val vecDF: DataFrame = assembler.transform(dflines);
  val lr1:LinearRegression = new LinearRegression();
  val lr2:LinearRegression = lr1.setFeaturesCol("features").setLabelCol("score").
    setPredictionCol("pre_score").setFitIntercept(true);
  val lr3:LinearRegression = lr2.setMaxIter(20).setRegParam(0.1).setElasticNetParam(0);
  val lr:LinearRegression = lr3;
  val lrModel:LinearRegressionModel = lr.fit(vecDF);
  val samplePredictions:DataFrame = lrModel.transform(vecDF);
  samplePredictions.createTempView("PreScore");
  val deviationDF:DataFrame=sqlContext.sql("select round(sum(pow(score-pre_score,2))/
    count(features),1) deviation from PreScore");
  deviationDF.show();
  val row1:Row=Row(14.0);
  val row2:Row=Row(16.0);
```

```
            val dataList:java.util.List[Row]=new java.util.ArrayList[Row]();
            dataList.add(row1)
            dataList.add(row2)
            val dataDF:DataFrame=sqlContext.createDataFrame(dataList,
            structType);
            val feaDF:DataFrame=assembler.transform(dataDF);
            val predictions:DataFrame = lrModel.transform(feaDF);
            val selDataFrame:DataFrame=predictions.selectExpr("features",
            "round(pre_score,1) as prediction");
            selDataFrame.show
            val predictResult:RDD[Row]=selDataFrame.toDF().rdd;
            val transRows:RDD[String]=predictResult.map((row:Row)=>{
              val features:DenseVector=row.getAs[DenseVector]("features");
              val prediction:Double=row.getAs[Double]("prediction");
              features+"\t"+prediction
            });
            val f:File=new File("/project/scala/SparkML/output/lr");
            if(f.exists()&&f.isDirectory())delDir(f);
            transRows.saveAsTextFile("/project/scala/SparkML/output/lr");
            sc.stop();
        }
    }
```

（4）在 Eclipse 中执行本地化测试，使用 Scala 运行本地 Spark 应用。
运行之后控制台打印如下：

```
+--------+----------+
|features|prediction|
+--------+----------+
|  [14.0]|      60.9|
|  [16.0]|      69.1|
+--------+----------+
```

运行之后本地文件中输出如下：

```
[hadoop@CloudDeskTop output]$ pwd
    /project/scala/SparkML/output
[hadoop@CloudDeskTop output]$ cat part-00000
    [14.0]  60.9
    [16.0]  69.1
```

18.2.4 基于集群模式的线性回归分析

1. 集群模式下线性回归的源码

```
package com.bnyw.bigdata.spark.ml.cluster
import java.io.File
```

```scala
import org.apache.spark.SparkConf
import org.apache.spark.SparkContext
import org.apache.spark.api.java.JavaRDD
import org.apache.spark.ml.feature.VectorAssembler
import org.apache.spark.ml.linalg.Vector
import org.apache.spark.ml.linalg.DenseVector
import org.apache.spark.ml.regression.LinearRegression
import org.apache.spark.ml.regression.LinearRegressionModel
import org.apache.spark.rdd.RDD
import org.apache.spark.sql.Row
import org.apache.spark.sql.DataFrame
import org.apache.spark.sql.SQLContext
import org.apache.spark.sql.types.DataType
import org.apache.spark.sql.types.DataTypes
import org.apache.spark.sql.types.StructField
import org.apache.spark.sql.types.StructType
import org.apache.hadoop.fs.FileSystem
import org.apache.hadoop.conf.Configuration
import java.net.URI
import org.apache.hadoop.fs.Path
object LinearRegressionML {
  var fs:FileSystem=null;
  {
        val hconf:Configuration=new Configuration();
        fs=FileSystem.get(new URI("hdfs://ns1/"),hconf,"hadoop");
  }
  def delDir(file:File):Unit={
    if(file.isFile||file.listFiles().length==0){
      file.delete();
      return;
    }
    val fileList:Array[File]=file.listFiles();
    fileList.foreach(file=>delDir(file));
    file.delete();
  }
  def getStructType(fieldNames:Array[String],fieldTypes:Array[DataType],
  isNUllS:Array[Boolean]):StructType={
    if(fieldNames.length!=fieldTypes.length||fieldTypes.length!=
    isNUllS.length)return null;
    val size:Int=fieldNames.length
    var i:Int=0;
    var featureField:StructField=null;
    val fieldList:java.util.List[StructField]=new java.util.ArrayList
    [StructField]();
    while(i<size){
        featureField=DataTypes.createStructField(fieldNames(i),
```

```scala
            fieldTypes(i),isNUllS(i));
            fieldList.add(featureField);
            i=i+1;
        }
        return DataTypes.createStructType(fieldList);
    }
    def mapStringToRowByRDD(rdd:RDD[String]):RDD[Row]={
        return rdd.map((line:String)=>{
            val fieldVals:Array[String]=line.split('\t');
            val d1:java.lang.Double=java.lang.Double.parseDouble(fieldVals
            (0));
            val d2:java.lang.Double=java.lang.Double.parseDouble(fieldVals
            (1));
            Row(d1,d2);
        });
    }
    def main(args: Array[String]): Unit = {
        val conf:SparkConf=new SparkConf();
        conf.setAppName("Cluster Scala Spark LinearRegressionML");
        val sc:SparkContext=new SparkContext(conf);
        val lines:RDD[String]=sc.textFile("/spark/ml/sample/lr", 1);
        val rows:RDD[Row]=mapStringToRowByRDD(lines);
        val structType:StructType=getStructType(Array[String]("eno",
        "score"),Array[DataType]
            (DataTypes.DoubleType,DataTypes.DoubleType),Array[Boolean]
            (false,false));
        val sqlContext:SQLContext=new SQLContext(sc);
        val dflines:DataFrame=sqlContext.createDataFrame(rows, structType);
        val inputColumns:Array[String]=Array[String]("eno");
        val assembler:VectorAssembler = new VectorAssembler().setInputCols
        (inputColumns).setOutputCol("features");
        val vecDF: DataFrame = assembler.transform(dflines);
        val lr1:LinearRegression = new LinearRegression();
        val lr2:LinearRegression = lr1.setFeaturesCol("features").
        setLabelCol("score").setPredictionCol("pre_score").
        setFitIntercept(true);
        val lr3:LinearRegression = lr2.setMaxIter(20);
        val lr:LinearRegression = lr3;
        val lrModel:LinearRegressionModel = lr.fit(vecDF);
        val srcRDD:RDD[String]=sc.textFile("/spark/ml/input/lr", 1);
        val srcRow:RDD[Row]=mapStringToRowByRDD(srcRDD);
        val dataDF:DataFrame=sqlContext.createDataFrame(srcRow, structType);
        val feaDF:DataFrame=assembler.transform(dataDF);
        val predictions:DataFrame = lrModel.transform(feaDF);
        val selDataFrame:DataFrame=predictions.selectExpr("features", "round
        (pre_score,1) as prediction");
```

```
    val predictResult:RDD[Row]=selDataFrame.toDF().rdd;
    val transRows:RDD[String]=predictResult.map((row:Row)=>{
      val features:DenseVector=row.getAs[DenseVector]("features");
      val prediction:Double=row.getAs[Double]("prediction");
      features+"\t"+prediction
    });
    val distDir:Path=new Path("/spark/ml/output/lr");
    if(fs.exists(distDir))fs.delete(distDir, true);
    transRows.saveAsTextFile("/spark/ml/output/lr");
    sc.stop();
  }
}
```

2. 启动集群

1）启动 Zookeeper 集群

```
cd zookeeper-3.4.10/bin/
./zkServer.sh start
cd -
jps
```

2）启动 HDFS 集群

```
start-dfs.sh
Starting namenodes on [master01 master02]
master01: starting namenode, logging to /software/hadoop-2.7.3/logs/
hadoop-hadoop-namenode-master01.out
master02: ssh: connect to host master02 port 22: No route to host
slave01: starting datanode, logging to /software/hadoop-2.7.3/logs/
hadoop-hadoop-datanode-slave01.out
slave03: starting datanode, logging to/software/hadoop-2.7.3/
logs/hadoop-hadoop-datanode-slave03.out
slave02: starting datanode, logging to/software/hadoop-2.7.3/
logs/hadoop-hadoop-datanode-slave02.out
Starting journal nodes [slave01 slave02 slave03]
slave02: starting journalnode, logging to/software/hadoop-2.7.3/logs/
hadoop-hadoop-journalnode-slave02.out
slave01: starting journalnode, logging to /software/hadoop-2.7.3/logs/
hadoop-hadoop-journalnode-slave01.out
slave03: starting journalnode, logging to/software/hadoop-2.7.3/logs/
hadoop-hadoop-journalnode-slave03.out
```

3）启动 Spark 集群

```
./start-master.sh
starting org.apache.spark.deploy.master.Master, logging to/software/
spark-2.1.1/logs/spark-hadoop-org.apache.spark.deploy.master.Master-1-
```

```
master01.out
 [hadoop@master01 sbin]$ ./start-slaves.sh
slave01: starting org.apache.spark.deploy.worker.Worker, logging to/
software/spark-2.1.1/logs/spark-hadoop-org.apache.spark.deploy.worker.
Worker-1slave01.out
slave02: starting org.apache.spark.deploy.worker.Worker, logging to/
software/spark-2.1.1/logs/spark-hadoop-org.apache.spark.deploy.worker.
Worker-1-slave02.out
slave03: starting org.apache.spark.deploy.worker.Worker, logging to/
software/spark-2.1.1/logs/spark-hadoop-org.apache.spark.deploy.worker.
Worker-1-slave03.out
```

3. 打包工程代码至工程目录下的 dist 目录

```
[hadoop@CloudDeskTop bin]$ pwd
    /project/scala/SparkML/bin
[hadoop@CloudDeskTop bin]$ ls
    com
[hadoop@CloudDeskTop bin]$ jar -cvfe ../dist/LinearRegressionML.jar
      com.bnyw.bigdata.spark.ml.cluster.LinearRegressionML com/
[hadoop@CloudDeskTop bin]$ cd ../dist/
[hadoop@CloudDeskTop dist]$ ls
    LinearRegressionML.jar
```

4. 准备样本数据

1）在集群上创建必要的输入输出目录

```
[hadoop@CloudDeskTop software]$ hdfs dfs -mkdir -p /spark/ml/sample/lr
[hadoop@CloudDeskTop software]$ hdfs dfs -mkdir -p /spark/ml/input/lr
[hadoop@CloudDeskTop software]$ hdfs dfs -mkdir -p /spark/ml/output/lr
[hadoop@CloudDeskTop software]$ hdfs dfs -ls /spark/ml/
    Found 3 items
    drwxr-xr-x   - hadoop supergroup    0 2017-08-24 00:05 /spark/ml/
      input/lr
    drwxr-xr-x   - hadoop supergroup    0 2017-08-24 00:05 /spark/ml/
      output/lr
    drwxr-xr-x   - hadoop supergroup    0 2017-08-24 00:05 /spark/ml/
      sample/lr
```

2）上传样本数据 HDFS 集群目录

```
[hadoop@CloudDeskTop lr]$ pwd
    /project/scala/SparkML/sample/lr
[hadoop@CloudDeskTop lr]$ hdfs dfs -put data /spark/ml/sample/lr/
[hadoop@CloudDeskTop lr]$ hdfs dfs -ls /spark/ml/sample/lr/
Found 1 items
```

```
-rw-r--r--   3 hadoop supergroup    50 2017-08-24 00:08
/spark/ml/sample/lr/data
[hadoop@CloudDeskTop lr]$ hdfs dfs -cat /spark/ml/sample/lr/data
   1    7
   2   12
   3   17
   4   18
   5   25
   6   27
   7   34
   8   37
   9   42
  10   42
```

3）上传需要处理的源数据 HDFS 集群目录

```
[hadoop@CloudDeskTop lr]$ pwd
    /project/scala/SparkML/input/lr
[hadoop@CloudDeskTop lr]$ ls
    data
[hadoop@CloudDeskTop lr]$ hdfs dfs -put data /spark/ml/input/lr/
[hadoop@CloudDeskTop lr]$ hdfs dfs -ls /spark/ml/input/lr/
    Found 1 items
    -rw-r--r--   3 hadoop supergroup    18 2017-08-24 00:16
    /spark/ml/input/lr/data
[hadoop@CloudDeskTop input]$ hdfs dfs -cat /spark/ml/input/lr/data
    14.0   0.0
    16.0   0.0
```

4）在工程目录下创建本地打包目录 dist

```
[hadoop@CloudDeskTop SparkML]$ pwd
    /project/scala/SparkML
[hadoop@CloudDeskTop SparkML]$ mkdir -p dist
[hadoop@CloudDeskTop SparkML]$ ls
   bin  dist  input  output  sample  spark-warehouse  src
```

5. 将线性回归应用提交至 Spark 集群

```
[hadoop@master01 bin]$ pwd
/software/spark-2.1.1/bin
[hadoop@CloudDeskTop bin]$ ./spark-submit --master spark://master01:7077
--class com.bnyw.bigdata.spark.ml.cluster.LinearRegressionML /project/
scala/SparkML/dist/LinearRegressionML.jar
17/08/24 00:35:01 WARN util.NativeCodeLoader: Unable to load native-hadoop
```

```
library for your platform... using builtin-java classes where applicable
17/08/24 00:35:02 INFO spark.SparkContext: Running Spark version 2.1.1
...
17/08/24 00:35:04 INFO spark.SparkEnv: Registering OutputCommitCoordinator
17/08/24 00:35:04 INFO util.log: Logging initialized @7614ms
17/08/24 00:35:04 INFO server.Server: jetty-9.2.z-SNAPSHOT
```

6. 查看集群运行的输出结果

```
[hadoop@CloudDeskTop ~]$ hdfs dfs -ls /spark/ml/output/lr
    Found 2 items
    -rw-r--r--   3 hadoop supergroup          0 2017-08-24 00:36 /spark/ml/
    output/lr/_SUCCESS
    -rw-r--r--   3 hadoop supergroup         24 2017-08-24 00:36 /spark/ml/
    output/lr/part-00000
[hadoop@CloudDeskTop ~]$ hdfs dfs -cat /spark/ml/output/lr/part-00000
    [14.0]  60.9
    [16.0]  69.1
```

18.3 聚类分析

聚类分析算法有很多，常用的聚类分析算法为 K-Means 聚类算法，是一种基于距离的迭代算法，即在一个 N 维的空间中将 n 个观察实例分类到 K 个聚类中，使每个观察实例距离它所在的聚类中心点比其他的聚类中心点距离更小。

18.3.1 K-Means 聚类算法原理

K-Means 聚类算法中的每组特征值（features）的组合决定了该组特征值所对应的类别（即聚类函数值 label），因此它的基础函数原理与线性回归一致，均为通过特征值计算函数值，此处函数值即聚类分析的结果。在线性回归分析中，几乎每组特征值的函数值均不同，而聚类分析中往往多组特征值的组合均对应到同一聚类函数值。

聚类分析的 API 中无须指定函数列，聚类分析仅基于特征值的组合推理各组特征所属的类别（聚类的质心），其预测出的类别（聚类的质心）可映射到实际的函数类别中，得出预测误差值。

18.3.2 聚类分析过程

（1）在 n 个观察实例中随机抽取 k 个点（因为需要分成 k 个类别）作为初始聚类中心点，然后计算所有的观察点距离各自最近的聚类中心点并将其划入对应的聚类中。

（2）求出各个聚类的中心点（质心，即聚类范围中距离各个观察点最近的中心点）作为新的聚类中心点，然后再次计算所有的观察点距离各自最近的新的聚类中心点并将其划入对应的聚类中。

（3）重复步骤（2），反复迭代，直到新的聚类中心点与前一个聚类中心点重合，或者距离满足设定条件为止，结束迭代。

18.3.3 基于本地模式的聚类算法分析

1. 创建 km 子文件夹

在 Scala 工程 SparkML 目录下的 input、output、sample 等子目录下分别创建 km 子文件夹。

```
[hadoop@CloudDeskTop SparkML]$ pwd
    /project/scala/SparkML
[hadoop@CloudDeskTop SparkML]$ mkdir -p {input,output,sample}/km
[hadoop@CloudDeskTop SparkML]$ tree {input,output,sample}
    input
    ├── km
    └── lr
    output
    ├── km
    └── lr
    sample
    ├── km
    └── lr
```

2. 构建样本数据

```
[hadoop@CloudDeskTop km]$ pwd
    /project/scala/SparkML/sample/km
[hadoop@CloudDeskTop km]$ cat data
    man    45    marry
    man    50    marry
    ...
    woman  36    marry
    woman  48    marry
```

3. 构建应用数据

```
[hadoop@CloudDeskTop km]$ pwd
    /project/scala/SparkML/input/km
[hadoop@CloudDeskTop km]$ ls
    data
[hadoop@CloudDeskTop km]$ cat data
    women    52
    man 34
    ...
    woman    25
    woman    20
```

4. 基于本地模式的聚类分析的源码

```
package com.bnyw.bigdata.spark.ml.local
```

```scala
import java.io.File
import org.apache.spark.rdd.RDD
import org.apache.spark.SparkConf
import org.apache.spark.SparkContext
import org.apache.spark.sql.SQLContext
import org.apache.spark.sql.types.DataTypes
import org.apache.spark.sql.Row
import org.apache.spark.sql.DataFrame
import org.apache.spark.sql.types.StructType
import org.apache.spark.sql.types.StructField
import org.apache.spark.sql.types.DataType
import org.apache.spark.ml.clustering.KMeans
import org.apache.spark.ml.feature.VectorAssembler
import org.apache.spark.ml.clustering.KMeansModel
import org.apache.spark.ml.feature.StringIndexer
import org.apache.spark.ml.Pipeline
import org.apache.spark.ml.PipelineModel
import org.apache.log4j.Level
import org.apache.log4j.Logger
import org.apache.spark.sql.Dataset
object KMeansML {
    def delDir(file:File):Unit={
        if(file.isFile||file.listFiles().length==0){
    file.delete();
    return;
  }
  val fileList:Array[File]=file.listFiles();
  fileList.foreach(file=>delDir(file));
  file.delete();
}
def getStructType(fieldNames:Array[String],fieldTypes:Array[DataType],
isNUllS:
  Array[Boolean]):StructType={
  if(fieldNames.length!=fieldTypes.length||fieldTypes.length!=isNUllS.
  length)return null;
  val size:Int=fieldNames.length
  var i:Int=0;
var featureField:StructField=null;
  val fieldList:java.util.List[StructField]=new java.util.ArrayList
  [StructField]();
  while(i<size){
    featureField=DataTypes.createStructField(fieldNames(i),
    fieldTypes(i),isNUllS(i));
    fieldList.add(featureFicld);
    i=i+1;
```

```scala
        }
        return DataTypes.createStructType(fieldList);
}
def rddStringToRow(lines:RDD[String]):RDD[Row]={
    val rows:RDD[Row]=lines.map((line:String)=>{
        val fieldVals:Array[String]=line.split('\t');
        var f1:Int=0;
        val f2:Int=fieldVals(1).toInt;
        if("man".equalsIgnoreCase(fieldVals(0)))f1=1;
        if(2==fieldVals.length){
            Row(f1,f2);
        }else{
            var fn:Int=0;
            if("marry".equalsIgnoreCase(fieldVals(2)))fn=1;
            Row(f1,f2,fn);
        }
    });
    return rows
}
def main(args: Array[String]): Unit = {
    Logger.getLogger("org.apache.spark").setLevel(Level.WARN)
    val conf:SparkConf=new SparkConf();
    conf.setAppName("Local Scala Spark KMeansML");
    conf.setMaster("local");

    val sc:SparkContext=new SparkContext(conf);
    val sqlContext:SQLContext=new SQLContext(sc);
    val structType:StructType=getStructType(Array[String]("gender","age",
    "marry"),
    Array[DataType](DataTypes.IntegerType,DataTypes.IntegerType,
    DataTypes.IntegerType),Array[Boolean](false,false,false));
    val lines:RDD[String]=sc.textFile("/project/scala/SparkML/sample/km", 1);
    val rows:RDD[Row]=rddStringToRow(lines);
    val dflines:DataFrame=sqlContext.createDataFrame(rows, structType);
    val inputColumns:Array[String]=Array[String]("gender","age");
    val assembler:VectorAssembler = new VectorAssembler().setInputCols
    (inputColumns)
    .setOutputCol("features");
    val vecDF: DataFrame = assembler.transform(dflines);
    val kmeans:KMeans=new KMeans();
    kmeans.setFeaturesCol("features");
    kmeans.setK(3);
    kmeans.setMaxIter(20);
    kmeans.setPredictionCol("pre_marry");
    kmeans.setInitMode("k-means||");
```

```
val kmeansModel:KMeansModel = kmeans.fit(vecDF);
val samplePredictions:DataFrame = kmeansModel.transform(vecDF);
samplePredictions.createTempView("SampleTab");
var selDF:DataFrame=sqlContext.sql("select marry,pre_marry,count
(features) level from SampleTab group by marry,pre_marry order by level
desc");
selDF.createTempView("mid1");
selDF.show(50);
selDF=sqlContext.sql("select pre_marry,max(level) maxLevel from mid1
group by pre_marry");
selDF.createTempView("mid2");
selDF.show(50);
selDF=sqlContext.sql("select mid1.marry,mid1.pre_marry,mid2.maxLevel
from mid1 inner join mid2 on mid1.pre_marry=mid2.pre_marry and mid1.
level=mid2.maxLevel");
selDF.createTempView("premodel");
selDF.show(50);
val cost:Double = kmeansModel.computeCost(vecDF);
println("cost:"+cost)
val requireLines:RDD[String]=sc.textFile("/project/scala/SparkML/
input/km", 1);
val requireRows:RDD[Row]=rddStringToRow(requireLines);
val requireStructType:StructType=getStructType(Array[String]
 ("gender","age"),
Array[DataType](DataTypes.IntegerType,DataTypes.IntegerType),Array
[Boolean](false,false));
val requireDFLines:DataFrame=sqlContext.createDataFrame(requireRows,
requireStructType);
val requireVecDF: DataFrame = assembler.transform(requireDFLines);
val requirePredictions:DataFrame = kmeansModel.transform(requireVecDF);
requirePredictions.createTempView("srcdata");
val resultDF:DataFrame=sqlContext.sql("select srcdata.
gender,srcdata.age,srcdata.pre_marry,premodel.marry from srcdata,
premodel where srcdata.pre_marry=premodel.pre_marry");
resultDF.show();
val genderMap=Map((1->"man"),(0->"women"));
val marryMap=Map((1->"marry"),(0->"unmarry"));
val resultRDD:RDD[Row]=resultDF.toDF().rdd;
val outRDD:RDD[String]=resultRDD.map((row:Row)=>{
  val age:Int=row.getAs[Int]("age");
  val gender:Int=row.getAs[Int]("gender");
  val marry:Int=row.getAs[Int]("marry");
  genderMap(gender)+"\t"+age+"\t"+marryMap(marry)
});
val fp:File=new File("/project/scala/SparkML/output/km");
```

```
    if(fp.exists())delDir(fp)
    outRDD.saveAsTextFile("/project/scala/SparkML/output/km");
    sc.stop();
 }
}
```

5. 直接在 Eclipse 中执行本地测试并查看结果

```
[hadoop@CloudDeskTop km]$ pwd
    /project/scala/SparkML/output/km
[hadoop@CloudDeskTop km]$ ls
    part-00000  part-00019  part-00038  part-00057  part-00076  part-00095
    part-00114  part-00133  part-00152  part-00171  part-00190
    part-00001  part-00020  part-00039  part-00058  part-00077  part-00096
    part-00115  part-00134  part-00153  part-00172  part-00191
    ...
    part-00017  part-00036  part-00055  part-00074  part-00093  part-00112
    part-00131  part-00150  part-00169  part-00188
    part-00018  part-00037  part-00056  part-00075  part-00094  part-00113
    part-00132  part-00151  part-00170  part-00189

[hadoop@CloudDeskTop km]$ cat *
    women   19  unmarry
    man  24  unmarry
    ...
    women   26  marry
    women   52  marry
```

18.3.4 基于集群模式的聚类算法分析

1. 基于集群模式的聚类算法分析的源码

```
            package com.bnyw.bigdata.spark.ml.cluster
            import java.io.File
            import org.apache.spark.rdd.RDD
            import org.apache.spark.SparkConf
            import org.apache.spark.SparkContext
            import org.apache.spark.sql.SQLContext
            import org.apache.spark.sql.types.DataTypes
            import org.apache.spark.sql.Row
            import org.apache.spark.sql.DataFrame
            import org.apache.spark.sql.types.StructType
            import org.apache.spark.sql.types.StructField
            import org.apache.spark.sql.types.DataType
            import org.apache.spark.ml.clustering.KMeans
            import org.apache.spark.ml.feature.VectorAssembler
            import org.apache.spark.ml.clustering.KMeansModel
```

```scala
import org.apache.spark.ml.feature.StringIndexer
import org.apache.spark.ml.Pipeline
import org.apache.spark.ml.PipelineModel
import org.apache.log4j.Level
import org.apache.log4j.Logger
import org.apache.spark.sql.Dataset
import org.apache.hadoop.fs.FileSystem
import java.net.URI
import org.apache.hadoop.conf.Configuration
import org.apache.hadoop.fs.Path
object KMeansML {
    var fs:FileSystem=null;
    {
    val hconf:Configuration=new Configuration();
    fs=FileSystem.get(new URI("hdfs://ns1/"),hconf,"hadoop");
    }
    def delDir(file:File):Unit={
        if(file.isFile||file.listFiles().length==0){
            file.delete();
            return;
        }
        val fileList:Array[File]=file.listFiles();
        fileList.foreach(file=>delDir(file));
        file.delete();
    }
    def getStructType(fieldNames:Array[String],fieldTypes:Array[DataType],isNUllS:Array[Boolean]):StructType={
        if(fieldNames.length!=fieldTypes.length||fieldTypes.length!=isNUllS.length) return            null;
        val size:Int=fieldNames.length
        var i:Int=0;
        var featureField:StructField=null;
        val fieldList:java.util.List[StructField]=new java.util.ArrayList[StructField]();
        while(i<size){
        featureField=DataTypes.createStructField(fieldNames(i),
        fieldTypes(i),
            isNUllS(i));
         fieldList.add(featureField);
        i=i+1;
         }
        return DataTypes.createStructType(fieldList);
    }
    def rddStringToRow(lines:RDD[String]):RDD[Row]={
```

```
        val rows:RDD[Row]=lines.map((line:String)=>{
            val fieldVals:Array[String]=line.split('\t');
            var f1:Int=0;
            val f2:Int=fieldVals(1).toInt;
            if("man".equalsIgnoreCase(fieldVals(0)))f1=1;
                if(2==fieldVals.length){
                Row(f1,f2);
                 }else{
        var fn:Int=0;
        if("marry".equalsIgnoreCase(fieldVals(2)))fn=1;
        Row(f1,f2,fn);
        }
            });
        return rows
        }
def main(args: Array[String]): Unit = {
   Logger.getLogger("org.apache.spark").setLevel(Level.WARN)
   val conf:SparkConf=new SparkConf();
   conf.setAppName("Cluster Scala Spark KMeansML");
   val sc:SparkContext=new SparkContext(conf);
   val sqlContext:SQLContext=new SQLContext(sc);
   val structType:StructType=getStructType(Array[String]("gender",
   "age","marry"),
   Array[DataType](DataTypes.IntegerType,DataTypes.IntegerType,DataTypes.
   IntegerType),Arra   y[Boolean](false,false,false));
   val lines:RDD[String]=sc.textFile("/spark/ml/sample/km", 1);
   val rows:RDD[Row]=rddStringToRow(lines);
   val dflines:DataFrame=sqlContext.createDataFrame(rows, structType);
   val inputColumns:Array[String]=Array[String]("gender","age");
   val assembler:VectorAssembler = new VectorAssembler().
   setInputCols(inputColumns).setOutputCol("features");
   val vecDF: DataFrame = assembler.transform(dflines);
   val kmeans:KMeans=new KMeans();
   kmeans.setFeaturesCol("features");
   kmeans.setK(3);
   kmeans.setMaxIter(20);
   kmeans.setPredictionCol("pre_marry");
   kmeans.setInitMode("k-means||");
   val kmeansModel:KMeansModel = kmeans.fit(vecDF);
   val samplePredictions:DataFrame = kmeansModel.transform(vecDF);
   samplePredictions.createTempView("SampleTab");
   var selDF:DataFrame=sqlContext.sql("select marry,pre_marry,count
   (features) level from SampleTab group by marry,pre_marry order by level
   desc");
   selDF.createTempView("mid1");
```

```
            selDF.show(50);
            selDF=sqlContext.sql("select pre_marry,max(level) maxLevel from mid1
            group by pre_marry");
            selDF.createTempView("mid2");
            selDF.show(50);
            selDF=sqlContext.sql("select mid1.marry,mid1.pre_marry,mid2.maxLevel
            from mid1 inner join mid2 on mid1.pre_marry=mid2.pre_marry and mid1.
            level=mid2.maxLevel");
            selDF.createTempView("premodel");
            selDF.show(50);
            val cost:Double = kmeansModel.computeCost(vecDF);
            println("cost:"+cost)
            val requireLines:RDD[String]=sc.textFile("/spark/ml/input/km", 1);
            val requireRows:RDD[Row]=rddStringToRow(requireLines);
            val requireStructType:StructType=getStructType(Array[String] ("gender",
            "age"),
            Array[DataType](DataTypes.IntegerType,DataTypes.IntegerType),Array[
            Boolean](false,false));
            val requireDFLines:DataFrame=sqlContext.createDataFrame(requireRows,
            requireStructType);
            val requireVecDF: DataFrame = assembler.transform(requireDFLines);
            val requirePredictions:DataFrame = kmeansModel.
            transform(requireVecDF);
            requirePredictions.createTempView("srcdata");
            val resultDF:DataFrame=sqlContext.sql("select srcdata.gender,
            srcdata.age,srcdata.pre_marry,premodel.marry from srcdata,premodel
            where srcdata.pre_marry=premodel.pre_marry");
            resultDF.show();

            val genderMap=Map((1->"man"),(0->"women"));
            val marryMap=Map((1->"marry"),(0->"unmarry"));
            val resultRDD:RDD[Row]=resultDF.toDF().rdd;
            val outRDD:RDD[String]=resultRDD.map((row:Row)=>{
              val age:Int=row.getAs[Int]("age");
              val gender:Int=row.getAs[Int]("gender");
              val marry:Int=row.getAs[Int]("marry");
              genderMap(gender)+"\t"+age+"\t"+marryMap(marry)
            });
            val output:Path=new Path("/spark/ml/output/km");
            if(fs.exists(output))fs.delete(output, true);
            outRDD.saveAsTextFile("/spark/ml/output/km");
            sc.stop();
        }
    }
```

2. 打包到 dist 目录

```
[hadoop@CloudDeskTop bin]$ pwd
    /project/scala/SparkML/bin
[hadoop@CloudDeskTop bin]$ jar -cvfe ../dist/KMeansML.jar
    com.bnyw.bigdata.spark.ml.cluster.KMeansML com/
[hadoop@CloudDeskTop bin]$ ls ../dist/
    KMeansML.jar  LinearRegressionML.jar
```

3. 启动集群

1）启动 Zookeeper 集群

```
cd zookeeper-3.4.10/bin/
./zkServer.sh start
cd -
jps
```

2）启动 HDFS 集群

```
start-dfs.sh
Starting namenodes on [master01 master02]
master01: starting namenode, logging to
/software/hadoop-2.7.3/logs/hadoop-hadoop-namenode-master01.out
master02: ssh: connect to host master02 port 22: No route to host
slave01: starting datanode, logging to /software/hadoop-2.7.3/logs/
hadoop-hadoop-datanode-slave01.out
slave03: starting datanode, logging to /software/hadoop-2.7.3/logs/
hadoop-hadoop-datanode-slave03.out
slave02: starting datanode, logging to /software/hadoop-2.7.3/logs/
hadoop-hadoop-datanode-slave02.out
Starting journal nodes [slave01 slave02 slave03]
slave02: starting journalnode, logging to
/software/hadoop-2.7.3/logs/hadoop-hadoop-journalnode-slave02.out
slave01: starting journalnode, logging to
/software/hadoop-2.7.3/logs/hadoop-hadoop-journalnode-slave01.out
slave03: starting journalnode, logging to
/software/hadoop-2.7.3/logs/hadoop-hadoop-journalnode-slave03.out
```

3）启动 Spark 集群

```
./start-master.sh
starting org.apache.spark.deploy.master.Master, logging to/software/
spark-2.1.1/logs/spark-hadoop-org.apache.spark.deploy.master.Master-1-
master01.out
[hadoop@master01 sbin]$ ./start-slaves.sh
slave01: starting org.apache.spark.deploy.worker.Worker, logging to /
```

```
software/spark-2.1.1/logs/spark-hadoop-org.apache.spark.deploy.worker.
Worker-1-slave01.out
slave02: starting org.apache.spark.deploy.worker.Worker, logging to /
software/spark-2.1.1/logs/spark-hadoop-org.apache.spark.deploy.worker.
Worker-1-slave02.out
slave03: starting org.apache.spark.deploy.worker.Worker, logging to /
software/spark-2.1.1/logs/spark-hadoop-org.apache.spark.deploy.worker.
Worker-1-slave03.out
```

4. 准备聚类样本数据

1）创建必要的 HDFS 路径

```
[hadoop@master01 software]$ hdfs dfs -ls /spark/
Found 2 items
drwxr-xr-x   - hadoop supergroup          0 2017-08-20 06:46 /spark/input
drwxr-xr-x   - hadoop supergroup          0 2017-08-21 00:27 /spark/streaming
[hadoop@master01 software]$ hdfs dfs -mkdir -p
/spark/ml/{sample,input,output}/{lr,km}
[hadoop@master01 software]$ hdfs dfs -ls /spark/
Found 3 items
drwxr-xr-x   - hadoop supergroup          0 2017-08-20 06:46 /spark/input
drwxr-xr-x   - hadoop supergroup          0 2017-08-27 04:44 /spark/ml
drwxr-xr-x   - hadoop supergroup          0 2017-08-21 00:27 /spark/streaming
[hadoop@master01 software]$ hdfs dfs -ls /spark/ml
Found 3 items
drwxr-xr-x   - hadoop supergroup          0 2017-08-27 04:44 /spark/ml/input
drwxr-xr-x   - hadoop supergroup          0 2017-08-27 04:44 /spark/ml/output
drwxr-xr-x   - hadoop supergroup          0 2017-08-27 04:44 /spark/ml/sample
```

2）上传聚类算法样本数据

```
[hadoop@CloudDeskTop SparkML]$ pwd
/project/scala/SparkML
[hadoop@CloudDeskTop SparkML]$ hdfs dfs -put sample/km/data /spark/ml/sample/km/
[hadoop@CloudDeskTop SparkML]$ hdfs dfs -put input/km/data /spark/ml/input/km/
```

3）提交聚类应用到 Spark 集群

```
[hadoop@CloudDeskTop bin]$ pwd
/software/spark-2.1.1/bin
[hadoop@CloudDeskTop bin]$ ./spark-submit --master spark://master01:7077
--class com.bnyw.bigdata.spark.ml.cluster.KMeansML /project/scala/
SparkML/dist/KMeansML.jar 1
```

4）查看聚类算法分析结果

```
[hadoop@CloudDeskTop dist]$ hdfs dfs -ls /spark/ml/output/km
```

```
Found 201 items
-rw-r--r--   3 hadoop supergroup          0 2017-08-27 05:08
/spark/ml/output/km/_SUCCESS
    -rw-r--r--   3 hadoop supergroup          0 2017-08-27 05:07
    /spark/ml/output/km/part-00000
    -rw-r--r--   3 hadoop supergroup          0 2017-08-27 05:07
    /spark/ml/output/km/part-00001
    -rw-r--r--   3 hadoop supergroup          0 2017-08-27 05:07
    /spark/ml/output/km/part-00002
    -rw-r--r--   3 hadoop supergroup          0 2017-08-27 05:07
    /spark/ml/output/km/part-00003
    -rw-r--r--   3 hadoop supergroup          0 2017-08-27 05:07
    /spark/ml/output/km/part-00004
    -rw-r--r--   3 hadoop supergroup          0 2017-08-27 05:07
    /spark/ml/output/km/part-00005
    -rw-r--r--   3 hadoop supergroup          0 2017-08-27 05:07
    /spark/ml/output/km/part-00006
    -rw-r--r--   3 hadoop supergroup          0 2017-08-27 05:07
    /spark/ml/output/km/part-00007
    -rw-r--r--   3 hadoop supergroup          0 2017-08-27 05:07
    /spark/ml/output/km/part-00008
    -rw-r--r--   3 hadoop supergroup          0 2017-08-27 05:07
    /spark/ml/output/km/part-00009
    -rw-r--r--   3 hadoop supergroup          0 2017-08-27 05:07
    /spark/ml/output/km/part-00010
    -rw-r--r--   3 hadoop supergroup          0 2017-08-27 05:07
    /spark/ml/output/km/part-00011
    -rw-r--r--   3 hadoop supergroup          0 2017-08-27 05:07
```

提示：由于篇幅所限，更多的日志打印输出在本书中略去。

```
[hadoop@CloudDeskTop dist]$ hdfs dfs -cat /spark/ml/output/km/*
women    19  unmarry
man  24  unmarry
man  25  unmarry
man  22  unmarry
man  23  unmarry
women    22  unmarry
women    23  unmarry
women    24  unmarry
women    25  unmarry
women    20  unmarry
man  34  marry
man  26  marry
women    26  marry
women    52  marry
```

18.4 协同过滤

协同过滤是推荐系统中应用最成功的一种算法，其宗旨是根据数据之间的相关性查找数据间潜在的相似度，基于相关性的描述在不同的业务领域中有不同的抽象。以下将电商领域中的推荐系统作为实例讲解协同过滤在推荐系统中的算法。

18.4.1 个性化推荐算法

（1）User-Based：基于用户特征相似度计算，具备相同特征的用户将被推荐相同的商品。

（2）Item-Based：基于商品特征相似度计算，具备相同特征的商品将会被推荐给同一个用户。

18.4.2 相关性推荐算法

1. 基于用户相似度的计算

若两个用户喜欢同一件商品，则认为这两个用户具备某种潜在特征上的相似度。当模型训练完毕，若其中某一用户喜欢 A 商品，系统将根据个性化相似度自动为另一用户推荐 A 商品。即不同的用户通过相同的商品作为桥梁建立起相似度。

2. 基于商品的相似度计算

若某一用户喜欢 A 商品的同时还喜欢 B 商品，那么系统将自动为喜欢 A 商品的另一用户推荐 B 商品，即不同的商品通过相同的用户作为桥梁建立起相似度。

提示：相关性推荐算法的思想出发点是"物以类聚、人以群分"。

18.4.3 基于本地的协同过滤算法分析

1. 创建 als 子文件夹

在 Scala 工程 SparkML 目录下的 input、output、sample 等子目录下分别创建 als 子文件夹。

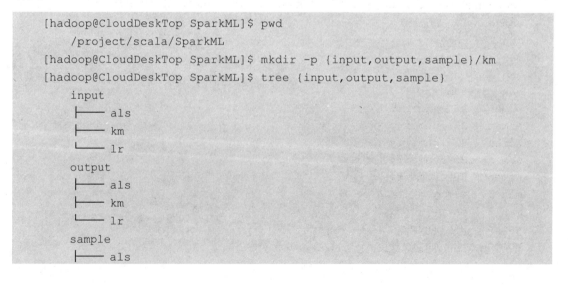

```
├── km
└── lr
```

2. 构建协同过滤样本数据

```
[hadoop@CloudDeskTop als]$ pwd
    /project/scala/SparkML/sample/als
[hadoop@CloudDeskTop als]$ cat data
    1   1   5.5
    1   2   10.5
    2   2   6.5
    2   3   4.5
    2   4   6.2
    3   1   6.5
    3   3   8.8
    3   5   7.8
    4   4   3.8
```

3. 构建协同过滤应用数据

```
[hadoop@CloudDeskTop als]$ pwd
    /project/scala/SparkML/input/als
[hadoop@CloudDeskTop als]$ cat data
    1   3
    1   4
    1   5
    2   1
    2   2
    2   3
    2   4
    2   5
    6   1
    6   2
```

4. 基于本地协同过滤分析的源码

```
package com.bnyw.bigdata.spark.ml.local
import org.apache.log4j.Level
import org.apache.log4j.Logger
import org.apache.spark.SparkConf
import org.apache.spark.sql.SQLContext
import org.apache.spark.SparkContext
import org.apache.spark.ml.recommendation.ALS
import java.io.File
import org.apache.spark.sql.types.DataTypes
import org.apache.spark.sql.types.DataType
import org.apache.spark.sql.types.StructType
```

```scala
import org.apache.spark.sql.types.StructField
import org.apache.spark.rdd.RDD
import org.apache.spark.sql.Row
import org.apache.spark.sql.DataFrame
import org.apache.spark.ml.recommendation.ALSModel
object ALSML {
    def delDir(file:File):Unit={
        if(file.isFile||file.listFiles().length==0){
            file.delete();
            return;
        }
        val fileList:Array[File]=file.listFiles();
        fileList.foreach(file=>delDir(file));
        file.delete();
    }
  def getStructType(fieldNames:Array[String],fieldTypes:Array
  [DataType],isNUllS:
  Array[Boolean]):StructType={
    if(fieldNames.length!=fieldTypes.length||fieldTypes.length!=
    isNUllS.length)return            null;
    val size:Int=fieldNames.length
    var i:Int=0;
    var featureField:StructField=null;
    val fieldList:java.util.List[StructField]=new java.util.
    ArrayList[StructField]();
    while(i<size){
        featureField=DataTypes.createStructField(fieldNames(i),
        fieldTypes(i),isNUllS(i));
        fieldList.add(featureField);
        i+=1;
    }
    return DataTypes.createStructType(fieldList);
  }
  def rddStringToRow(lines:RDD[String]):RDD[Row]={
    val rows:RDD[Row]=lines.map((line:String)=>{
        val fieldVals:Array[String]=line.split('\t');
        var f1:Int=fieldVals(0).toInt;
        val f2:Int=fieldVals(1).toInt;
        if(2==fieldVals.length){
          Row(f1,f2);
        }else{
          var fn:Double=fieldVals(2).toDouble;
          Row(f1,f2,fn);
        }
    });
```

```scala
    return rows;
}
def main(args: Array[String]): Unit = {
  Logger.getLogger("org.apache.spark").setLevel(Level.WARN);
  val conf:SparkConf=new SparkConf();
  conf.setAppName("Local Scala Spark ALSML");
  conf.setMaster("local");
  val sc:SparkContext=new SparkContext(conf);
  val sqlContext:SQLContext=new SQLContext(sc);
  val structType:StructType=getStructType(Array[String]("userId",
  "itemId","rating"),
  Array[DataType](DataTypes.IntegerType,DataTypes.IntegerType,
  DataTypes.DoubleType),Array[Boolean](false,false,false));
  val lines:RDD[String]=sc.textFile("/project/scala/SparkML/sample/als", 1);
  val rows:RDD[Row]=rddStringToRow(lines);
  val dflines:DataFrame=sqlContext.createDataFrame(rows, structType);
  val als:ALS=new ALS();
  als.setMaxIter(10);
  als.setUserCol("userId");
  als.setItemCol("itemId");
  als.setRatingCol("rating");
  als.setPredictionCol("pre_rating");
  val alsModel:ALSModel=als.fit(dflines);
  val samplePredictions:DataFrame = alsModel.transform(dflines);
  samplePredictions.show();
  samplePredictions.createTempView("samplePreView");
  val deviationDF:DataFrame=sqlContext.sql("select round(sum(
  pow(rating-pre_rating,2))/count(userId),5) deviation from samplePreView");
  deviationDF.show()
  val appLines:RDD[String]=sc.textFile("/project/scala/SparkML/input/
  als", 1);
  val appRows:RDD[Row]=rddStringToRow(appLines);
  val appStructType:StructType=getStructType(Array[String]("userId",
  "itemId"),Array[DataType](DataTypes.IntegerType,DataTypes.
  IntegerType),Array[Boolean](false,false));
  val appDflines:DataFrame=sqlContext.createDataFrame(appRows,
  appStructType);
  val appPredictions:DataFrame = alsModel.transform(appDflines);
  appPredictions.show();
  appPredictions.createTempView("appPreView");
  val appResult:DataFrame=sqlContext.sql("select mid.userId,ap.itemId,
  mid.maxRating fro
  (select userId,max(pre_rating) maxRating from appPreView group by
  userId)
  mid,appPreView ap where mid.maxRating=ap.pre_rating order by userId");
```

```
        appResult.show();
    }
}
```

5. 本地运行聚类分析的应用
1)预测样本数据结果

```
|userId|itemId|rating|pre_rating|
+------+------+------+----------+
|     1|     1|   5.5|  5.505122|
|     3|     1|   6.5| 6.4941416|
|     3|     3|   8.8|  8.627493|
|     2|     3|   4.5| 4.5032167|
|     3|     5|   7.8|  7.844314|
|     4|     4|   3.8| 3.7673109|
|     2|     4|   6.2| 6.0356483|
|     1|     2|  10.5| 10.412292|
|     2|     2|   6.5|  6.504319|
+------+------+------+----------+
```

2)预测样本数据方差评估

```
|deviation|
+---------+
|  0.00751|
+---------+
```

3)预测实际应用数据结果

```
|userId|itemId|pre_rating|
+------+------+----------+
|     6|     1|       NaN|
|     2|     1|  4.511438|
|     1|     3|  4.049105|
|     2|     3| 4.5032167|
|     1|     5|  3.627529|
|     2|     5| 4.0324507|
|     1|     4|  5.348643|
|     2|     4| 6.0356483|
|     6|     2|       NaN|
|     2|     2|  6.504319|
+------+------+----------+
```

4)预测实际应用数据分析统计推荐结果

```
|userId|itemId|maxRating|
+------+------+---------+
|     1|     4| 5.348643|
```

```
|    2|    2| 6.504319|
|    6|    1|      NaN|
|    6|    2|      NaN|
+-----+-----+---------+
```

若实际应用数据中出现的 userId 在样本中从未出现，则无法进行预测，显示为 NaN。协同过滤的基本思想是统计分析不同人群的偏好，即通过相同的 itemId 建立不同 userId 之间的关联或通过相同的 userId 建立不同 itemId 之间的关联。如果在建模训练时没有覆盖到某个 userId，则无法实现对该 userId 的预测和推荐。

18.4.4 基于集群的协同过滤算法分析

1. 启动集群

1）启动 Zookeeper 集群

```
cd zookeeper-3.4.10/bin/
./zkServer.sh start
cd -
jps
```

2）启动 HDFS 集群

```
start-dfs.sh
Starting namenodes on [master01 master02]
master01: starting namenode, logging to
/software/hadoop-2.7.3/logs/hadoop-hadoop-namenode-master01.out
master02: ssh: connect to host master02 port 22: No route to host
slave01: starting datanode, logging to
/software/hadoop-2.7.3/logs/hadoop-hadoop-datanode-slave01.out
slave03: starting datanode, logging to
/software/hadoop-2.7.3/logs/hadoop-hadoop-datanode-slave03.out
slave02: starting datanode, logging to
/software/hadoop-2.7.3/logs/hadoop-hadoop-datanode-slave02.out
Starting journal nodes [slave01 slave02 slave03]
slave02: starting journalnode, logging to
/software/hadoop-2.7.3/logs/hadoop-hadoop-journalnode-slave02.out
slave01: starting journalnode, logging to
/software/hadoop-2.7.3/logs/hadoop-hadoop-journalnode-slave01.out
slave03: starting journalnode, logging to
/software/hadoop-2.7.3/logs/hadoop-hadoop-journalnode-slave03.out
```

3）启动 Spark 集群

```
./start-master.sh
starting org.apache.spark.deploy.master.Master, logging to/software/
spark-2.1.1/logs/spark-hadoop-org.apache.spark.deploy.master.Master-1-
```

```
master01.out
[hadoop@master01 sbin]$ ./start-slaves.sh
slave01: starting org.apache.spark.deploy.worker.Worker, logging to /
software/
spark-2.1.1/logs/spark-hadoop-org.apache.spark.deploy.worker.Worker-1-
slave01.out
slave02: starting org.apache.spark.deploy.worker.Worker, logging to /
software/
spark-2.1.1/logs/spark-hadoop-org.apache.spark.deploy.worker.Worker-1-
slave02.out
slave03: starting org.apache.spark.deploy.worker.Worker, logging to /
software/
spark-2.1.1/logs/spark-hadoop-org.apache.spark.deploy.worker.Worker-1-
slave03.out
```

2. 准备协同过滤样本数据

1）创建必要的集群目录

```
[hadoop@CloudDeskTop software]$ hdfs dfs -mkdir -p /spark/
ml/{sample,input}/als
```

2）上传本地测试数据到 HDFS 集群

```
[hadoop@CloudDeskTop software]$ cd /project/scala/SparkML/sample/als/
[hadoop@CloudDeskTop als]$ hdfs dfs -put data /spark/ml/input/als/
[hadoop@CloudDeskTop als]$ hdfs dfs -cat /spark/ml/sample/als/data
    1   1   5.5
    1   2   10.5
    2   2   6.5
    2   3   4.5
    2   4   6.2
    3   1   6.5
    3   3   8.8
    3   5   7.8
    4   4   3.8

[hadoop@CloudDeskTop als]$ hdfs dfs -cat /spark/ml/input/als/data
    1   3
    1   4
    1   5
    2   1
    2   2
    2   3
    2   4
    2   5
    6   1
    6   2
```

3）基于集群模式的协同过滤的源码

```
package com.bnyw.bigdata.spark.ml.cluster
import org.apache.log4j.Level
import org.apache.log4j.Logger
import org.apache.spark.SparkConf
import org.apache.spark.sql.SQLContext
import org.apache.spark.SparkContext
import org.apache.spark.ml.recommendation.ALS
import java.io.File
import org.apache.spark.sql.types.DataTypes
import org.apache.spark.sql.types.DataType
import org.apache.spark.sql.types.StructType
import org.apache.spark.sql.types.StructField
import org.apache.spark.rdd.RDD
import org.apache.spark.sql.Row
import org.apache.spark.sql.DataFrame
import org.apache.spark.ml.recommendation.ALSModel
import org.apache.hadoop.fs.FileSystem
import java.net.URI
import org.apache.hadoop.conf.Configuration
import org.apache.hadoop.fs.Path
object ALSML {
  var fs:FileSystem=null;
  {
    val hconf:Configuration=new Configuration();
    fs=FileSystem.get(new URI("hdfs://ns1/"),hconf,"hadoop");
  }
  def getStructType(fieldNames:Array[String],fieldTypes:Array[DataType],isNUllS:
  Array[Boolean]):StructType={
    if(fieldNames.length!=fieldTypes.length||fieldTypes.length!=
    isNUllS.length)return null;
    val size:Int=fieldNames.length
    var i:Int=0;
    var featureField:StructField=null;
    val fieldList:java.util.List[StructField]=new java.util.ArrayList
    [StructField]();
    while(i<size){
      featureField=DataTypes.createStructField(fieldNames(i),
      fieldTypes(i),isNUllS(i));
      fieldList.add(featureField);
      i+=1;
    }
```

```
        return DataTypes.createStructType(fieldList);
    }
    def rddStringToRow(lines:RDD[String]):RDD[Row]={
        val rows:RDD[Row]=lines.map((line:String)=>{
            val fieldVals:Array[String]=line.split('\t');
            var f1:Int=fieldVals(0).toInt;
            val f2:Int=fieldVals(1).toInt;
            if(2==fieldVals.length){
                Row(f1,f2);
            }else{
                var fn:Double=fieldVals(2).toDouble;
                Row(f1,f2,fn);
            }
        });
        return rows
    }
    def main(args: Array[String]): Unit = {
        Logger.getLogger("org.apache.spark").setLevel(Level.WARN);
        val conf:SparkConf=new SparkConf();
        conf.setAppName("Cluster Scala Spark ALSML");
        val sc:SparkContext=new SparkContext(conf);
        val sqlContext:SQLContext=new SQLContext(sc);
        val structType:StructType=getStructType(Array[String]("userId",
        "itemId","rating"),
        Array[DataType](DataTypes.IntegerType,DataTypes.IntegerType,
        DataTypes.DoubleType),Array[Boolean](false,false,false));
        val lines:RDD[String]=sc.textFile("/spark/ml/sample/als", 1);
        val rows:RDD[Row]=rddStringToRow(lines);
        val dflines:DataFrame=sqlContext.createDataFrame(rows, structType);
        val als:ALS=new ALS();
        als.setMaxIter(10);
        //als.setRegParam(0.16);
        als.setUserCol("userId");
        als.setItemCol("itemId");
        als.setRatingCol("rating");
        als.setPredictionCol("pre_rating");
        val alsModel:ALSModel=als.fit(dflines);
        val samplePredictions:DataFrame = alsModel.transform(dflines);
        samplePredictions.createTempView("samplePreView");
        val deviationDF:DataFrame=sqlContext.sql("select round(sum(pow
        (rating-pre_rating,2))
        /count(userId),5) deviation from samplePreView");
        deviationDF.show()
        val appLines:RDD[String]=sc.textFile("/spark/ml/input/als", 1);
        val appRows:RDD[Row]=rddStringToRow(appLines);
```

```
    val appStructType:StructType=getStructType(Array[String]("userId",
    "itemId"),
    Array[DataType](DataTypes.IntegerType,DataTypes.IntegerType),
    Array[Boolean](false,false));
    val appDflines:DataFrame=sqlContext.createDataFrame(appRows,
    appStructType);
    val appPredictions:DataFrame = alsModel.transform(appDflines);
    appPredictions.createTempView("appPreView");
    val appResult:DataFrame=sqlContext.sql("select mid.userId,ap.
    itemId,mid.maxRating from (select userId,max(pre_rating) maxRating from
 appPreView group by userId) mid,appPreView ap where mid.maxRating=
 ap.pre_rating order by userId");
    val resultRDD:RDD[Row]=appResult.rdd;
    val outputRDD:RDD[String]=resultRDD.map((row:Row)=>{
      val userId:Int=row.getAs[Int]("userId");
      val itemId:Int=row.getAs[Int]("itemId");
      val maxRating:Float=row.getAs[Float]("maxRating");
      userId+"\t"+itemId+"\t"+maxRating
    });
    val dist:Path=new Path("/spark/ml/output/als");
    if(fs.exists(dist))fs.delete(dist, true);
    outputRDD.saveAsTextFile("/spark/ml/output/als");
  }
}
```

3. 打包工程到 dist 目录

```
[hadoop@CloudDeskTop bin]$ pwd
    /project/scala/SparkML/bin
[hadoop@CloudDeskTop bin]$ jar -cvfe ../dist/ALSML.jar com.bnyw.bigdata.
spark.ml.cluster.ALSML com/
[hadoop@CloudDeskTop bin]$ ls ../dist/
    ALSML.jar  KMeansML.jar  LinearRegressionML.jar
```

4. 提交协同过滤应用到 Spark 集群

```
[hadoop@CloudDeskTop bin]$ ./spark-submit --master spark://master01:7077
--class com.bnyw.bigdata.spark.ml.cluster.ALSML /project/scala/SparkML/
dist/ALSML.jar
17/08/30 09:09:14 WARN util.NativeCodeLoader: Unable to load native-hadoop
library for your platform... using builtin-java classes where applicable
17/08/30 09:09:15 WARN spark.SparkContext: Support for Java 7 is deprecated
as of Spark 2.0.0
……
17/08/30 09:09:36 INFO handler.ContextHandler: Started o.s.j.s.
ServletContextHandler@de28c5e{/static/sql,null,AVAILABLE,@Spark}
```

```
17/08/30 09:09:45 INFO mapred.FileInputFormat: Total input paths to process : 1
        +---------+
        |deviation|
        +---------+
        |  0.00751|
        +---------+

17/08/30 09:11:34 INFO mapred.FileInputFormat: Total input paths to process : 1
17/08/30 09:12:26 INFO Configuration.deprecation: mapred.tip.id is
deprecated. Instead, use mapreduce.task.id
...
17/08/30 09:12:32 INFO handler.ContextHandler: Stopped o.s.j.s.
ServletContextHandler@29669b01{/jobs/json,null,UNAVAILABLE,@Spark}
17/08/30 09:12:32 INFO handler.ContextHandler: Stopped o.s.j.s.
ServletContextHandler@21ec6eed{/jobs,null,UNAVAILABLE,@Spark}
```

5. 查看协同过滤算法的推荐结果

```
[hadoop@CloudDeskTop bin]$ hdfs dfs -cat /spark/ml/output/als/*
    1   4   5.348643
    2   2   6.504319
    6   1   NaN
    6   2   NaN
```

第4篇 项目实战

第 19 章 基于电力能源的大数据实战

19.1 需求分析

电网属国家级垄断行业，该行业需求遍及城乡企业、集体组织和个体用户，覆盖率几乎达 100%。由于在电网线路传送过程中，电流流经具有电阻的电网线路时会产生热量而造成电能损耗，将进一步导致线路老化、终端电压过低、部分线路发生热损故障、用电成本持续上升等负面影响。解决这些问题，需通过大量数据进行统计和分析。

电力大数据主要服务应用于以下几个方面：首先是规划，通过大数据分析利用数据挖掘技术提升负荷预测能力；其次是检修，通过研究消缺、检修、运行工况、气象条件等因素对设备状态的影响，利用大数据计算等技术实现检修策略化，提升状态检修管理能力；最后是运监，利用聚类、模式识别、可视化、并行处理技术，实现全方位的在线监测、分析、计算，提高数据质量监控和治理，对设备故障做出预判断，提前消除设备故障对生产带来的不良因素，提升连续运行能力。

大数据具有 4V 特点：Volume（大量）、Velocity（高速）、Variety（多样）、Value（价值）。2012 年 7 月 10 日，联合国发布政务白皮书《大数据促发展：挑战和机遇》建议联合国成员国开发大数据潜在价值。李克强总理在 2014 年两会政府工作报告中，多次提到一个热词——"大数据"。此外，百度、新浪、京东等著名企业已在大数据方面展开相关研究，发布了若干大数据研究报告。

电网公司基于现有各种监测指标建立了较为完善的供电质量运行体系，信息系统建设方面已实现基地试验设备数据采集。由于采集数据量庞大，传统的分析技术难以进行细化和量化分析。为充分发挥采集数据的作用，满足对数据深化应用和数据存储、查询、统计、分析及对数据价值深入挖掘的需求，通过领先的数据融合、数据清洗、数据治理以及大数据挖掘等相关技术手段提升数据应用价值成为重要手段。

19.2 项目设计

基于电力大数据采集产生监测数据构建的供电设备分析模型，用于工作站对辖区线路设备的运行情况进行综合评估，有利于从分析结果中预测线路设备是否会发现异常，从而做到节约能源和提前预防以减少临时故障带来的损失。电力数据分析平台主要分为三部分：数据采集、数据处理和数据呈现。

19.2.1 数据采集

采用分布数据采集系统对监测点采集站进行数据收集，通过传输电缆将收集到的数据传送至中央控制站，其中地面采集站探测器按线路相线布置，负责采集一个或几个测点的物理数据，中央控制站则完成其所在站点的数据记录和质量监控。最后将中央控制站的数据发送至大数据集群，流程如图19.1所示。

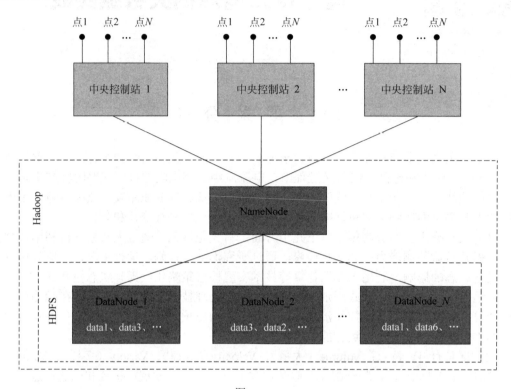

图 19.1

19.2.2 数据处理

基于大数据集群中的生产数据，按照业务流程对数据进行统计分析，通过季节、气候和专家经验评价等因素，综合计算得出分析结果，再将分析结果上传至 RDBMS 使用 Web 页面进行展示，展示流程如图 19.2 所示。

19.2.3 数据呈现

基于 RDBMS 和 Web 的业务展示分为柱状对比图、趋势图、热力图、饼图和列表。

柱状图用于展示同一时间内各个单位的电能质量、各个单位的上月（年）电能质量和上级单位的电能质量柱状对比。通过对比可发现在当前时间哪个单位的电能质量下降，以判断该单位所管的辖区内是否出现线路异常。也可通过单击柱状条进一步获取该单位的哪些线路出现了异常。

通过趋势图可展示出同一时间内各单位的电能质量历史趋势线和预测趋势线。决策者可以根据预测趋势线作出决策，为风险防范提供依据。

通过地图热力图显示某块区域内的电能质量的变化，以颜色的从浅到深变化来展示电能质量的危险级别。决策者可根据危险级别查看该区域的详细信息，并生成报表以向下级单位下达处理命令。

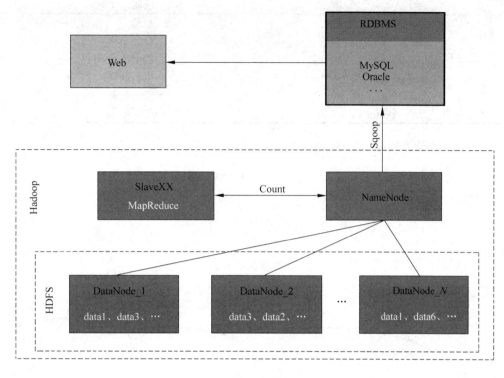

图 19.2

饼图展示各单位的电能质量合格的占比。通过对占比数据的分析可判断电能质量过低的区域，进一步获取该区域可查看区域内线路的详细信息及不达标情况。

通过 Web 页面展示数据，登录页面将进入到图 19.3 页面，默认为省会城市近一年的电压波动范围超标的曲线，可单选或多选不同城市并查看该城市电压波动范围超标情况的对比。单击曲线可获取该城市该月电压波动范围超标的详细情况。

下面介绍项目的数据模型。

电压波动趋势的表结构如表 19.1 所示。

表 19.1

字段	说明
city_name	城市名称
power_error_number	电压超标数量
month_time	月度时间

超标电压的详细表结构如表 19.2 所示。

表 19.2

字段	说明
city_name	城市名称
month_time	月度时间

续表

字段	说明
user_name	户号
dq_name	台区名称
power_org	供电单位
power_type	电压级别
time	监测时间
error_power	超标电压

超标电压波动趋势展示如图19.3所示。

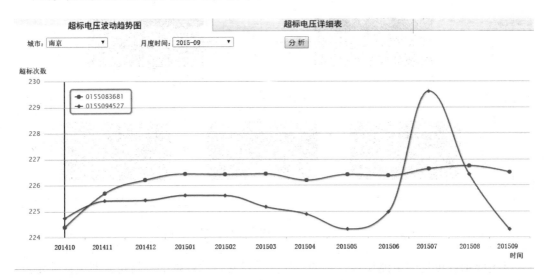

图 19.3

超标电压的详细表展示如图19.4所示。

图 19.4

19.3　数据收集与处理

19.3.1　数据收集

电力监测数据是采用分布数据采集系统对监测点采集站进行数据收集，并通过传输电缆传送至中央控制站，其中地面采集站探测器按线路相线布置，负责采集一个或几个测点的物理数据。由于电力大数据分析项目以数据处理和呈现讲解为主，并无真实数据，因此我们通过人为造数假设已有数据进行分析处理。

监测数据（monitor_data）表如图 19.5 所示。

接线方式	计量点电压等级	户号	台区编号	数据日期	U1	U2	…	U95	U96
三相四线	0.4kV380	0152405348	0190000207675	2015/1/16 0:00	224.40	224.90		226.90	225.00	224.80
三相四线	0.4kV380	0732036159	0190000182806	2015/1/16 0:00	221.60	221.10		226.20	226.50	221.70
三相四线	0.4kV380	0702353810	3090102301575	2015/1/16 0:00	221.60	222.90		221.40	222.30	222.40
三相四线	0.4kV380	0155062744	90100685501	2015/1/16 0:00	222.80	222.80		223.80	222.60	222.80
三相四线	0.4kV380	0738019173	0190000246986	2015/1/16 0:00	224.40	223.90		224.90	226.80	227.10
三相四线	0.4kV380	0155317687	3090102145727	2015/1/16 0:00	223.00			224.80	225.60	225.30
三相四线	0.4kV380	0702356766	0190000190359	2015/1/16 0:00	226.50	226.40		227.10	226.60	226.30
三相四线	0.4kV380	0155484499	3090102058657	2015/1/16 0:00	222.00	220.90		224.80	225.80	224.30
三相四线	0.4kV380	0734014427	0190000232752	2015/1/16 0:00	232.50	232.20			231.40	230.90
三相四线	0.4kV380	0736016692	0190000232723	2015/1/16 0:00	228.20	227.40		226.20	226.70	226.40
三相四线	0.4kV380	0155141277	0190000185079	2015/1/16 0:00	230.10	229.90		230.40	231.00	231.20

图　19.5

监测数据记录台区（一条线路范围）某个点的电压波动范围，每 15 分钟监测一次，每天监测 96 次。

单位信息数据（city_data）表如图 19.6 所示。

台区编号	台区名称	供电单位	城市
0190000207675	PMS_万福城11#	南京市高淳区供电公司	南京
0190000182806	PMS_阳江新正新圩	阳江供电所	南京
3090102301575	PMS_春栋村2#	南京市高淳区供电公司	南京
90100685501	PMS_阳江王家圩#1	阳江供电所	南京
0190000246986	PMS_桠溪村永兴集贸小区2#	桠溪供电所	南京
3090102145727	PMS_漆桥双望路2#	漆桥供电所	南京

图　19.6

单位信息记录台区编号、台区名称和该台区属于哪个供电单位的结构信息。

19.3.2　数据处理

1. 转换数据（Dataconvert）

通过观察，一天为 1 条记录数据。若统计电压超标的波动次数，首先需将一天 1 条的统计数据转变成一天 96 条，每条数据均有接线方式、计量点电压等级、户号、台区编号、数据日期和电压字段。监测时间每 15 分钟一次，U1 为 00:15，U2 为 00:30，…，U96 为 23:45 分，将时间精确到分钟并对应值，一次监测点为一条记录，如图 19.7 所示。

接线方式	计量点电压等级	户号	台区编号	数据日期	U1	U2	...	U95	U96
三相四线	0.4kV380	0152405348	0190000207675	2015/1/16 0:00	224.40	224.90		226.90	225.00	224.80
三相四线	0.4kV380	0732036159	0190000182806	2015/1/16 0:00	221.60			226.20	226.50	221.70
三相四线	0.4kV380	0702353810	3090102301575	2015/1/16 0:00	221.60	222.90		221.40	222.30	222.40
三相四线	0.4kV380	0155062744	90100685501	2015/1/16 0:00	222.80	222.80		223.80	222.60	222.80

↓ 转换

	接线方式	计量点电压等级	户号	台区编号	数据日期	电压
U1	三相四线	0.4kV380	0152405348	0190000207675	2015/1/16 00:15	224.40
U2	三相四线	0.4kV380	0152405348	0190000207675	2015/1/16 00:30	224.90
...	三相四线	0.4kV380	0152405348	0190000207675	2015/1/16 ...	226.90
U95	三相四线	0.4kV380	0152405348	0190000207675	2015/1/16 23:30	225.00
U96	三相四线	0.4kV380	0152405348	0190000207675	2015/1/16 23:45	224.80

图 19.7

配置类：

```java
package com.bunfly.dataconvert;
import org.apache.hadoop.conf.Configuration;
import org.apache.hadoop.fs.Path;
import org.apache.hadoop.io.Text;
import org.apache.hadoop.mapreduce.Job;
import org.apache.hadoop.mapreduce.lib.input.FileInputFormat;
import org.apache.hadoop.mapreduce.lib.output.FileOutputFormat;

public class DataConvert_Conf {
    public void Config() throws Exception {
        //获取配置文件中的参数
        Configuration conf = new Configuration();
        //通过参数配置获取一个Job对象
        Job job = Job.getInstance(conf,"DataConvert_Conf");
        //启动Job需要知道的入口
        job.setJarByClass(DataConvert_Conf.class);
        //启动Map入口
        job.setMapperClass(DataConvert_Mapper.class);
        //设置Reduce输出的数量
        job.setNumReduceTasks(0);
        //设置key值的输出类型
        job.setOutputKeyClass(Text.class);
        //设置value值的输出类型
        job.setOutputValueClass(Text.class);
        //设置输入源
        String in1 = "/input/monitor_data";
        //设置输出源
        String out = "/output/convert_result";
        //将输入、输出字符路径转为Path类型
        Path input1 = new Path(in1);
        Path output = new Path(out);
        //将路径加入到文件格式中
```

```java
        FileInputFormat.addInputPath(job, input1);
        FileOutputFormat.setOutputPath(job, output);
        System.out.println(job.waitForCompletion(true) ? 0 : 1);
    }
}
```

Mapper 类:

```java
package com.bunfly.dataconvert;
import java.io.IOException;
import java.text.DateFormat;
import java.text.SimpleDateFormat;
import java.util.Calendar;
import java.util.Date;
import org.apache.hadoop.io.Text;
import org.apache.hadoop.mapreduce.Mapper;
public class DataConvert_Mapper extends Mapper<Object, Text, Text, Text> {
    public void map (Object key, Text value, Context context)
            throws IOException, InterruptedException {
        String[] q = value.toString().split("\t");
        String ymd = DataConvert_Mapper.YearConvert(q[4]);
        String hms=null;
        String time=null;
        int j=0;
        for (int i=5; i<q.length; i++) {
            hms = DataConvert_Mapper.HourConvert(j);
            time = ymd +"/"+ hms;
            //q[0]接线方式、q[1]计量点电压等级、q[2]户号、q[3]台区编号、q[4]数据
            //日期、q[5]电压
            context.write(new Text(q[0] +"\t"+ q[1] +"\t"+ q[2] +"\t"+ q[3]
            +"\t"+ time),   new Text(q[i]));
            j+=15;
        }
    }
    //提取时分
    public static String HourConvert (int i) {
        String time = "2010-01-01 00:00:00";
         DateFormat dateFormat2 = new SimpleDateFormat("yyyy-MM-dd HH:mm:ss");

        Calendar cal = Calendar.getInstance();
        String ss = null;
        try {
            cal.setTime(dateFormat2.parse(time));
            //当前时间加多少分钟
            cal.add(Calendar.MINUTE,i);
            Date a = cal.getTime ();
```

```
            ss = dateFormat2.format(a).substring(11, 19);
        } catch (Exception e) {
            e.printStackTrace();
        }
        return ss;
    }
    //提取年月月
    public static String YearConvert (String time) {
        DateFormat dateFormat2 = new SimpleDateFormat("yyyy/MM/dd");
        String year = null;
        try {
            String tiem = dateFormat2.format(dateFormat2.parse(time));
            year = tiem.substring(0, 10);
        } catch (Exception e) {
            e.printStackTrace();
        }
        return year;
    }
}
```

2. 清理数据（Cleardata）

根据业务要求，需找出电压波动超标的电压。按电网电压标准，家用正常电压的范围是 220−0.1V～220+0.07V，须通过 MapReduce 程序筛选出非正常的电压。在电压中如存在空、0 等特殊电压，不能作为路线正常与否的判断依据，因此须剔除该类记录。

配置类：

```
package com.bunfly.cleardata;
import org.apache.hadoop.conf.Configuration;
import org.apache.hadoop.fs.Path;
import org.apache.hadoop.io.Text;
import org.apache.hadoop.mapreduce.Job;
import org.apache.hadoop.mapreduce.lib.input.FileInputFormat;
import org.apache.hadoop.mapreduce.lib.output.FileOutputFormat;
public class ClearData_Conf {
    public static void main(String[] str) throws Exception {
        ClearData_Conf a = new ClearData_Conf();
        a.Config();
    }
    public void Config() throws Exception {
        //获取配置文件中的参数
        Configuration conf = new Configuration();
        //通过参数配置获取一个Job对象
        Job Job = Job.getInstance(conf,"ClearData_Conf");
        //启动Job需要知道的入口
        job.setJarByClass(ClearData_Conf.class);
```

```java
        //启动Map入口
        job.setMapperClass(ClearData_Mapper.class);
        //启动Reduce入口
        //设置Reduce输出的数量
        job.setNumReduceTasks(0);
        //设置key值的输出类型
        job.setOutputKeyClass(Text.class);
        //设置value值的输出类型
        job.setOutputValueClass(Text.class);
        //设置输入源
        String in1 = "/output/convert_result/p*";
        //设置输出源
        String out = "/output/cleardata";
        //将输入、输出字符路径转为Path类型
        Path input1 = new Path(in1);
        Path output = new Path(out);
        //将路径加入到文件格式中
        FileInputFormat.addInputPath(job, input1);
        FileOutputFormat.setOutputPath(job, output);
        System.out.println(job.waitForCompletion(true) ? 0 : 1);
    }
}
```

Mapper 类：

```java
package com.bunfly.cleardata;
import java.io.IOException;
import org.apache.hadoop.io.Text;
import org.apache.hadoop.mapreduce.Mapper;
public class ClearData_Mapper  extends Mapper<Object, Text, Text, Text> {
    public void map (Object key, Text value, Context context)
            throws IOException, InterruptedException {
        String[] q = value.toString().split("\t");
        //剔除值为空的记录
        if (q.length < 6) {
            return;
        }
        //剔除值为0和值为空的无效数据
        if (q[5].equals("0") || q[5].equals(null) || q[5].equals("")) {
            return;
        }
        try {
            Double d = Double.parseDouble(q[5]);
            // 判断电压是否超标
            if (d < 198 || d > 235.4) {
                context.write(new Text(q[0] +"\t"+q[1] +"\t"+q[2] +
```

```
                "\t"+q[3] +"\t"+q[4] +"\t"+q[5]),
                    new Text());
            }
        } catch(Exception e) {
            e.printStackTrace();
        }
    }
}
```

3. 关联数据（Relation）

根据业务要求，关联台区编码、台区名称、供电单位、城市、接线类型、电压类型、用户号、监测时间、月度时间和超标电压字段。

配置类：

```
package com.bunfly.Relation;
import org.apache.hadoop.conf.Configuration;
import org.apache.hadoop.fs.Path;
import org.apache.hadoop.io.Text;
import org.apache.hadoop.mapreduce.Job;
import org.apache.hadoop.mapreduce.lib.input.FileInputFormat;
import org.apache.hadoop.mapreduce.lib.output.FileOutputFormat;
public class Relation_Conf {
    //合并所需要的字段
    public static void Config() throws Exception {
        //获取配置文件中的参数
        Configuration conf = new Configuration();
        //通过参数配置获取一个Job对象
        Job job = Job.getInstance(conf,"CountPower_Conf");
        //启动Job需要知道的入口
        job.setJarByClass(Relation_Conf.class);
        //启动Map入口
        job.setMapperClass(Relation_Mapper.class);
        //启动Reduce入口
        job.setReducerClass(Relation_Reduce.class);

        //设置Reduce输出的数量
        job.setNumReduceTasks(1);
        //设置key值的输出类型
        job.setOutputKeyClass(Text.class);
        //设置value值的输出类型
        job.setOutputValueClass(Text.class);
        //设置输入源
        String in1 = "/output/cleardata/part*";
        String in2 = "/input/city_data";
        //设置输出源
```

```java
        String out = "/output/countpower";
        //将输入、输出字符路径转为Path类型
        Path input1 = new Path(in1);
        Path input2 = new Path(in2);
        Path output = new Path(out);
        //将路径加入到文件格式中
        FileInputFormat.addInputPath(job, input1);
        FileInputFormat.addInputPath(job, input2);
        FileOutputFormat.setOutputPath(job, output);
        System.out.println(job.waitForCompletion(true) ? 0 : 1);
    }
}
```

Mapper 类：

```java
package com.bunfly.Relation;
import java.io.IOException;
import org.apache.hadoop.io.Text;
import org.apache.hadoop.mapreduce.InputSplit;
import org.apache.hadoop.mapreduce.Mapper;
import org.apache.hadoop.mapreduce.lib.input.FileSplit;
public class Relation_Mapper extends Mapper<Object, Text, Text, Text> {
    public void map (Object key, Text value, Context context)
            throws IOException, InterruptedException {
        InputSplit t = context.getInputSplit();
        String path = ((FileSplit) t).getPath().getName();
        System.out.println(path);
        if (path.contains("part")) {
            String[] q = value.toString().split("\t");
            String month_thime = q[4].substring(0,7);
            //以台区编码为key，并给记录信息作标记（AAAAA）
            //q[0]接线类型、q[1]电压等级、q[2]户号、q[4]监测时间、q[5]超标值、
            //时度时间、超标次数为1
            context.write(new Text(q[3]),
                    new Text("AAAAA" +"\t"+ q[0] +"\t"+ q[1] +"\t"+ q[2]
                            +"\t"+ q[4]
                            +"\t"+ q[5] +"\t"+ month_thime +"\t"+ "1"));
        } else if (path.contains("city_data")) {
            String[] q = value.toString().split("\t");
            //将县组区域编码作为key,value：BBBBB标识，q[1]台区名称，q[2]供电所
            //q[3]城市
            context.write(new Text(q[0]),
                    new Text("BBBBB" +"\t"+ q[1] +"\t"+ q[2] +"\t"+ q[3]));
        }
    }
}
```

Reduce 类:

```java
package com.bunfly.Relation;
import java.io.IOException;
import java.util.ArrayList;
import java.util.List;
import org.apache.hadoop.io.Text;
import org.apache.hadoop.mapreduce.Reducer;
public class Relation_Reduce extends Reducer<Text, Text, Text, Text> {
    public void  reduce (Text key, Iterable<Text> values, Context context)
    throws IOException, InterruptedException {
        //用来装houses_number表内容
        List<String> list = new ArrayList<String>();
        //用来装org_number表内容
        List<String> list2 = new ArrayList<String>();
        for (Text val : values) {
            String[] q = val.toString().split("\t");
            if (q[0].equals("AAAAA")) {
                list.add(q[1] +"\t"+ q[2] +"\t"+ q[3] +"\t"+ q[4] +"\t"+ q[5]
                +"\t"+ q[6] +"\t"+ q[7]);
            } else if (q[0].equals("BBBBB")) {
                list2.add(q[1] +"\t"+ q[2] +"\t"+ q[3]);
            }
        }
        for(String i : list) {
            for(String j : list2) {
                context.write(key, new Text(i +"\t"+ j));
            }
        }
    }
}
```

4. 统计超标数量（Count_Error_Number）

由分析可知，电压波动趋势页面需要展示电压波动范围超标的同一城市在同一时间内的数量，满足页面要求需城市名称、电压超标数据和月度时间字段。

配置类:

```java
package com.bunfly.count_error_number;
import org.apache.hadoop.conf.Configuration;
import org.apache.hadoop.fs.Path;
import org.apache.hadoop.io.Text;
import org.apache.hadoop.mapreduce.Job;
import org.apache.hadoop.mapreduce.lib.input.FileInputFormat;
import org.apache.hadoop.mapreduce.lib.output.FileOutputFormat;
public class Count_Error_Number {
```

```java
    //统计超标数量
    public static void Config() throws Exception {
        //获取配置文件中的参数
        Configuration conf = new Configuration();
        //通过参数配置获取一个Job对象
        Job job = Job.getInstance(conf,"Count_Error_Number");
        //启动Job需要知道的入口
        job.setJarByClass(Count_Error_Number.class);
        //启动Map入口
        job.setMapperClass(Count_Error_Number_Mapper.class);
        //启动Reduce入口
        job.setReducerClass(Count_Error_Number_Reduce.class);
        //设置Reduce输出的数量
        job.setNumReduceTasks(1);
        //设置key值的输出类型
        job.setOutputKeyClass(Text.class);
        //设置value值的输出类型
        job.setOutputValueClass(Text.class);
        //设置输入源
        String in1 = "/output/countpower/part*";
        //设置输出源
        String out = "/output/count_error_number";
        //将输入、输出字符路径转为Path类型
        Path input1 = new Path(in1);
        Path output = new Path(out);
        //将路径加入到文件格式中
        FileInputFormat.addInputPath(job, input1);
        FileOutputFormat.setOutputPath(job, output);
        System.out.println(job.waitForCompletion(true) ? 0 : 1);
    }
}
```

Mapper 类：

```java
package com.bunfly.count_error_number;
import java.io.IOException;
import org.apache.hadoop.io.Text;
import org.apache.hadoop.mapreduce.Mapper;
public class Count_Error_Number_Mapper extends Mapper<Object, Text, Text, Text> {
    public void map (Object key, Text value, Context context)
            throws IOException, InterruptedException {
        String[] q = value.toString().split("\t");
        //以台区编码为key，并给记录信息作标记（AAAAA）
        //q[10]城市 、q[6]月度时间
        context.write(new Text(q[10] +"\t"+ q[6]),
```

```
                          new Text("1"));
    }
}
```

Reduce 类:

```
package com.bunfly.count_error_number;
import java.io.IOException;
import org.apache.hadoop.io.Text;
import org.apache.hadoop.mapreduce.Reducer;
public class Count_Error_Number_Reduce extends Reducer<Text, Text, Text, Text> {
    public void reduce (Text key, Iterable<Text> values, Context context)
    throws IOException, InterruptedException {
        int i=0;
        int sum=0;
        for(Text val : values) {
            i = Integer.valueOf(val.toString());
            sum = sum + i;
        }
        context.write(key, new Text(sum+""));
    }
}
```

5. 列出超标详情（List_Error_Data）

通过异常电压趋势图可单击某点获取该单位在选取时间内超标电压的详细情况，超标电压详细情况表由城市、日度时间、户号、台区名称、供电单位、电压级别、监测时间和超标电压字段组成。在超标电压详细情况表中可根据供电单位和台区名称进一步筛选出超标电压的详细信息。

配置类:

```
package com.bunfly.list_error_power;
import org.apache.hadoop.conf.Configuration;
import org.apache.hadoop.fs.Path;
import org.apache.hadoop.io.Text;
import org.apache.hadoop.mapreduce.Job;
import org.apache.hadoop.mapreduce.lib.input.FileInputFormat;
import org.apache.hadoop.mapreduce.lib.output.FileOutputFormat;
public class List_Error_Data {
    //列出超标详情
    public static void Config() throws Exception {
        //获取配置文件中的参数
        Configuration conf = new Configuration();
        //通过参数配置获取一个Job对象
        Job job = Job.getInstance(conf,"List_Error_Data");
        //启动Job需要知道的入口
```

```java
            job.setJarByClass(List_Error_Data.class);
            //启动Map入口
            job.setMapperClass(List_Error_Data_Mapper.class);
            //启动Reduce入口
            //设置reduce输出的数量
            job.setNumReduceTasks(0);
            //设置key值的输出类型
            job.setOutputKeyClass(Text.class);
            //设置value值的输出类型
            job.setOutputValueClass(Text.class);
            //设置输入源
            String in1 = "/output/countpower/part*";
            //设置输出源
            String out = "/output/list_error_data";
            //将输入、输出字符路径转为Path类型
            Path input1 = new Path(in1);
            Path output = new Path(out);
            //将路径加入到文件格式中
            FileInputFormat.addInputPath(job, input1);
            FileOutputFormat.setOutputPath(job, output);
            System.out.println(job.waitForCompletion(true) ? 0 : 1);
    }
}
```

Mapper 类：

```java
package com.bunfly.list_error_power;
import java.io.IOException;
import org.apache.hadoop.io.Text;
import org.apache.hadoop.mapreduce.Mapper;
public class List_Error_Data_Mapper extends Mapper<Object, Text, Text, Text> {
    public void map (Object key, Text value, Context context)
            throws IOException, InterruptedException {
        String[] q = value.toString().split("\t");
        //以台区编码为key,并给记录信息做标记（AAAAA）
        /* q[10]城市  q[6]月度时间  q[3]户号  q[8]台区名称  q[9]供电单位
         * q[1]电压级别  q[4]监测时间 q[5]值
         */
        context.write(new Text(q[10]),
                new Text(q[6] +"\t"+ q[3] +"\t"+ q[8] +"\t"+ q[9]
                    +"\t"+ q[1] +"\t"+ q[4] +"\t"+ q[5]));
    }
}
```

6. 入口（MainMethod）

MainMethod 类是执行整个离线程序的主入口。

```java
package com.bunlfy.main;
import com.bunfly.Relation.Relation_Conf;
import com.bunfly.cleardata.ClearData_Conf;
import com.bunfly.count_error_number.Count_Error_Number;
import com.bunfly.dataconvert.DataConvert_Conf;
import com.bunfly.list_error_power.List_Error_Data;
public class MainMethod {
    public static void main(String[] args) throws Exception {
        // TODO Auto-generated method stub
        //转换数据
        DataConvert_Conf.Config();
        //数据清理
        ClearData_Conf.Config();
        //合并所需要的字段
        Relation_Conf.Config();
        //统计超标数量
        Count_Error_Number.Config();
        //列出超标电压详情
        List_Error_Data.Config();
    }
}
```

7. 程序运行流程图

将程序生成的 PowerProject.jar 包上传至大数据集群，执行 hadoop jar PowerProject.jar 命令开始运行程序。首先找到程序入口 MainMethod 类，并依次执行 MainMethod 类中的语句。DataCaonvert_Conf 类将 HDFS 中的 HDFS:/input/monitor_data 文件中的数据转换成一次监测时间为一条记录，并输出至 HDFS:/output/convert_result 目录。ClearDate_Conf 类在 HDFS:/output/convert_result 目录中读取数据，对无效数据进行清洗，并输出数据至 HDFS:/output/cleardata。Relation_Conf 从 HDFS:/output/cleardata 和 HDFS:/input/city_data 中读取数据，根据业务要求从 HDFS:/input/city_data 文件中关联出城市、台区名称等信息后将数据输出至 HDFS:/output/countpower 目录。Count_Error_Number 类用于实现电压波动超标趋势图页面，根据业务分析，得知该页面需要城市、月度时间和电压超标数量三个字段。Count_Error_Number 类读取 HDFS:/output/countpower/part* 文件，通过对数据统计分析得出结果，并输出至 HDFS:/output/count_error_number 文件。List_Error_Data 类用于实现超标电压详细列表，用户可在电压波动超标趋势图页面中单击某个城市获取某个时间电压超标的详细情况，可在详细页面中通过选择供电单位和电压名称进行筛选电压超标详细情况，根据对业务的分析，得知该页面需要城市名称、月度时间、户号、台区名称、供电单位、电压级别、监测时间和超标电压字段。List_Error_Data 类读取 HDFS:/output/countpower/part* 文件，通过对数据筛选得出结果，并输出至 HDFS:/output/list_error_data 目录。

程序运行流程如图 19.8 所示。

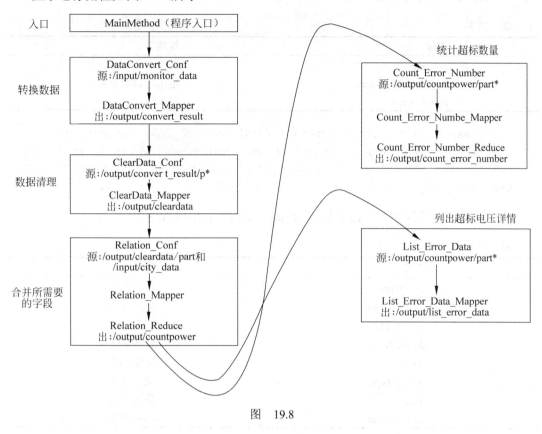

图 19.8

19.4 大数据呈现

19.4.1 数据传输

大数据呈现是以 B/S 方式在 Web 页面呈现出来。由于无法在 Hadoop 大数据集群中进行实时响应，所以需要把分布文件系统中的数据传输至关系型数据库，再编写 Java 代码，在关系型数据库（MySQL）中读取数据。

Sqoop 是一款开源工具，主要用于在 Hadoop 与传统的数据库（MySQL、PostgreSQL…）间进行数据的传递，可将一个关系型数据库（例如 MySQL、Oracle、PostgreSQL 等）中的数据导进到 Hadoop 的 HDFS 中，也可将 HDFS 的数据导进到关系型数据库中。若从 HDFS 中导入数据到关系数据库中，关系数据库中需先有对应的表来存储导入的数据。具体操作如下：

```
sqoop export --connect jdbc:mysql://192.168.153.205:3306/bunfly?
characterEncoding=UTF-8 --username root -password 123456 --table count_
error_eonf --export-dir '/output/count_error_number/p*' --fields-
terminated-by '\t'
```

导入的电压波动趋势表（Count_Error_Conf）的表结构如表 19.3 所示。

表 19.3

字段	说明
city_name	城市名称
power_error_number	电压超标数量
month_time	月度时间

```
sqoop export --connect jdbc:mysql://192.168.153.205:3306/bunfly?
characterEncoding=UTF-8 --username root -password 123456 --table list_
error_data --export-dir '/output/list_error_data/p*' --fields-
terminated-by '\t'
```

导入的超标电压详细表（List_Error_Data）结构如表 19.4 所示。

表 19.4

字段	说明
city_name	城市名称
month_time	月度时间
user_name	户号
dq_name	台区名称
power_org	供电单位
power_type	电压级别
time	监测时间
error_power	超标电压

19.4.2 数据呈现

超标电压波动趋势页面如图 19.9 所示。

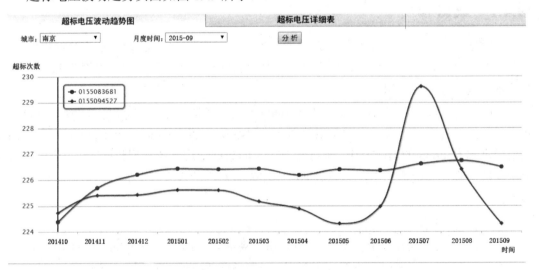

图 19.9

超标电压详细表页面如图 19.10 所示。

城市名称	月度时间	户号	台区名称	供电单位	电压级别	监测时间	超标电压
南京	2015/06	738019173	PMS_姚家	东坝供电所	三相四线	2015/06/01 23:45	236.83
南京	2015/06	155317687	PMS_阳江孙家庄	阳江供电所	三相四线	2015/06/11 23:45	55.51
南京	2015/06	702356766	PMS_浮兴路2	南京市高淳区供电公司	三相四线	2015/06/21 23:45	74.65
南京	2015/06	155484499	PMS_化肥厂宿舍1#	南京市高淳区供电公司	三相四线	2015/06/31 20:45	164.19
南京	2015/06	734014427	PMS_北漪小区1	南京市高淳区供电公司	三相四线	2015/06/23 23:45	308.78
南京	2015/06	736016692	PMS_固城湖北路3#	南京市高淳区供电公司	三相四线	2015/06/18 23:45	273.00
南京	2015/06	155141277	PMS_镇南村平圩	桠溪供电所	三相四线	2015/06/01 23:45	25.52
南京	2015/06	702354302	PMS_镇南村荷花塘	桠溪供电所	三相四线	2015/06/01 23:45	359.72
南京	2015/06	739011318	PMS_花义村广寺里2#	桠溪供电所	三相四线	2015/06/01 03:15	307.81
南京	2015/06	732033197	PMS_纺机厂宿舍	南京市高淳区供电公司	三相四线	2015/06/01 23:45	194.86
南京	2015/06	702354033	PMS_荆山村王马村	桠溪供电所	三相四线	2015/06/01 23:45	248.61
南京	2015/06	155490557	PMS_赵village长村	桠溪供电所	三相四线	2015/06/01 23:45	342.85
南京	2015/06	702366211	PMS_永庆村地山大	桠溪供电所	三相四线	2015/06/01 23:45	253.00
南京	2015/06	702354354	PMS_观溪村新村	桠溪供电所	三相四线	2015/06/01 23:45	33.68
南京	2015/06	702359584	PMS_跃进村东边庄	桠溪供电所	三相四线	2015/06/01 23:45	20.66

当前是第:1/124页 首页 上一页 下一页 末页 跳转到 □ 页

图 19.10

19.5 项 目 总 结

建设节约型社会是我国重大发展战略之一。我国的现实国情要求必须把电力节能放在节约能源的重要位置，电力节能在能源节约、减少环境污染等方面有着巨大的潜力，是建设节约型社会的突破口。本成果可根据各市实际情况进行实际运营，通过对采集线路的电压数据收集、存储、挖掘、分析和展示，将各单位管辖的线路电能运行情况和风险趋势以图表形式展现，为职能部门风险评估提供依据。若本成果能够上线，通过大数据分析预测风险进行预警，可下发风险检修通知做出预防工作，将减少异常耗能和意外故障对企业的能源浪费、时间浪费和经济损失。

本项目从数据收集、数据处理和数据呈现三方面讲解大数据项目开发流程，重点讲解数据处理。为实现业务要求将数据处理部分分为入口、数据转换、数据清洗、关联、数据统计和数据详细 6 部分，通过对数据的转换、清洗、处理将结果数据使用 Sqoop 工具导入到 MySQL 数据库，通过 B/S 的方式在 Web 页面进行展示。

本项目从实际项目中提取部分业务讲解大数据编程的流程，由浅到深的讲解将帮助读者理解项目开发流程和代码现实，增加学习大数据编程的信心。

图书资源支持

感谢您一直以来对清华版图书的支持和爱护。为了配合本书的使用,本书提供配套的资源,有需求的读者请扫描下方的"书圈"微信公众号二维码,在图书专区下载,也可以拨打电话或发送电子邮件咨询。

如果您在使用本书的过程中遇到了什么问题,或者有相关图书出版计划,也请您发邮件告诉我们,以便我们更好地为您服务。

我们的联系方式:

地　　址:北京海淀区双清路学研大厦 A 座 707

邮　　编:100084

电　　话:010-62770175-4604

资源下载:http://www.tup.com.cn

电子邮件:weijj@tup.tsinghua.edu.cn

QQ:883604(请写明您的单位和姓名)

用微信扫一扫右边的二维码,即可关注清华大学出版社公众号"书圈"。

书圈